fórmula bruta	валовая формула	sumární vzorec
peso molecular	молекулярный вес	molekulová váha
energía de formación	теплота образования	slučovací teplo
balance de oxígeno	кислородный баланс	kyslíková bilance
porcentaje de nitrógeno	содержание азота	obsah dusíku
volumen de gases de detonación	обьем газообразных продуктов взрыва при нормальных условиях	objem plynných zplodin výbuchu za norm. podmínek
calor de explosión	теплота взрыва	výbuchové teplo
energía específica	удельная энергия	specifická energie
densidad	плотность	hustota
punto de fusión	точка плавления	bod tání
ensayo del bloque de plomo	расширение канала свинцового блока трауцля	váduť v olověném bloku
velocidad de detonación	скорость детонации	det. rychlost
punto de deflagración	температура вспышки	bod výbuchu
sensibilidad al choque	чувствительность к удару	citlivost k nárazu
sensibilidad a fricción	чувствительность к трению	citlivost k tření
diámetro crítico en el ensayo con vaina de acero	критический диаметр при испытании в стальных гильзах	kritický průměr pro stanovení v ocelov. trub.

R. Meyer
J. Köhler
A. Homburg

Explosives

Rudolf Meyer
Josef Köhler
Axel Homburg

Explosives

Sixth, Completely Revised Edition

WILEY-VCH Verlag GmbH & Co. KGaA

Dr. Rudolf Meyer (†)
(formerly: WASAG Chemie AG, Essen, Germany)

Josef Köhler
Fronweg 1
A-4784 Schardenberg

Dr.-Ing. Axel Homburg
c/o Dynamit Nobel GmbH
Kaiserstr. 1
D-53839 Troisdorf

> This book was carefully produced. Nevertheless, authors and publisher do not warrant the information contained therein to be free of errors. Readers are advised to keep in mind that statements, data, illustrations, procedural details or other items may inadvertently be inaccurate.

First Edition 1977
Second, Revised and Extended Edition 1981
Third, Revised and Extended Edition 1987
Fourth, Revised and Extended Edition 1994
Fifth, Completely Revised Edition, 2002
Sixth, Completely Revised Edition, 2007

Library of Congress Card No.: Applied for.

British Library Cataloguing-in-Publication Data
A catalogue record for this book is available from the British Library.

Bibliographic information published by the Deutsche Nationalbibliothek
Die Deutsche Nationalbibliothek lists this publication in the Deutsche Nationalbibliografie; detailed bibliographic data are available in the Internet at <**http://dnb.d-nb.de**>

© Wiley-VCH Verlag GmbH, Weinheim, 2007

ISBN 978-3-527-31656-4

Printed on acid-free paper.

All rights reserved (including those of translation in other languages). No part of this book may be reproduced in any form – by photoprinting, microfilm, or any other means – nor transmitted or translated into machine language without written permission from the publishers. Registered names, trademarks, etc. used in this book, even when not specifically marked as such, are not to be considered unprotected by law.

Composition: Typomedia, Ostfildern
Printing: betz-druck GmbH, Darmstadt
Bookbinding: Litges & Dopf Buchbinderei GmbH, Heppenheim
Cover Design: Wolfgang Scheffler, Mainz
Wiley Bicentennial Logo: Richard J. Pacifico

Printed in the Federal Republic of Germany.

Preface

The sixth English edition of "Explosives" which is now available can look back over a history of 30 years since it was first published.

On the initiative of its main author Dr. Rudolf Meyer (who regrettably died in 2000), the first copy of the German version of "Explosivstoffe", which had become known far beyond the borders of German-speaking countries, was first translated in 1977. It was also Dr. Rudolf Meyer, who in 1961, in his function of Technical Director of the German company WASAG Chemie AG, gave new impetus to the company brochure to achieve its present form.

The preparation of all the current German and English editions have a closely-knit history with the Fraunhofer Institut für chemische Technologie (ICT, formerly the Institut für Chemie der Treib- und Explosivstoffe with its head offices in Pfinztal/Berghausen near Karlsruhe, Germany. This institute was initially founded by Dr. Karl Meyer, Dr. Rudolf Meyer's elder brother in 1957, as part of the Technische Hochschule, Karlsruhe, and was extended later on.

We regret to say that while this sixth edition was under preparation, the Assistant Technical Director of ICT, Dr. Fred Volk, (1930–2005) passed away.

Immediately after obtaining his doctorate in chemistry in 1960, Dr. Fred Volk joined the ICT, which had then just been built in Pfinztal, where he worked closely with Dr. Karl Meyer. The main focus of his research within the area of explosive analysis was on the use of thin-layer chromatography and mass spectrometry as well as on calculating thermodynamic energies used in explosive and combustion processes. The many key words and the related articles in "Explosives" dealing with theoretical and thermodynamic performances were painstakingly checked, or written by Dr. F. Volk himself, each time before a new edition was printed.

As an addition to the previous edition of "Explosives", this book includes even more information on new explosive formulations.

Dr. Alfred Kappl has provided information and text materials on Fuel Air and Thermobaric explosives, and Dr. Robert Bickes (Sandia National Labs, USA) has contributed an article about Semiconductor Bridge Igniters (SCB). The special department II.3 (Dr. Thomas Lehmann, Dr. Silke Schwarz and Dr. Dietrich Eckhardt) of the Deutsche Bundesanstalt für Materialforschung und -prüfung (German Federal Institute for Material Testing and Research, BAM) has also contributed important notes and references on the subjects of hazardous substances and testing procedures.

Prof. Dr. Charles L. Mader has provided relevant information on the subject area of detonation physics. The Fraunhofer ICT has once again provided valuable help towards the publication of this volume,

and Prof. Dr. P. Elsner, Dr. S. Kelzenberg, Dr. Th. Keicher, Dr. N. Eisenreich, Dr. K. Menke, Dr. H. Krause, Dr. M. Hermann, and Dr. P.B. Kempa deserve particular mention for their generous contributions.
The authors would particularly like to thank their colleagues mentioned above for their contributions and for providing valuable comments.

Due to reductions and restructuring in the explosives industry, which has sometimes resulted in a complete loss of former product and company names used over decades, a range of names used in the past has not been included in this edition.

As an addition a CD containing a demo version of the ICT-Database of Thermochemical Values and information about the ICT-Thermodynamic-Code ist attached to the book. The full version of the database contains detailed information of more than 14,000 substances, including structure formulae, oxygen balance, densities and enthalpies of formation. The Code may be used for calculating properties of formulations like the heat of explosion or specific impulse of explosives, propellants or pyrotechnics. Both programs, updated regularly, are available by the Fraunhofer ICT.

Among these may be mentioned Dr. B. Eulering (WASAG Chemie, Essen), Dipl.-Ing. W. Franke (BAM, Berlin), Dipl.-Ing. H. Grosse † (WASAG Chemie), Dr. E. Häusler † (BICT), Dr. R. Hagel (Ruag Ammotec, Fürth), Dr. H. Hornberg † (ICT), Dr. H. Krebs (BAM, Berlin), Dr. G. Kistner (ICT), Prof. Dr H. Köhler † (Austron), Dr. A. Kratsch (Rheinmetall Industrie GmbH), Dip.-Ing. H. Krätschmer, Dr. K. Meyer † (ICT), Prof. Dr.-Ing. K. Nixdorff (BW University, Hamburg), Dr. K. Redecker, Dr. H. J. Rodner (BAM, Berlin), Dr. J. F. Roth † (DNAG, Troisdorf), Prof. Dr. H. Schubert (ICT), Prof. Dr. M. Steidinger (BAM, Berlin), Dipl.-Ing. G. Stockmann (WNC-Nitrochemie), Dr. G. Traxler (ORS Wien), Mr. R. Varosh (RISI, USA), Mr. J. Wraige (Solar Pyrotechnics, GB), Mrs. Christine Westermaier and Dr. R. Zimmermann (BVS, Dortmund).

We hope that the large number of people who remain unmentioned will also feel that they share in this expression of thanks.

The authors also wish to thank the publishers, the WILEY-VCH Verlag GmbH Company, and in particular Mrs. K. Sora, Mrs. R. Dötzer and Mrs. D. Kleemann, for the most pleasant co-operation in the production and printing of this book.

The publishers and authors continue to welcome suggestions and communications of any kind. We hope that our book will remain an important reference work and a quick source of information in this edition as well.

Schardenberg, April 2007   Josef Köhler   Axel Homburg

From the preface of previous editions:

"Explosives" is a concise handbook covering the entire field of explosives. It was preceded by the booklet "Explosivstoffe" published in 1932 by WASAG, Berlin, and by the handbook of industrial and military explosives published by WASAG-CHEMIE in 1961 under the same name.

The book contains about 500 entries arranged in alphabetical order. These include formulas and descriptions of about 120 explosive chemicals, about 60 additives, fuels, and oxidizing agents, and a 1500-entry subject index.

The objective of the book is to provide fundamental information on the subject of explosives not only to experts but also to the general public. The book will therefore, apart from industrial companies and research facilities concerned, be found useful in documentary centers, translation bureaus, editorial offices, patent and lawyer offices, and other institutions of this nature.

The properties, manufacturing methods, and applications of each substance are briefly described. In the case of key explosives and raw materials, the standard purity specifications are also listed.

The asymmetric margins are provided for entries and marginal notes of the reader.

Instructions for the thermodynamic calculations of the performance parameters of high explosives, gun propellants, and rocket propellants are given in somewhat greater detail. The basic thermodynamic data will be found in the extensive synoptic tables. They are based on the metric system; conversion from the English or the U.S. system can be made using the conversion tables on the back flyleaf. The front flyleaf contains a glossary of the terms denoting the characteristics of explosive materials in six languages.

The standard temperature selected for the energy of formation and enthalpy of formation data is 25 °C − 298.15 K. The elementary form of carbon was taken to be graphite (and not diamond, as before). The numerical values of the energies of formation (which, as known, appear both in the relevant entry and in the tables) are the optimum molar values found in the enthalpy tables of *Volk, Bathelt* and *Kuthe*: "Thermochemische Daten von Raketentreibstoffen, Treibladungspulvern sowie deren Komponenten", published by the Institut für Chemische Technologie (ICT), D-76327 Pfinztal-Berghausen 1972.

The US experts in rocket-techniques* and the Institute of Makers of Explosives** published glossaries on the definition and explanations of technical terms. Parts of them have been incorporated in the text.

---

* Published as appendix 4 of the Aerospace Ordnance Handbook by *Pollard, F. B.* and Arnold, J. H. Prentice Hall Inc., 1966
** Published as Publication No. 12 by the Institute of Makers of Explosives.

From the preface of previous editions VIII

The book is not intended as a systematic presentation of the science of explosives. Interested readers are referred to the many excellent publications on the subject, which are available in English (see, for example, the books by *M. A. Cook*) and the now nearly complete encyclopedia covering the whole explosive field, edited by *Seymour M. Kaye* (formerly by Basil *T. Fedoroff* †: "Encyclopedia of Explosives and Related Items"). Users of explosives should consult the "Blaster's Handbook" of DU PONT Inc., which is by far the best book on the subject.

A comprehensive list of literature references will be found at the end of the book.

Abel Test

This test on chemical stability was proposed by *Abel* in 1875. The test parameter determined is the time after which a moist potassium iodide starch paper turns violet or blue when exposed to gases evolved by one gram of the explosive at 180°F (82.2 °C).

In commercial nitroclycerine explosives, for example, this coloration only develops after 10 minutes or more. In a more sensitive variant of the method, Zinc iodide – starch paper is employed.

The *Abel* test is still used in quality control of commercial nitrocellulose, nitroclycerine and nitroglycol, but is currently no longer employed in stability testing of propellants.

Acceptor*)

Empfängerladung; charge réceptrice

A charge of explosives or blasting agent receiving an impulse from an exploding → *Donor* charge.

Acremite

This is the name given by the U.S. inventor *Acre* to his mixture of about 94 % ammonium nitrate with 6 % fuel oil. This mixture was at first prepared in a primitive manner by the users themselves to obtain a very cheap explosive for open pit mining under dry conditions. As → *ANFO* the material has widely displaced the conventional cartridged explosives.

Actuator

Mechanical device operated by a solid propellant.

Adiabatic

Processes or phenomena assumed to occur in a closed system without energy exchange with the surroundings.

adiabatic flame temperature

The temperature obtained by thermodynamics calculations for the products of combustion of energetic materials neglecting energy loss to the surroundings.

---

* Text quoted from glossary.

isobaric adiabatic flame temperature

Adiabatic flame temperature attained under constant pressure conditions.

isochoric adiabatic flame temperature

Adiabatic flame temperature attained under constant volume conditions

adiabatic temperature

The temperature attained by a system undergoing a volume or pressure change in which no heat enters or leaves the system.

Adobe Charge

Auflegerladung; pétardage

Synonymous with → *Mud Cap*

ADR

Abbreviation for "Accord Européen Relatif au Transport des Marchandises Dangereuses par Route" (European Agreement Concerning the international Carriage of Dangerous Goods by Road). It is based on the Recommendations on the Transport of Dangerous Goods Model Regulations (United Nations).

Aerozin

A liquid fuel for rocket engines, which is composed of 50 % anhydrous hydrazine and 50 % *asym*-dimethylhydrazine.

AGARD

Abbreviation for the NATO Advisory Group for Aeronautical Research and Development.

Airbag

Gasgenerator

The basic idea of the airbag as a passive restraint system in a motor vehicle was already patented fort he first time in 1951 in Germany. However, it takes nearly 20 years for start of development on two basic types of generators. Both types are manufactured nearly exclusively in series production and were built in cars starting 1975. Mainstream applications of restraint systems in almost every car started in 1990.

The first type is based on pure *pyrotechnic* for gas generation to fill the bag with hot gas. The second type is also known as *hybrid* design, where a gas is stored under high pressure. It acts as cold gas source since there is no need for this gas to be generated pyrotechnically. Most types of *hybrid* generators have a pyrotechnic heating charge to prevent excess cooling to undesired low temperatures. Both basic types of gas generator for airbags are used for driver, passenger, side and head airbags. Their diagrammatic construction is shown in Fig. 1 and Fig. 2.

In the *hybrid* system the pre-pressurized gas (nitrogen, argon/helium or pure helium) is stored in high pressure containers fitted with a bursting membrane. Opening this membrane by pyrotechnic means allows the gas to flow out into the airbag. The cooling of the expanding gas is compensated or even over-compensated by a pyrotechnic charge. Since the total amount of pyrotechnic mixture is small in quantitative terms, the prescribed threshold values of the toxic impurities contained in the working gas can be adhered to relatively easily. This fact in addition to the ideal temperature of the working gas is the main advantage of *hybrid* gas generators. The disadvantages are the large weight, the subjection to the Pressure Vessel Regulation and the high noise level that occurs when the sealing disk opens, because initially the full gas pressure is present.

The unique feature of almost all *pyrotechnical gas generators* is the concentric assembly of three different chambers with different designs corresponding to their pressure conditions and functions. The innermost chamber with the highest pressure resistance contains the igniter unit consisting of a plug, squib and booster charge. Depending on the generator construction a pre-ignition unit may also be installed, whose task is to ignite the pyrotechnic mixture without electric current in case of high temperatures, which could occur in case of a fire. During normal electrical ignition the thin resistance wire of the igniter is heated and the ignition train started. The booster charge normally used is boron / potassium nitrate. The hot gases and particles generated by this charge enter the concentrically arranged combustion chamber and ignite the pyrotechnic main charge. Both chambers are designed for pressures up to 40 MPa. The pyrotechnic main charge consists generally of compressed pellets which generate the working gas and slag residues by a combustion process. The products leave the combustion chamber through nozzles and enter the low pressure region of the filter compartment, where the slag is removed from the gas flow. The filter compartment is equipped with various steel filters and deflector plates. The resulting gas flows through the filter compartment apertures into the bag.

The basic task of a gas generator is to provide sufficient non-toxic gas within approximately 25 ms to inflate the airbag to the specification pressure. The first pyrotechnic mixture used in airbag gas generators was based on sodium azide. Sodium azide reacts with oxidising agents, which bond chemically the remaining sodium as the nitrogen is liberated. Established oxidisers are the alkali and alkaline earth nitrates, metal oxides (e.g. CuO, Fe2O3), metal sulfides (MoS2) and sulphur. If necessary slag forming agents (e.g. SiO_2, aluminosilicates) were also added.

The consequence of advance in environmental awareness was that the toxic sodium azide has to be replaced despite pure nitrogen generation, lower reaction temperatures and greater long-term stability. Another factor against sodium azide was the relative low specific gas yield and the unsolved disposal procedure for this type of pyrotechnic mixture.

The unique feature of almost all *pyrotechnical* gas generators (specifically on the driver side) is the concentric assembly of three different chambers with different designs corresponding to their pressure conditions and functions. The innermost chamber with the highest pressure resistance contains the igniter unit consisting of a plug, electrical igniter matchhead and the igniter mixture. Depending on the generator construction, a pre-ignition unit may also be installed, whose task is to ignite the gas mixture without an electric current in the event of exposure to elevated external temperature – for example during a fire. During normal electrical ignition, the thin resistance wire of the igniter matchhead is heated to melting point and the ignition train started. As the ignition mixture burns away – usually a boron/potassium nitrate mixture – the resulting hot gases and particles flow through the peripheral holes and into the combustion chamber filled with the gas mixture, which is arranged concentrically around the igniter chamber and is designed for an operating pressure of 100–180 bar. The gas mixture consists of compressed tablets which, after ignition, burn to form the working gas and slag. The combustion products leave the combustion chamber through the nozzle holes. The low pressure region of the filter compartment is arranged around the combustion chamber. The filter compartment is fitted with various steel filters and deflector plates. In the filter compartment the hot gases are cooled down and freed from liquid/solid slag. The resulting working gas flows through the filter compartment apertures towards the gas bag. The liquid slag constituents must be cooled down to solidification in the filter compartment so that they can also be filtered out there. It is clear that the nature of the gas mixture – formerly called the propellant or propellant mixture – is exceptionally important with regard to providing the gas (fume) cloud during burn-up. The basic task of a gas generator is, when necessary, to supply sufficient non-toxic gas within approx.

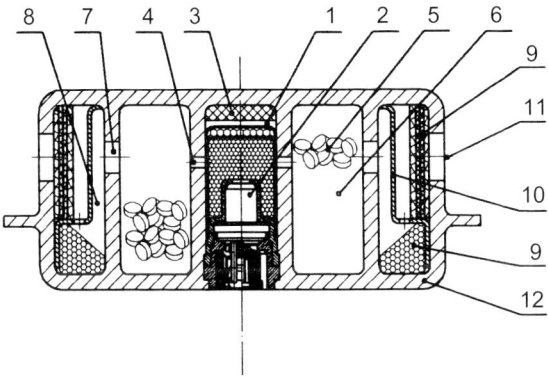

1. Ignition chamber
2. Igniter unit
3. Pre-ignition unit
4. Nozzle holes
5. Gas mixture
6. Combustion chamber
7. Nozzle holes
8. Filter chamber
9. Filter
10. Deflector plate
11. Filter chamber apertures
12. Gas generator housing

Fig. 1. Sectional diagram of a pyrotechnical gas generator for airbags

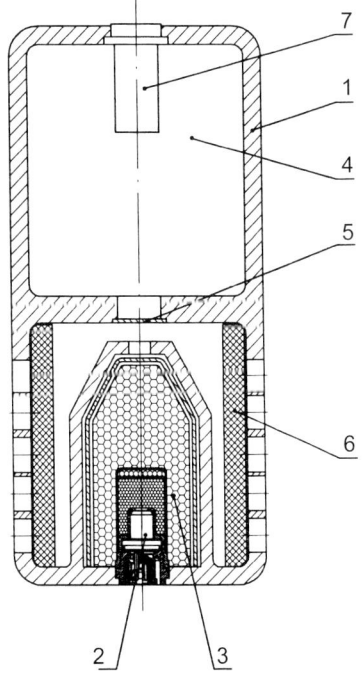

1. Hybrid generator housing
2. Igniter
3. Pyrotechnic mixture
4. High-pressure vessel
5. Sealing disk
6. Filter pack
7. Pressure measurement device

Fig. 2. Sectional diagram of a hybrid gas generator for airbags

40 ms to inflate the airbag to the specification pressure. From the mid-seventies to the mid-nineties the vast majority of gas mixtures in pyrotechnic gas generators were based on sodium azide. Sodium azide reacts with oxidising agents that bond chemically to the resulting sodium as the nitrogen is liberated. Established oxidisers include the alkali and alkaline earth nitrates, metal oxides (e.g. CuO, Fe2O3), metal sulphides (e.g. MoS2) and sulphur. If necessary, slag forming agents (e.g. SiO2, aluminosilicates) are also added.

The consequence of advances in environmental awareness is that gas mixtures containing azide are to be replaced because of the toxicity of their sodium azide, and this in spite of lower reaction temperature, purer nitrogen yield and greater long-term stability. However, one factor against sodium azide is that the correct disposal of unused gas mixtures throughout the world, which arise on a scale of thousands of tons per year, has not yet been guaranteed.

With regard to azide-free gas mixtures, there have been numerous patents and initial applications since the early nineties. These new gas mixtures generate more gas per gram (gas yields from gas mixtures containing NaN3: 0.30–0.35 l/g) and thus enable smaller and to some extent a more lightweight construction of the gas generators.

They can be classified into three categories:
1. High-nitrogen organic compounds (C, H, O, N) are combined with inorganic oxidisers:
   The fuels are, for example, 5-aminotetrazole, azodicarbonamide, → *Guanidine nitrate*, → *Nitroguanidine*, dicyandiamide, → *Triaminoguanidine nitrate* and similar compounds, as well as salts of, for example, 5-nitrobarbituric acid, urea derivatives and also nitramines and similar compounds. The oxidisers are, for example, alkali or alkaline earth nitrates, → *Ammonium*, alkali or alkaline earth perchlorates and metal oxides.
   Gas yield of these mixtures: 0.50–0.65 l/g.

2. Cellulose nitrate in combination (gelation) with nitrate esters of polyols (plus → *Stabilisers* and plasticizers), e.g. NC/NGL (→ *Nitroglycerine*) or NC/EDDN (→ *Ethylenediamine dinitrate*).
   Because of the unfavourable oxygen balance, it is necessary to secondary oxidise (e.g. with Hopcalite) to avoid excess CO formation. Despite favourable raw materials costs, the unfavourable storage stability, see below, must be noted here.
   Gas yield of the mixture: 0.8–0.9 g/l (not including the secondary oxidation).

3. High-oxygen, nitrogen-free organic compounds (C, H, O) are blended with inorganic oxidisers. The fuels used are, for example, tri or dicarboxylic acids (e.g. citric acid, tartaric acid, fumaric acid) or similar compounds. The oxidisers used are especially perchlor-

ates and chlorates with additional assistance from metal oxides. This enables any formation of NOx to be excluded.

Gas yield of the mixture: 0.5–0.6 l/g.

The gas mixtures are usually manufactured by grinding and blending the raw materials, which after a pre-compacting step are pressed into pellets or disks on (rotary table) presses, after which they are weighed out. Gas mixtures containing → Nitrocellulose are moulded after gelatinising in the usual way.

The fact that the transition from gas mixtures containing azide to ones free from azide is not simple is attributable to the following problems
(a) The considerably higher combustion temperatures impose higher demands on both the gas generator housing and on the airbag.
(b) The cooling curve of the combustion gases is steeper and must be taken into account.
(c) Condensation/filtration of the liquid/solid slag components is more difficult because of the temperature (fine dust problem).
(d) Gas mixtures containing nitrocellulose can cause difficulties in the long-term temperature test (400 hours at 107 °C, specification weight loss: < 3 %) and during temperature cycling storage (→ exudation).
(e) The long-term stability of the various azide-free gas mixtures is not yet sufficiently known.
(f) Despite an equilibrated oxygen balance, there is a tendency during the combustion of organic substances for toxic gases to be formed as by-products, although these are limited as follows:

Effluent Gas	Vehicle Level Limit	Driverside Limit
Chlorine (Cl_2)	5 ppm	1.7 ppm
Carbon monoxide (CO)	600 ppm	200 ppm
Carbon dioxide (CO_2)	20,000 ppm	6,700 ppm
Phosgene ($CoCl_2$)	1 ppm	0.33 ppm
Nitric Oxide (NO)	50 ppm	16.7 ppm
Nitrogen Dioxide (NO_2)	20 ppm	60.7 ppm
Ammonia (NH_3)	150 ppm	50 ppm
Hydrogen Chloride (HCl)	25 ppm	8.3 ppm
Sulphur Dioxide (SO_2)	50 ppm	16.7 ppm
Hydrogen Sulphide (H_2S)	50 ppm	16.7 ppm
Benzene (C_6H_6)	250 ppm	83.3 ppm
Hydrogen Cyanide (HCN)	25 ppm	8.3 ppm
Formaldehyde (HCHO)	10 ppm	3.3 ppm

In the case of the azide-free gas mixtures, there is currently no recognisable trend towards any particular fuel, since the size of the market entails a large range of variants with different requirements.

For example liquid gas generators are described in which carbon-free compounds are used and which can also be reacted to form working gases without any slag, e.g. systems consisting of hydrazine/hydrazine nitrate.

Air Blast

Druckwelle; onde de choc

The airborne acoustic or shock wave generated by an explosion
→ *Detonation*, → Fuel Air Explosives, → Thermobaric Explosives.

Air Loaders

Blasgeräte; chargeurs pneumatiques

Air loaders serve to charge prilled → *ANFO* blasting agents into boreholes. If the free-running prills cannot be charged by pouring, e.g. horizontal boreholes, boreholes with neglectable slope or boreholes with small diameters, they can be introduced by air loaders. This is done by loading the charge into a pressurized vessel and applying an air pressure of about 0,4 MPa (4 atm); a valve at the lowest point of the machine, which can be controlled from the borehole to be filled, leads to a long hose; when the valve is opened, a stream of air containing the explosive charge in suspension, is sent through it into the borehole. Other, portable machines work on the injector principle.

Akardite I

diphenylurea; Diphenylharnstoff; diphénylurée

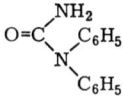

colorless crystals
empirical formula: $C_{13}H_{12}N_2O$
energy of formation: -117.3 kcal/kg -490.6 kJ/kg
enthalpy of formation: -138.2 kcal/kg -578.2 kJ/kg
oxygen balance: -233.7
nitrogen content: 13.21 %
density: 1.276 g/cm^3

Akardite I serves as → *Stabilizer* for gunpowders, in particular for → *Double Base Propellants*.

Specifications

melting point: at least	183 °C = 361°F
moisture: not more than	0.2%
ashes: not more than	0.1%
chlorides: not more than	0.02%
pH value: at least	5.0
acid, n/10 NaOH/100 g: not more than	2.0 cm^3

Akardite II

methyldiphenylurea; Methyldiphenylharnstoff; N-méthyl-N', N'diphénylurée

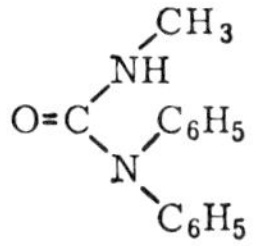

$$O=C \begin{matrix} NH-CH_3 \\ \\ N-C_6H_5 \\ C_6H_5 \end{matrix}$$

colorless crystals
empirical formula: $C_{14}H_{14}N_2O$
molecular weight: 226.3
energy of formation: -90.5 kcal/kg = -378.5 kJ/kg
enthalpy of formation: -112.7 kcal/kg = -471.5 kJ/kg
oxygen balance: -240.4%
nitrogen content: 12.38%

Akardite II is an effective → *Stabilizer* for double base gunpowders

Specifications

same as for Akardite I, except
melting point: at least 170–172 °C = 338–342°F

Akardite III

ethyldiphenylurea; Ethyldiphenylharnstoff; N-éthyl-N', N'-diphénylurée

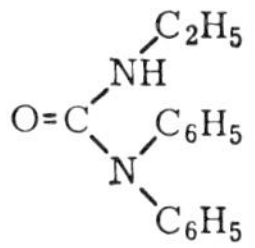

$$O=C \begin{matrix} NH-C_2H_5 \\ \\ N-C_6H_5 \\ C_6H_5 \end{matrix}$$

colorless crystals
empirical formula: $C_{15}H_{16}N_2O$
molecular weight: 240.3
energy of formation: -128.5 kcal/kg = -537.7 kJ/kg

enhalpy of formation: −151.9 kcal/kg = −635.5 kJ/kg
oxygen balance: −246.3%
nitrogen content: 11.65%

Akardite III is an effective → *Stabilizer* for double base propellants. Both Akardite II and Akardite III are gelatinizers as well as → *Stabilizers*.

Specifications

same as for Akardite I, exept
melting point: at least   89 °C = 192°F

Alginates

Salts of alginic acid which are capable of binding 200–300 times their own volume of water. They are added as swelling or gelling agents to explosive mixtures in order to improve their resistance to moisture, and to → *Slurries* to increase viscosity.

Alex

Alex is an → *aluminum powder* formed by explosion of electrically heated aluminum wires in inert atmospheres with particle sizes between 50 and 200 nm. Due to a passivation layer of a thickness between 2 to 4 nm a substantial amount of the particles is already converted to alumina the formation of which should be avoided by in-situ coating. In addition to the diffusion controlled oxidation at lower temperatures, a partial oxidation of the particles can occur by a fast chemically controlled reaction. Alex can increase the burning rate of solid composite rocket propellants up to a factor of 2. An increase of detonation velocity is not confirmed but Alex might improve → *air blast* or fragment velocities of some high explosives.

All Fire

Mindestzündstrom; ampèrage minime d'amorcage

Minimum current that must be applied to an igniter circuit for reliable ignition of the primer-charge without regard to time of operation.

Aluminum Powder

Aluminiumpulver; poudre d'aluminum

Aluminum powder is frequently added to explosives and propellants to improve their performance. The addition of aluminium results in considerable gain in heat of explosion because of the high heat of formation of aluminia (1658 kJ/mole, 16260 kJ/kg) leading to higher temperatures of the fumes. Aluminium not reacted in the detonation front might be oxidized atmospheric oxygen to induce "post-heating" in the fume zone and to increase the → *air blast* or even to initiate a delayed secondary explosion.

Widely used mixtures of explosives with aluminum powder include → *Ammonals,* → DBX, → HBX-1, → Hexal, → Minex, → Minol, → Torpex, → Trialenes, → Tritonal and *Hexotonal*. In addition underwater explosives often contain aluminium powders.

The performance effect produced by aluminum powder is frequently utilized in → *Slurries*, also in → *Composite Propellants*.

Important characteristics of aluminum powders are shape and grain size of the powder granules. Waxed and unwaxed qualities are marketed. Propellant formulations often prescribe systematically varied grain sizes for obtaining optimal densities.

Amatex

A pourable mixture of trinitrotoluene, ammonium nitrate and RDX

Amatols

Pourable mixtures of ammonium nitrate and trinitrotoluene of widely varying compositions (40:60, 50:50, 80:20). The composition 80:20 may be loaded e.g. into grenades using a screw press (extruder).

Ammonals

Compressible or pourable mixtures containing ammonium nitrate and aluminum powder; the pourable mixtures contain → *TNT*

Ammongelit 2; 3

Trade names of ammonium nitrate – nitroglycol-based gelatinous explosives distributed in Germany and exported by ORICA and WASAGCHEMIE.

Ammongelit	Density g/cm³	Weight Strength %
2	1.45	88

Ammon-Gelit TDF

Trade names of ammonium nitrate – nitroglycol-based gelatinous explosives distributed in Germany and exported by WASAGCHEMIE.

Ammon-Gelit TDF, a safehandling, gelatinous ammonium nitrate explosive, is, due to its similar properties, an alternative to → Wasag-Gelit 2 and it is manufactured without any nitrous aromatic components (→ DNT, → TNT).

Ammonium Azide

Ammoniumazid; azoture d'ammonium

$(NH_4)N_3$

colorless crystals
molecular weight: 60.1
energy of formation: +499.0 kcal/kg = +2087.9 kJ/kg
enthalpy of formation: +459.6 kcal/kg = +1922.8 kJ/kg
oxygen balance: −53.28
nitrogen content: 93.23 %
density: 1.346 g/cm³

Ammonium azide is prepared by introducing a solution of ammonium chloride and sodium azide into dimethylformamide at 100 °C.

The solvent is then drawn off in vacuum.

Owing to its high vapor pressure, this compound has not yet found any practical application.

Vapor pressure:

Pressure millibar	Temperature °C	°F
1.3	29.2	84.6
7	49.4	121.0
13	59.2	138.6
27	69.4	157.0
54	80.1	176.2
80	86.7	188.1
135	95.2	203.4
260	107.7	225.9
530	120.4	248.7
1010	133.8	272.8

Ammonium Chloride

Ammoniumchlorid; chlorure d'ammonium

NH_4Cl

colorless crystals
molecular weight: 53.49
energy of formation: −1371.6 kcal/kg = −5738.9 kJ/kg
enthalpy of formation: −1404.9 kcal/kg = −5878.1 kJ/kg
oxygen balance: −44.9 %
nitrogen content: 26.19 %
sublimation point: 335 °C = 635°F

Ammonium chloride serves as a partner component to alkali nitrates in the so-called inverse salt-pair (ion-exchanged) explosives (→ *Permitted Explosives*).

Specifications

net content: at least	99.5 %
moisture: not more than	0.04 %
glow residue: not more than	0.5 %
Ca; Fe; SO_4: NO_3: not more than	traces
pH value:	4.6−4.9

Ammonium Dichromate

Ammoniumdichromat; dichromate d'ammonium

$(NH_4)_2Cr_2O_7$

orange red crystals
molecular weight: 252.1
energy of formation: −1693.1 kcal/kg = −7083.9 kJ/kg
enthalpy of formation: −1713.1 kcal/kg = −7167.4 kJ/kg
oxygen balance: ± 0 %
nitrogen content: 11.11 %
density: 2.15 g/cm^3

Ammonium dichromate decomposes on heating, but is not an explosive. It is a component of pyrotechnical charges, and is an effective additive which is introduced into ammonium nitrate-based propellants in order to catalyze the decomposition reaction.

Ammonium dinitramide

Ammoniumdinitramid; ADN

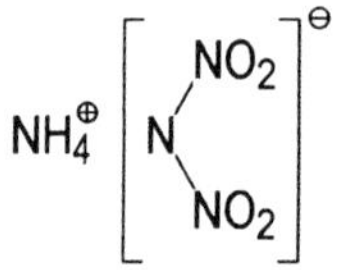

$$NH_4^{\oplus} \left[N \begin{matrix} NO_2 \\ NO_2 \end{matrix} \right]^{\ominus}$$

empirical formula: H$_4$N$_4$O$_4$
molecular weight: 124.06
energy of formation: −259.96 kcal/kg = −1086.6 kJ/kg
enthalpy of formation: −288.58 kcal/kg = −1207.4 kJ/kg
oxygen balance: +25.8 %
nitrogen content: 45.1 %
volume of explosion gases: 1084 l/kg
heat of explosion
  (H$_2$O liq.): 3337 kJ/kg
  (H$_2$O gas): 2668 kJ/kg
specific energy: 843 kJ/kg
density: 1.812 g/cm3 at 20 °C
melting point: 93.5 °C (decomposition at 135 °C and above)
impact sensitivity: 4 Nm
friction sensitivity: 64 N

Ammonium dinitramide is obtained by ammonolysis of dinitroamines, which are formed by the step-wise nitration of urethanes, β,β-iminodipropionitrile or nitramide. The last nitration step in each case requires the most powerful nitration reagents such as nitronium tetrafluoroborate or dinitrogen pentoxide. Other methods pass via the direct nitration of ammonia with dinitrogen pentoxide to a mixture of ADN and → *Ammonium Nitrate* or the nitration of ammonium sulfamate with nitric acid to a mixture of ADN and ammomium hydrogensulfate. On the basis of its good → *Oxygen Balance* and high → *Enthalpy of Formation*, ADN appears to be attractive as a halogen-free oxidising agent for solid rocket propellants and is currently the subject of intensive studies.

Ammonium Nitrate

Ammoniumnitrat; nitrate d'ammonium; AN

NH_4NO_3

colorless crystals
molecular weight: 80.0
energy of formation: −1058.3 kcal/kg = −4428.0 kJ/kg
enthalpy of formation: −1091.5 kcal/kg = −4567.0 kJ/kg
oxygen balance: +19.99 %
nitrogen content: 34.98 %
volume of explosion gases: 980 l/kg
heat of explosion
   (H_2O liq.): 593 kcal/kg = 2479 kJ/kg
   (H_2O gas): 345 kcal/kg = 1441 kJ/kg
melting point: 169.6 °C = 337.3°F
lead block test: 180 cm^3/10 g
deflagration point:
begins decomposition at melting point, complete at
   210 °C = 410°F
impact sensitivity: up to 5 kp m = 50 Nm no reaction
   friction sensitivity:
   up to 36 kp = 353 N pistil load no reaction
critical diameter of steel sleeve test: 1 mm

Ammonium nitrate is hygroscopic and readily soluble in water (the saturated solution contains about 65 % NH_4NO_3). Transitions from one polymorph to another take place at 125.2 °C, 84.2 °C, 32.3 °C and -16.9 °C. The product shows a great tendency to cake. The difficulties therefore involved are avoided by transformation into → *Prills*. Ammonium nitrate is marketed as dense prills and as porous prills. Both can be introduced in industrial explosives after milling except → *ANFO* blasting agents, which need unmilled porous prills.

Ammonium nitrate is the most important raw material in the manufacture of industrial explosives. It also serves as a totally gasifiable oxygen carrier in rocket propellants.

Phase Stabilized Ammonium Nitrate PSAN and Spray Crystallized Ammonium Nitrate SCAN are special qualities provided by ICT.

Specifications

net content (e.g. by N-determination): at least	98.5%
glow residue: not sandy, and not more than	0.3%
chlorides, as NH_4Cl: not more than	0.02%
nitrites:	none
moisture: not more than	0.15%
Ca; Fe; Mg: not more than	traces
reaction:	neutral
Abel test at 82.2 °C = 180°F: at least	30 min.
pH:	5.9+0.2
solubles in ether: not more than	0.05%
unsolubles in water: not more than	0.01%
acidity, as HNO_3: not more than	0.02%

Specifications for prills

boric acid	0.14+0.03%
density of grain: at least	1.50 g/cm^3
bulk density: at least	0.8 g/cm^3

Ammonium Nitrate Explosives

Ammonsalpeter-Sprengstoffe; explosifs au nitrate d'ammonium

Ammonium nitrate explosives are mixtures of ammonium nitrate with carbon carriers such as wood meal, oils or coal and sensitizers such as → *Nitroglycol* or → *TNT* and → *Dinitrotoluene*. They also may contain → *Aluminum Powder to* improve the → *Strength*. Such mixtures can be cap-sensitive. The non-cap-sensitive ones are classed as → *Blasting agents*.

Mixtures of porous ammonium nitrate prills with liquid hydrocarbons, loaded uncartridged by free pouring or by means of → *Air Loaders* are extensively used under the name → ANFO blasting agents.

The resistance to moisture of powder-form ammonium nitrate explosives and blasting agents is low, but can be improved by addition of hydrophobic agents (e.g. calcium stearate). The densities of the powders are about 0.9–1.05 g/cm^3.

Higher density and better water resistance are obtained using gelatinous ammonium nitrate explosives. They are based on ammonium nitrate and 20–40% gelatinized nitroglycol or a nitroclycerine-nitroglycol mixture. The German Ammongelites also contain low-melting TNT-dinitrololuene mixtures. Ammonium nitrate gelatins have widely replaced the elder sodium nitratenitroclycerine gelignites. The density of the gelatinous explosives is about 1.5–1.6 g/cm^3.

Water-containing ammonium nitrate mixtures with fuels are known as → *Slurries* and → *Emulsion Slurries*

Many permitted explosives are ammonium nitrate in powder form or gelatinous explosives with added inert salts such as sodium chloride or potassium chloride which reduce their explosion temperature.

Ammonium Nitrate Emulsion

ANE

Intermediate for → Emulsion (blasting) explosives. These emulsions are non sensitized and are intended to produce an emulsion (a blasting) explosive only after further processing prior to use. Emulsions typicallyconsist of ammonium nitrate (partly replaced by other inorganic nitrate salts), water, fuel and emulsifier agents.

Ammonium Perchlorate

Ammoniumperchlorat; perchlorate d'ammonium; APC

NH_4ClO_4

colorless crystals
molecular weight: 117.5
energy of formation: −576.5 kcal/kg = −2412.0 kJ/kg
enthalpy of formation: −601.7 kcal/kg = −2517.4 kJ/kg
oxygen balance: +34.04%
nitrogen content: 11.04%
volume of explosion gases: 799 l/kg
heat of explosion (H$_2$O liq.): 471 kcal/kg = 1972 kJ/kg
density: 1.95 g/cm^3
melting point: decomposition on heating
lead block test: 195 cm^3/10 g
deflagration point: 350 °C = 662°F
impact sensitivity: 1.5 kp m = 15 N m

Ammonium perchlorate is prepared by neutralizing ammonia by perchloric acid. It is purified by crystallization.

Ammonium perchlorate is the most important oxygen carrier for → *Composite Propellants*. Unlike alkali metal perchlorates, it has the advantage of being completely convertible to gaseous reaction products.

Table 1. Specifications

	Grade A	Grade B	Grade C
net content: at least	99.0%	99.0%	98.8%
water-insolubles:			
not more than	0.03%	0.01%	0.25%
bromates, as NH_4BrO_3:			
not more than	0.002%	0.002%	0.002%
chlorides, as NH_4Cl:			
not more than	0.15%	0.10%	0.15%
chromates, as K_2CrO_4:			
not more than	0.015%	0.015%	0.015%
iron, as Fe:			
not more than	0.003%	0.003%	0.003%
residue from sulfuric acid fuming:			
not more than	0.3%	0.3%	0.3%
moisture: not more than	0.08%	0.05%	0.08%
surface moisture:			
not more than	0.020%	0.015%	0.020%
ash, sulfated:			
not more than	0.25%	0.15%	0.45%
chlorate as NH_4ClO_3:			
not more than	0.02%	0.02%	0.02%
Na and K: not more than	0.08%	0.05%	0.08%
$Ca_3(PO_4)_2$:	none	none	0.15-0.22%
pH:	4.3–5.3	4.3–5.3	5.5–6.5

Granulation classes

Class 1 – Through 420 and 297 micron sieve, retained on 74 micron sieve.
Class 2 – Through 297 micron sieve.
Class 3 – Through 149 micron sieve.
Class 4 – 50 to 70% through 210 micron sieve.
Class 5 – Through 297 micron sieve, retained an 105 micron sieve.
Class 6 – 89 to 97% through 297 micron sieve.
Class 7 – 45 to 65% through 420 micron sieve.

Ammonium Picrate

ammonium-2,4,6-trinitrophenolate; Ammonpikrat; picrate d'ammonium; explosive D

$$\underset{NO_2}{\underset{|}{O_2N\overset{ONH_4}{\underset{|}{-C_6H_2-}}NO_2}}$$

yellow crystals
empirical formula: $C_6H_6N_4O_7$
molecular weight: 246.1
energy of formation: -355.0 kcal/kg = -1485.2 kJ/kg
enthalpy of formation: -375.4 kcal/kg = -1570.7 kJ/kg
oxygen balance: -52.0%
nitrogen content: 22.77%
volume of explosion gases: 909 l/kg
heat of explosion
 (H_2O liq.): 686 kcal/kg = 2871 kJ/kg
 (H_2O gas): 653 kcal/kg = 2732 kJ/kg
density: 1.72 g/cm^3
melting point: 280 °C = 536°F (decomposition)
lead block test: 280 cm^3/10 g
detonation velocity:
 7150 m/s = 23 500 ft/s at ρ = 1.6 g/cm^3
deflagration point: 320 °C = 608°F
impact sensitivity: at 2 kp m = 19 N m no reaction

Ammonium picrate is soluble in water, alcohol and acetone, and is practically insoluble in ether. It is prepared by saturating an aqueous solution of picric acid with ammonia; a red form is formed first which passes into the stable yellow form in the presence of water vapor, on prolonged storage or by recrystallization from water. Ammonium picrate has been employed as an explosive in military charges.

Amorces

This term denotes very small priming plates utilized in children's toys. They contain an impact-sensitive mixture of potassium chlorate and red phosphorus.

The French word "amorce" means all of initiating or inflaming devices.

Andex 1, ML

Trade names of → *ANFO* explosives marketed in Germany by MSW-CHEMIE, and ORICA Germany GmbH (formerly DYNAMIT NOBEL) in 25-kg carton packs or in containers of about 900 kg capacity. Andex 2 contains a small percentage of rock salt.

>Andex 1:
>bulk density: 0.9 g/cm^3
>weight strength: 75 %

ANFO

An abbreviation for ammonium nitrate fuel oil, a blasting agent composed of ammonium nitrate and liquid hydrocarbons. The application technique of these mixtures has now become very much easier owing to the fact that the material, which has a strong tendency to agglomeration, is commercially produced as porous prills. These are granules solidified from the liquid melt, sufficiently porous to take up about 6 % of the oil, which is the amount needed to produce oxygen balance. The nitrate, and the explosive produced from it retain their free flowing capacity. (see also → Acremite)

The explosive must be utilized in the form of a continuous column, and must be ignited by a powerful primer. This means that it must be poured loose (not as cartridges) into the borehole, or else blown into it with an → *Air Loader.*

Its manufacture is very cheap, and may even take place on continuous mixers an wheels. The material has now almost completely replaced conventional explosives in cartridge form in open-pit mining and in potash mining.

>density: 0.9 g/cm^3
>weight strength: 75 %

"Heavy Anfo" is a 50/50-mixture of Anfo and → *Emulsion Slurries* – which has higher loading densities than poured Anfo alone.

APU

(Auxiliary Power Unit) – Propellant-powered device used to generate electric or fluid power.

Aquarium Test

The parameter which is measured in this test is the pressure of underwater explosion. Lead or copper membranes are employed, and the membrane deformation as a function of the performance of the explosive and of the distance from the explosion site is estimated. The measuring apparatus, consisting of piston and anvil, resembles the *Kast* brisance meter. An alternative technique is to measure the deformation of diaphragms or copper discs accommodated inside an air-containing vessel such as a can.

In addition to the mechanical method described there are also electro-mechanical measuring techniques in which the impact pressure is recorded by an oscillograph with the aid of a piezoquartz crystal.

The measurements can be carried out in natural waters. A basin, made of steel concrete and bulkhead steel, has a buffering floor made of foamed polystyrene. Air is blown in along the bulkhead walls for damping purposes, so that an "air curtain" is formed.

→ *Underwater Detonations.*

Argon Flash

Argon-Blitz; éclair par détonation dans l'argon

The intensity of the light appearing during a detonation is caused, primarily by compression of the surrounding air.

If the air is replaced by a noble gas such as argon, the light intensity increases considerably. The duration of the flash is only as long as that of the explosion, i.e., of the order of a few μs.

The recently developed ultra-short flash lamps work on the principle of detonation of an explosive in an argon medium. They are particularly suited to the illumination of detonation processes, since the detonation moment of the lamp can be accurately adjusted to the course of the detonation of the test specimen.

The intensity of the method can be considerably increased and the explosion time considerably reduced if the shock wave travelling from the explosive through the argon medium is reflected. This can be achieved by interposing a mass, which may be very small, such as an 0.2 mm-thick acetate foil, as an obstacle. The same effect can also be produced by using curved glass, such as a watch glass.

Armor Plate Impact Test

This is a test developed in the USA to study the behavior of a given explosive, employed as charge in a projectile, on impact against hard, solid targets. The explosive is charged into the test projectile and is fired from a "gun" against a steel plate. The impact velocity which causes the charge to detonate is determined. The test description: → *Susan Test*.

Armstrong Blasting Process

This is an extraction method in the USA in coal mining. The highly Compressed (700–800 atm) air in the borehole is suddenly released by means of so-called blasting tubes equipped with bursting discs. The compressed air is generated underground by special compressors (→ also *Gas Generators*).

A similar method has received the name *Airdox*. The bursting elements in the blasting tubes have a different construction; the compressed air utilized in the method can be generated overground and distributed over a network of ducts.

ARRADCOM

US Army Armament Research and Development Command; Picatinny Arsenal Dover, New Jersey, USA

Center for research, development, approval and documentation on weapons and military materials.

ASTROLITE

Stochiometric mixture from hydrazine and ammonium nitrate. It was a spin off from the US-rocket program in the 60's. Like in 19th century Sprengel (→ *Sprengel Explosives*) already suggested, two non explosive components become cap sensitive after being mixed. Astrolite G and Astrolite A/A-1-5 (with 20% aluminium additive) are called also "liquid land mine" or "binary explosive". The explosive strength of such mixtures is very high (see also → *Hydan*) and exceeds in certain aspects even Nitroglycol. Detonation velocity of Astrolite G: 8600 m/s and Astrolite A/A-1-5: 7800 m/s.

Audibert Tube

Audibert-Rohr

This testing apparatus, which was first proposed by *Audibert* and *Delmas*, measures the tendency to → *Deflagration of* a permitted explosive. A cartridge containing the test sample is placed, with its front face open, in the tube and is packed tightly on all sides with coal dust. An incandescent spiral is placed in the cartridge opening; if the material is difficult to ignite (e.g. inverse salt-pair permissibles) the spiral is covered with a flammable igniter mixture. The tube is then closed by a perforated plate. The parameter measured is the minimum hole diameter at which the initiated deflagration arrives at the bottom of the cartridge.

In a modification of the method two cartridges placed coaxially one an top of the other are tested.

Aurol

T-Stoff; Ingolin

Concentrated (88–86%) hydrogen peroxide. It is employed in liquid fuel rocket engines as → *Oxidizer* or, after catalytic decomposition, as → *Monergol*. For its explosive properties, see *Haeuseler*, Explosivstoffe 1, pp. 6–68 (1953).

AUSTROGEL G1/G2

Austrogel G1 and Austrogel G2 are a blasting cap sensitive gelatinous explosives. They do not contain any nitroaromatic compounds (→ *DNT* or → *TNT*). The main ingredients are ammonium nitrate, nitroglycole and combustibles. Austrogel G1 and G2 may be used above and under ground. Austrogel G1 and G2 is also qualified very well as booster charge for low sensitive explosives such as → *ANFO* and other not cap sensitive explosives, and its manufactured by the Austin Powder GmbH, Austria (formerly Dynamit Nobel Wien)

	AUSTROGEL G1:	AUSTROGEL G1:
density:	1,5 g/cm^3	1,45 g/cm^3
oxygen balance:	+4,0%	+3,5%
gas volume:	891 l/kg	881 l/kg
specific energy:	1020 kJ/kg	1064 kJ/kg
velocity of detonation: (steel tube confinement 52/60/500 mm)	6000 m/s	6200 m/s

Average Burning Rate

Mittlere Abbrandgeschwindigkeit; vitesse moyenne de combustion

The arithmetic mean (statistical average) burning rate of pyrotechnic or propellants at specific pressures and temperatures. Dimension – length/time or mass/time.

Azides

Azide; azotures

Azides are salts of hydrazoic acid (N_3H). Alkali metal azides are the most important intermediates in the production of → Lead Azide.

Sodium azide is formed by the reaction between sodium amide ($NaNH_2$) and nitrous oxide (N_2O). Sodium amide is prepared by introducing gaseous ammonia into molten sodium.

Ballistic Bomb

closed vessel; ballistische Bombe; bombe pour essais ballistiques
(→ Burning Rate)

The ballistic bomb (pressure bomb, manometric bomb) is used to study the burning behavior of a → Gunpowder or → Propellant charge powder. It consists of a pressure-resistant (dynamic loading up to about 1000 MPa (10000 bar) hollow steel body that can be bolted together and has a hole to adapt a piezoelectric pressure transducer. The pressure p in the bomb is measured as a function of time t.

As a rule, studies of powder in the pressure bomb are carried out in comparison with a powder of known ballistic performance. They are very useful both in the development of powders and in production monitoring.

If the dynamic liveliness L ($= 1/p_{max} * d\ln p/dt$) is determined as a function of p/p_{max} from the primary measured signal, then for a defined powder geometry the parameters characterising its burn-up, the linear burning rate $\dot{e}$ (→ Burning Rate) and the pressure exponent α can be determined. Pressure bomb shots of the same powder at different charge densities δ (= mass m_c of powder/volume V_B of the pressure vessel) enable the specific covolume η of the combustion gases from the powder and the force f (powder force) of the powder to be determined in addition. From these, if the → *Heat of Explosion* Q_{Ex} of the powder is known, the value of the average adiabatic coefficient $æ$ ($= 1 + f/Q_{Ex}$) of the combustion gases, which is of interest for the ballistic performance, can be derived.

Since the combustion gases of powders satisfy Abel's equation of state to a good approximation, it is possible by using the auxiliary parameters (ρ_c) density of the powder)

$\Delta := m_c/(V_B * \rho_c)$ 'normalised charge density'   (1)

$\chi := (1 - \eta\rho_c) * \Delta/(1 - \Delta)$ 'real gas correction term'   (2)

$\Phi := f\rho_c\Delta/(1 - \Delta)$ 'characteristic pressure'   (3)

to write the relationship between the pressure p in the manometric bomb and the burnt volume proportion z of the powder as

$$z(p/p_{max}) = p/p_{max}/\{1 + \chi(1 - p/p_{max})\} \tag{4}$$

and

$$p(z) = \Phi * z / (1 + \chi z). \tag{5}$$

Accordingly, the maximum gas pressure achieved at the end of burn-up ($z = 1$) is calculated as

$$p_{max} = \Phi / (1 + \chi). \tag{6}$$

The dynamic liveliness L is calculated from

$$L = \frac{S(0)}{V(0)} * \varphi(z) * \frac{\dot{e}(p_{ref})}{p_{ref}} * \left[\frac{p}{p_{ref}}\right]^{\alpha-1} * \frac{1 + \chi}{(1 + \chi z)^2} \tag{7}$$

$S(0)/V(0)$ is the ratio of the initial surface area to the initial volume of the powder,

$\varphi(z)$ is the shape function of the powder, which takes account of the geometrical conditions (sphere, flake, cylinder, N-hole powder) during the burn-up ($\varphi(z)$ = current surface area / initial surface area)

$\dot{e}(p_{ref})$ is the linear burning rate at the reference gas pressure p_{ref}

p_{ref} is the reference gas pressure and

α is the pressure exponent, which for many powders is close to 1.

To evaluate Eq. (7), z should be replaced by p/p_{max} using Eq. (4).

Figure 3 shows the time profile of the pressure in the manometric bomb for a typical 7-hole powder. Initially the pressure is increasingly steep, since burn-up takes place more quickly the higher the pressure and in addition the burning surface of the powder becomes greater as

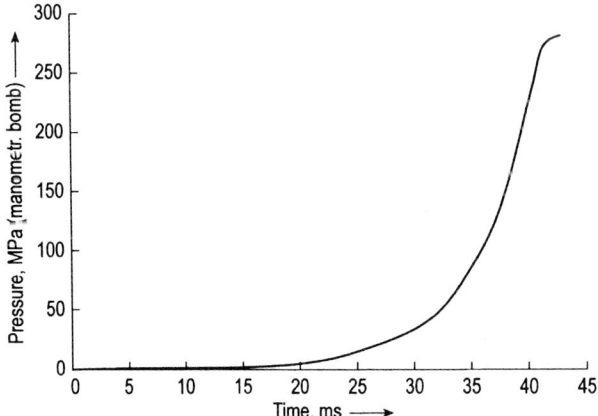

Fig. 3. Pressure-time graph p = f(t)

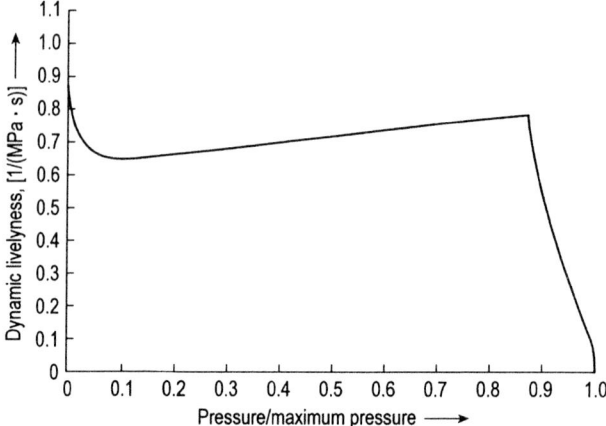

Fig. 4. Dynamic liveliness as a function of p/p_{max}

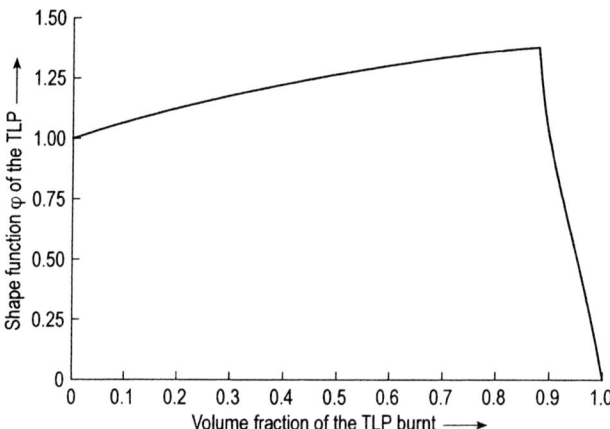

Fig. 5. Shape function pf the powder as a function of the current surface area relative to the initial surface area

the burn-up progresses (progressive burn-up). Towards the end of the burn-up the pressure profile levels out rapidly because the burning surface area of the powder becomes drastically smaller as soon as approx. 88% of the powder has been burnt.

Figure 4, which shows the calculated profile of the dynamic liveliness as a function of p/p_{max}, again reflects essentially the shape of the form function for $p/p_{max} > 0.2$ (see Fig. 5). On the other hand for small values of p/p_{max}, the dependence on $p^{\alpha-1}$ resulting for $\alpha = 0.9$ is dominant. The kink in the shapes of the form function and the dynamic *liveliness at* p/p_{max} = 0.87 (disintegration of the powder granules into slivers) is greatly rounded off in the measured curves, because not all of the granules burn up at exactly the same time and small differences in geometry always arise (manufacturing tolerances)

Ballistic Mortar

ballistischer Mörser, mortier ballistique

An instrument for comparative determinations of the performance of different explosives. A mortar, provided with a borehole, into which a snugly fitting solid steel projectile has been inserted, is suspended at the end of a 10 ft long pendulum rod. Ten grams of the explosive to be tested are detonated in the combustion chamber. The projectile is driven out of the mortar by the fumes, and the recoil of the mortar is a measure of the energy of the projectile; the magnitude determined is the deflection of the pendulum. This deflection, which is also known as weight strength, is expressed as a percentage of the deflection produced by blasting gelatine, arbitrarily taken as 100. Also, relative values referring to the deflection produced by TNT are listed, especially for explosives of military interest.

This method, which is commonly employed in English-speaking countries, and which is suited for the experimental determination of the

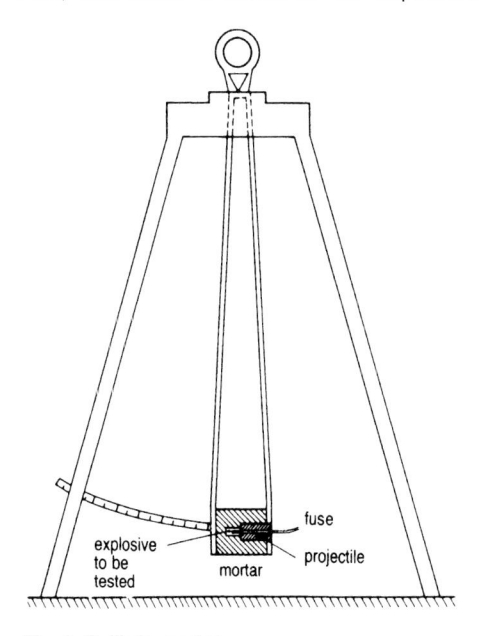

Fig. 6. Ballistic mortar.

work performed by the explosive, has now been included in the list of standard tests recommended by the European Commission for the Standardization of Explosive Testing.

An older comparison scale is "grade strength", which determines the particular explosive in standard "straight" dynamite mixtures (the mixtures contain ungelatinized nitroglycerine in different proportions, sodium nitrate and wood or vegetable flour (→ *Dynamites*) which gives a pendulum deflection equal to that given by the test material. The percentage of nitroglycerine contained in the comparitive explosive is reported as grade strength.

The grade strength percentage is not a linear indicator of the performance of the explosive; the performance of a 30% dynamite is more than half of the performance of a 60% dynamite, because the fuel-oxidizer mixtures as well as nitroglycerine also contribute to the gas- and heat-generating explosive reaction.

For comparison of weight strength values with other performance tests and calculations → *Strength*.

Ball Powder

Kugelpulver, Globularpulver; poudre sphérique

Ball powder is a propellant with ball-shaped particles, produced by a special method developed by *Mathieson* (USA). A concentrated solution of nitrocellulose in a solvent which is immiscible with water (e.g., ethyl acetate) is suspended in water by careful stirring, so that floating spheres are formed. The solution is warmed at a temperature below the boiling point of the solvent, and the latter gradually evaporates and the floating spheres solidify.

Since the spherical shape is unfavorable from internal ballistical considerations (very degressive), follows, a thorough → *Surface Treatment*, the purpose of which is to sheathe the faster-burning core by a slower-burning shell.

BAM

Bundesanstalt für Materialforschung und -prüfung

Unter den Eichen 87, D-12200 Berlin (www.bam.de)

Federal Institute for Materials Research and Testing (including explosives). BAM sensitivity tests: → *Friction Sensitivity* → *Heat Sensitivity* and → *Impact Sensitivity.*

BAM is the German competent Authority for the dangerous goods of Class 1 and Notified Body in Compliance with EU Directive 93/15.

Baratols

Pourable TNT mixtures with 10–20% barium nitrate.

Barium Chlorate

Bariumchlorat; chlorate de barium

$Ba(ClO_3)_2 \cdot H_2O$

colorless crystals
molecular weight: 322.3
energy of formation: −789.3 kcal/kg = −3302.3 kJ/kg
enthalpy of formation: −799.4 kcal/kg = −3344.6 kJ/kg
oxygen balance: +29.8%
density: 3.18 g/cm^3
melting point: 414 °C = 779°F

Barium chlorate and → *Barium Perchlorate* are used in pyrotechnical mixtures using green flames.

Barium Nitrate

Bariumnitrat; nitrate de barium: BN

$Ba(NO_3)_2$

colorless crystals
molecular weight: 261.4
energy of formation: −898.2 kcal/kg = −3758.1 kJ/kg
enthalpy of formation: −907.3 kcal/kg = −3796.1 kJ/kg
oxygen balance: +30.6%
nitrogen content: 10.72%
density: 3.24 g/cm^3
melting point: 592 °C = 1098°F

component in green flame pyrotechnicals and In ignition mixtures (with → *Lead Styphnate*). For the specification see Table 1.

Barium Perchlorate

Bariumperchlorat; perchlorate de barium

$Ba(ClO_4)_2 \cdot 3H_2O$

colorless crystals
molecular weight: 390.3
oxygen balance: +32.8%
density: 2.74 g/cm^3
melting point: 505 °C = 941°F

Table 2. Specifications

	Class 1	Class 2	Class 3	Class 4	Class 5	Class 6
net content by nitrogen analysis: at least	99.7%	99.0%	99.5%	99.5%	98.5%	99.5%
Sr: not more than	0.6%	–	0.6%	–	–	0.6%
Ca: not more than	0.05%	–	0.05%	–	–	0.05%
$Al_2O_3 + Fe_2O_3$: not more than	–	0.50%	–	–	–	–
Na, as Na_2O: not more than	–	0.15	–	–	–	0.15
Chloride, as $BaCl_2$, not more than	0.0075%	0.0075%	0.0075%	0.0075%	–	0.0075%
grit: not more than	0.05%	0.05%	0.05%	0.05%	–	0.05%
Fe and other metals	none	none	none	none	none	none
moisture: not more than	0.20%	0.10%	0.20%	0.20%	0.05%	0.10%
pH:	5.0–8.0	5.0–8.0	5.0–8.0	–	5.0–8.0	5.0–9.0
insoluble matter: not more than	0.1%	0.1%	0.1%	0.1%	0.1%	0.1%

An oxidizer in propellant formulations and for → *Pyrotechnical Compositions*.

Barricade

Schutzwall; merlon, écran

Barricades are grown-over earth embankments erected for the protection of buildings which may be endangered by an explosion. The overgrown height of the barricade must be at least one meter above the building to be protected. The required safety distances between explosive manufacture buildings or storage houses can be halved if the houses are barricaded.

Base-Bleed Propellants

Gas generating elements inserted in the bottom of projectiles. The generated gas fills the subathmosperic pressure behind the projectile.

Base Charge

Sekundärladung; charge de base de détonateur

The main explosive charge in the base of a blasting cap, an electric blasting cap, or a non-electric delay cap.

Bazooka

A shaped-charge anti-tank weapon first used by the Americans in the Second World War; → *Shaped Charges*. Its operating method is identical with that of the "Panzerfaust" developed in Germany at that time.

B-Black Powder

Sprengsalpeter; poudre noir au nitrate de soude

is a → *Black Powder* mixture which contains sodium nitrate instead of potassium nitrate. It is marketed and utilized in the form of compressed cylinder-shaped grains, 25 and 30 mm of diameter, with a central hole 5 mm in diameter.

Bengal Fireworks

→ *Pyrotechnical Compositions*.

Benzoyl Peroxide

Benzoylperoxid; peroxyde de benzoyle

colorless crystals
empirical formula: $C_{14}H_{10}O_4$
molecular weight: 242.1
oxygen balance: -191.6%
melting and deflagration point: 107 °C = 225°F
impact sensitivity: 0.5 kp m = 5 N m
friction sensitivity: at 12 kp = 120 N pistil load
decomposition: at 24 kp = 240 N pistil load crackling
critical diameter of steel sleeve test: 10 mm

Benzoyl peroxide is sparingly soluble in water and alcohol, but soluble in ether, benzene and chloroform. It can be prepared by reaction of benzoyl chloride with sodium peroxide.

The explosion strength of the product is low, but its sensitivity relatively high.

The organic peroxides serve as catalysts for polymerization reactions. They must be wetted or phlegmatized (→ *Phlegmatization*) for transportation and handling.

Benzoyl peroxide can also be used as a bleaching agent for oils and fats.

Bergmann-Junk Test

A method, developed by *Bergmann* and *Junk in* 1904, for testing the chemical stability of nitrocellulose; it was also subsequently employed for testing single-base powders. The test tube, which contains the specimen being tested, and which is equipped with a cup attachment, is heated at 132 °C = 270.4°F for two hours (nitrocellulose) or five hours (single base powders). At the end of the heating period the sample is extracted with water, and the test tube filled to the 50-ml mark with the water in the cup. The solution is filtered, and the content of nitrous oxides is determined by the *Schulze-Tiemann* method on an aliquot of the filtrate.

The main disadvantage of the method is that nitrous compounds are only incompletely absorbed in water, especially since the atmospheric oxygen which has remained behind in the tube is expelled during heating or is displaced by the carbon dioxide evolved at the powder surface. Moreover, the results vary with the volume of the specimen employed, since differing volumes of water are required to fill the tube up to the mark in gelled and porous powders.

Siebert suggested the use of H_2O_2 rather than water as the absorption medium in 1942. He also suggested that the employed apparatus should be redesigned, to avoid gas losses which occur when the cup attachment is taken off. In the new design, the cup is replaced by a large (over 50 ml) attachment resembling a fermentation tube, which need not be taken off during the extraction of the sample. In this way quantitative determination of the liberated No, even in large amounts, becomes possible.

Siebert also suggested that the total acidity be determined by titration against N/100 NaOH, in the presence of Tashiro's indicator. In this manner → *Double Base Propellants* can also be tested as well; the test is carried out at 115 °C, the duration of heating being 8 or 16 hours depending on the nitroglycerine content of the sample (or of similar products, e.g. → *Diethyleneglycol Dinitrate*).

Bichel Bomb

Bichel-Bombe; bombe Bichel

Used to study the composition and → *Volume of Explosion Gases*. It consists of a heavy steel case sealed by a screw cap. The construction withstands the dynamic shock of a detonating explosive sample. The gas developed can be vented by a valve in the screw cap for measurement of volume and for gas analysis.

The → *Lead Block Test* has been used for the same purpose: the block is sealed hermetically by a plug, and held in position by a steel construction. After detonation of the explosive sample in the block, the gas content has been vented by a special sealed drilling tool.

For the evaluation of the specific gas volume by computing → *Thormo dynamic Calculation of Decomposition Reactions*.

BICT

Bundesinstitut für Chemisch-Technische Untersuchungen

German Federal institute for testing of and research on propellants and explosives for military purposes.

Since 01. 04. 97 the name of this institute has been changed "Wehrwissenschaftliches Institut für Werk-, Explosiv- und Betriebsstoffe (WIWEB)" (= Defence Scientific for Construction Materials, Explosives and Operating Materials).

Billet

Monolithic charge of solid propellant of any geometry; term usually applied to a formed propellant prior to final shaping (→ *Grain*).

Binder

Compositions that hold together a charge of finely divided particles and increase the mechanical strength of the resulting propellant grain when it is consolidated under pressure. Binders are usually resins, plastics, or asphaltics, used dry or in solution (→ *Energetic Binders*).

BITA

Abbreviation denotation for an aziridine curing agent in → *Composite Propellants;* it has the following structure:

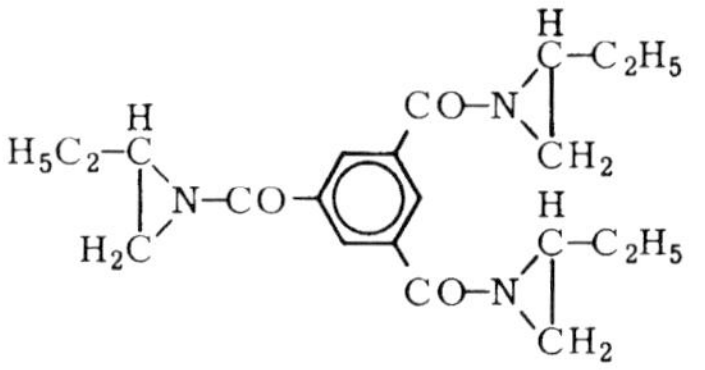

empirical formula: $C_{21}H_{27}O_3N_3$
molecular weight: 369.24
density: 1.00 g/cm^3

Bi-trinitroethylnitramine

Di (2,2,2-trinitroethyl)nitramin; di-trinitroéthylnitramine; BTNENA, HOX = High Oxygen Explosive

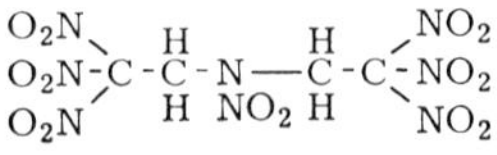

empirical formula: $C_4H_4N_8O_{14}$
molecular weight: 388.1
energy of formation: +2.8 kcal/kg = +11.9 kJ/kg
enthalpy of formation: −17.0 kcal/kg = −71.2 kJ/kg
oxygen balance: +16.5 %
nitrogen content: 28.80 %
volume of explosion gases: 693 l/kg

heat of explosion
(H₂O liq.): 1299 kcal/kg = 5436 kJ/kg
(H₂O gas): 1248 kcal/kg = 5222 kJ/kg

and

Bi-trinitroethylurea

Di (2,2,2-trinitroethyl)-Harnstoff; di-trinitroéthylurée; BTNEU

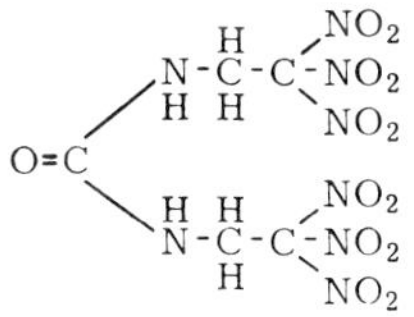

$$O=C \begin{cases} N-\underset{H}{\overset{H}{C}}-C-NO_2 \\ \overset{H}{}NO_2 \\ N-\underset{H}{\overset{H}{C}}-\overset{H}{C}-NO_2 \\ NO_2 \end{cases}$$

empirical formula: $C_5H_6N_8O_{13}$
molecular weight: 386.1
energy of formation: −178.5 kcal/kg = −746.7 kJ/kg
enthalpy of formation: −199.2 kcal/kg = −833.2 kJ/kg
oxygen balance: ±0%
nitrogen content: 29.02%
volume of explosion gases: 697 l/kg
heat of explosion
(H₂O liq.): 1543 kcal/kg = 6454 kJ/kg
(H₂O gas): 1465 kcal/kg = 6131 kJ/kg
specific energy: 114 mt/kg = 1119 kJ/kg

are derivatives of trinitroethylalcohol, addition product of → *Trinitromethane* and formaldehyde.

Black Powder

Schwarzpulver; poudre noire

Black powder is a mechanical mixture of potassium nitrate, sulfur and charcoal, which is mostly pressed, granulated and classified into definite grain fractions. It faster deflagrates than it detonates; it is thus classified as a "low" explosive, compared to the detonating "high" explosives.

The standard composition is: 75% potassium nitrate, 10% sulfur and 15% charcoal. There are also graded compositions containing 74, 70, 68 or 64% potassium nitrate. Corresponding compositions based on sodium nitrate are known as → *B-Black Powder*.

The starting components are finely ground, mixed and compacted in rolling mills and then pressed into cakes in hydraulic presses. The cakes are then broken and grain-classified; the resulting granules are polished with the application of graphite.

When in granulated form, black powder can be freely poured into boreholes.

Black powder is sensitive to impact, friction, and sparks. It is suitable for controlled blastings in which the treatment of stone must be mild – e.g., in the manufacture of roofing slates, and in quarrying for paving stones.

It is employed in safety fuses, in pyrotechnics and in priming charges for smokeless powders. It is still the only suitable explosive for many purposes. It rapidly builds up pressure in relatively wear confinement. It does not detonate under normal conditions; the maximum rate of the explosion is about 500 m/s.

Blast Area

Sprengbereich (Absperrzone); chantier de tir

The area of direct blast impact, in addition adjacent areas are added which could be hit by flying debris (rocks etc.).

Blaster

Sprengmeister; boutefeu

That qualified person in charge of, and responsible for, the loading and firing of a blast (same as *shot firer*).

Blasting Accessories

Sprengzubehör; accessoires pour sautage

Non-explosive devices and materials used in blasting, such as, but not limited to, cap crimpers, tamping bags, → *Blasting Machines, Blasting Galvanometers*, and cartridge punches.

Blasting Agents

The notion of a blasting agent was conceived in the USA. Contrary to high explosives, which may contain, say, nitroglycerine, and which are sensitive to blasting caps, the term "blasting agents" denotes relatively low-sensitive explosives, usually based on ammonium nitrate, which are insensitive to blasting caps and do not contain any high explosives such as nitroglycerine or TNT. In many countries (not Germany) the safety regulations governing the transport and storage of blasting agents are considerably less severe than those applicable to high explosives. NCN (nitricarbonitrate) is designated in the USA as an ammonium nitrate non-cap-sensitive explosive. The components are named by *nitro:* dinitrotoluene; by *carbo:* solid carbon carriers as fuel; by *nitrate:* ammonium nitrate. Meanwhile, NCN as a shipping name has been removed by the US Department of Transportation and replaced by the shipping name "Blasting Agent". A blasting agent has to be non-capsensitive (→ *Cap Sensitivity*). → *ANFO* explosives and most of → *Slurries* have to be classified as blasting agents.

Blasting Caps

Sprengkapseln; détonateurs

Blasting caps serve as initiators of explosive charges. They consist of a cylindrical copper or aluminum capsule containing a primary charge of an initiating explosive or a mixture of initiating explosives (e.g. lead azide with lead trinitroresorcinate); in order to achieve a higher brisance, they also contain a secondary charge of a highbrisance explosive (e.g. → *Tetryl;* → *PETN;* → *Hexogen*).

A blasting cap can be ignited by the flame of a safety fuse or electrically. In the past, 10 standard types of blasting caps were marketed; these differed from each other by the quantity of the explosive in the charge and by their size. Currently, No. 8 blasting cap (0.3 g primary charge. 0.8 g secondary charge, 4–50 mm in length and 7.0 mm in external diameter) is, for all practical purposes, the main type of blasting cap on the market.

Blasting Galvanometer

→ *Circuit Tester.*

Blasting Gelatin

Sprenggelatine; dynamite-gomme

This product is one of the strongest commercial explosives. It consists of 92–94 % nitroglycerine, gelatinized with 6–8 % soluble guncotton.

Since such a high explosive strength is rarely required, blasting gelatin is scarcely ever used in practice.

Blasting gelatin is used as a comparitive explosive in determinations of relative weight strength (→ *Ballistic Mortar*).

Blasting Machines

Zündmaschinen; exploseurs

Blasting machines are used for electric firing of explosive charges by sending an electric pulse (indicated in mW·s/ohm) through the firing circuit to the round of electric detonators connected in series. Except during the moment of actuation of the blasting machine, the entire electrical system is tensionless (unlike: → *Blasting Switch*).

In mines endangered by a potential firedamp explosion, the duration of the electric pulse must be limited to 4 ms with the aid of a triggering switch in the blasting machine, so that flying fragments cannot strike the firing circuit while the latter is still live, and then generate a short-circuit spark. Also, the housing must withstand an internal pressure of 1 MPa (10 atm), so that it cannot be destroyed by a burst due to intruded methane. These special conditions are only requested in blasting areas endangered by firedamp.

Two types of blasting machines are used:

1) blasting machines with direct energy supply, equipped with a self-induction or a permanent magnet generator, which are made to rotate with the aid of a twist knob, impact knob or a spring extension, and

2) blasting machines with an indirect energy supply, in which the generated electrical energy is stored in a capacitor and, after the discharge voltage has been attained, the breakthrough pulse is sent to a blasting train ("CD Type"). A misfire due to incorrect handling is impossible.

Capacitor machines have now superseded direct-generation machines. In order to set off → *Bridgewire Detonators*, which are connected in parallel, the output of the machines must be particularly high, since more than 95% of the electric energy becomes lost in the blasting circuit. Special powerful machines are required to set off "HU"-(highly unsensitive) detonators for blastings carried out in high mountain areas and in other locations endangered by high-voltage induction; a very strong (3000 mW · s/ohm) priming pulse must be applied in such cases.

→ *Bridgewire Detonators.*

Blasting Mat

Sprengmatte; réseau de fils d'acier

A mat of woven steel wire, rope, scrap tires, or other suitable material or construction to cover blast holes for the purpose of preventing rock missiles against flying debris.

Blasting Switch

Zündschalter; commande de tir; ignition switch

Device which actuates electric primers by using main voltage – in openpit and potash mining, for example. The switch can be located in a surface stand(shelter outside a mine) e.g. if the danger of gas outbursts exists.

Blastmeter

Blastmeters are simple devices which are used to determine the maximum pressure of a shock wave (→ p. 78) They consist of steel bodies into which holes of different diameters are drilled and convered with aluminum foil.

The smallest diameter is determinted at whitch the foil covering is penetrated. The device can be calibrated by static pressure.

Bomb Drop Test

Serves to test the sensitivity of military explosives as bomb fillers. Bomb drops are made using bombs assembled in the conventional manner, as for service usage, but containing either inert or simulated fuzes. The target is usually reinforced concrete.

Boom Powder

A pyrotechnic ignition mixture designed to produce incandescent particles. A typical boom composition is:

Ingredient	Parts by Weight %
Iron Oxide	50
Titanium (powdered)	32.5
Zirconium (powdered)	17,5

plus about 1 part of cellulose nitrate as a binder.

Booster

Verstärkungsladung; relais

A device to ensure → *Initiation*. A booster can be a cap-sensitive cartridge or press molded cylinder for the initiation of non-cap-sensitive charges, e.g. blasting agents or cast TNT. A booster is, in rocketry, a rocket device that accelerates the missile to attain the required speed after the start.

Booster Sensitivity Test

The booster sensitivity test procedure is a scaled-up modification of the *Bruceton Test* (unconfined charge). The source of the shock consists of two Tetryl pellets, each 1.57 inches in diameter and 1.60 inches long, of approximately 100 g total weight. The initial shock is degraded through wax spacers of cast Acrawax B, 1–5/8 inches in diameter. The test charges are 1–5/8 inches diameter by 5 inches long. The value given is the thickness of wax in inches at the 50 % detonation point. The weight of the Tetryl pellet noted is the minimum which will produce detonation with the spacer indicated.

Bootleg

Bohrlochpfeife; trou ayant fait canon

That part of a drilled blast hole that remains when the force of the explosion does not break the rock completely to the bottom of the hole (→ *Large Hole Blasting*).

Boss

Messanschlüsse; raccords de mesurage

Outlets provided in the generator case for hot gas flow, igniter, pressure measurement, and safety diaphragm.

Break

Spalt; fente

Cleft in the rock formation, especially in coal mines, which endangers blasting in fire damp areas; → *Permitted Explosives*.

Breech

Patronenkammer; chambre pour cartouche

Reloadable pressure vessel used to contain a propellant cartridge.

Bridgewire Detonator

Brückenzünder; amorce à pont

Bridgewire detonators are used in industrial blasting operations for the initiation of explosive charges. They contain an incandescent bridge made of thin resistance wire, which is made to glow by application of an electric pulse. An igniting pill is built around the wire by repeated immersion in a solution of a pyrotechnical material followed by drying. The igniting flash acts directly onto the detonating surface in the case of instantaneous detonators; in delayed-action detonators it is sent over a delay device onto the detonating surface of a blasting cap which has been pressed onto the detonating pill so as to produce a water-tight bond with it. Non-armed bridgewire detonators have an open casing, into which a blasting cap may be inserted.

The "U"-detonators which are now employed in mining in the Germany need a pulse of 16 mW·s/ohm; the earlier detonators required only 3 mW·s/ohm. Thus new detonators afford much better protection against stray currents. Locations exposed to electrostatic stray charges (thunderstorms) and which are therefore particularly dangerous, are equipped with low-sensitivity detonators, which require as much as 2500 mW·s/ohm for actuation and may therefore be considered safe ("HU"-detonators).

The delayed-action detonators may be set for a delay of half a second (half-second detonators) or for a delay of 2–34 ms (millisecond detonators). Blasting with the latter type of detonators results in a larger yield of blasted stone fragments; moreover, a smaller shock will be imparted to the ground around the explosion site.

In coal mining only copper casings rather than the conventional aluminum casings are permitted because of the danger of firedamp. Explosive charges equipped with bridgewire detonators are fired by wire-connected → *Blasting Machines* from a safe location. If several charges have to be initiated at the same time, the detonators are connected in series with the connecting wire. Parallel connection of the detonators is used only in special cases (extremely wet conditions with danger of shunting); special blasting machines must be employed for this purpose.

Brisance

Brisanz

The performance of an explosive cannot be expressed by means of a single characteristic parameter. Brisance is the destructive fragmentation effect of a charge on its immediate vicinity. The relevant parameters are the detonation rate and the loading density (compactness) of the explosive, as well as the gas yield and the heat of explosion. The higher the loading density of the explosive (molding or pressing density), the higher its volume specific performance; also, the faster the reaction rate, the stronger the impact effect of the detonation. Moreover, an increase in density is accompanied by an increase in the detonation rate of the explosive, while the shock wave pressure in the detonation front (→ *Detonation*) varies with the square of the detonation rate. Thus it is very important to have the loading density as high as possible.

This is particulary true for → *Shaped Charges*.

Kast introduced the concept of "brisance value", which is the product of loading density, specific energy and detonation rate.

Brisance tests are upsetting tests according to *Kast* and *Heß*; the compression of a copper cylinder is determined by actuating a piston instrument; alternatively, a free-standing lead cylinder is compressed by the application of a definite cylindrical load of the explosive being tested: → *Upsetting Tests*.

Bulk Density

Schüttdichte; densité apparente

The mass per unit volume of a bulk material such as grain, cement, coal. Used in connection with packaging, storage or transportation.

Bulk Mix

Sprengstoffmischung für unpatronierte Anwendung; explosif en vrac

A mass of explosive material prepared for use in bulk form without packaging.

Bulk Mix Delivery Equipment; Misch-Lade-Fahrzeug; véhicule mélangeur-chargeur

Equipment (usually a motor vehicle with or without a mechanical delivery device) that transports explosives, blasting agents or ingredients for explosive materials in bulk form for mixing and/or loading directly into blast holes.

Bulk Strength

Cartridge Strength: Volume Strength

The strength per unit volume of an explosive calculated from its → *Weight Strength* and → *Density*.

Bulldoze

Auflegeladung; pétardage

A mud covered or unconfined explosive charge fired in contact with a rock surface without the use of a bore hole. Synonymous with *Adobe Charge* and → *Mud Cap*.

Bullet Hit Squib

Filmeffektzünder; Squib

Bullet hit Squibs are used in motion pictures and television to simulate ballistic impact of fired projectiles.

What is refered to here are small, pyrotechnic, electrical devices with varying charges and containing several milligrams of a compound consisting of → *Lead Azide,* → *Lead Styphnate,* → *Diazodinitrophenol* and Tetrazole Derivatives.

The initiating explosive material must be specially treated and phlegmatized to avoid the undesired byproduct of smoke and flash. One method achieves this by using an admixture of alkaline earth sulfates or by means of micro-encapsulation of the explosive crystals.

These special electrical igniters are produced by the company J. Köhler Pyrotechnik in Schardenberg/Austria (www.pyrochemie.at).

Bullet-resistant

Kugelsicher; résistant au balles

Magazine walls or doors of construction resistant to penetration of a bullet of 150-grain M2 ball ammunition having a nominal muzzle velocity of 2700 feet per second fired from a .30 caliber rifle from a distance of 100 feet perpendicular to the wall or door.

When a magazine ceiling or roof is required to be *Bullet-Resistant*, the ceiling or roof shall be constructed of materials comparable to the side walls or of other materials which will withstand penetration of the bullet above described when fired at an angle of 45 degrees from the perpendicular.

Tests to determine bullet resistance shall be conducted on test boards or empty magazines which shall resist penetration of 5 out of 5 shots placed independently of each other in an area at least 3 feet by 3 feet. If hardwood or softwood is used, the water content of the wood must not exceed 15%.

Bullet-sensitive Explosive Material

Beschussempfindlicher Sprengstoff; explosif sensible a l'impact de balles

Explosive material that can be detonated by 150-grain M2 ball ammunition having a nominal muzzle velocity of 2700 feet per second when the bullet is fired from a .30 caliber rifle at a distance of not more than 100 feet and the test material, at a temperature of 70 ° to 75°F., is placed against a backing material of 1/2-inch steel plate.

(→ *Impact Sensitivity.*)

Burden

Vorgabe; distance entre 1a charge et la surface du massif

That dimension of a medium to be blasted measured from the borehole to the face at right angles to the spacing. It means also the total amount of material to be blasted by a given hole, usually measured in cubic yards or in tons.

Bureau of Alcohol, Tobacco and Firearms (BATF)

A bureau of the (US-)Department of the Treasury having responsibility for the enactment and enforcement of regulations related to commerce in explosives under Part 181 of Title 26 of the Code of Federal Regulations.

Bureau of Explosives

A bureau of the Association of American Railroads which the U.S. Department of Transportation may consult to classify explosive material for the purposes of interstate transportation.

Bureau of Mines

→ *U.S. Bureau of Mines.*

Bureau of Mines Test

→ *Impact Sensitivity.*

Burning Rate

Abbrandgeschwindigkeit; velocity of combustion; vitesse de combustion

The linear burning rate of a propellant is the velocity with which a chemical reaction progresses as a result of thermal conduction and radiation (at right angles to the current surface of the propellant). It depends on the chemical composition, the pressure, temperature and physical state of the propellant (porosity; particle size distribution of the components; compression). The gas (fume) cloud that is formed flows in a direction opposite to the direction of burning.

The burning rate describes the velocity with which the volume of the burning propellant changes. It is proportional to the linear burning rate and in addition it depends on the specific shape of the propellant (size of the powder elements and conformation, e. g. flakes, spheres, tubes, multi-perforated tubes etc. extending to the most complicated shapes of rocket propellant charges).

In rocket engineering, "Burning rate" means specifically the stationary progress of burning rate in the rocket chamber.

The following relationship exists between the burning rate dz/dt and the linear burning rate $\dot{e}$:

$$\frac{dz}{dt} = \frac{S(0)}{V(0)} \times \varphi(z) \times \dot{e}$$

where $\dot{e}$ is given by

$$\dot{e} = \dot{e}(p_{ref}) \times \left(\frac{p(z)}{p_{ref}}\right)^{\alpha}$$

z means the ratio of the volume burnt to that originally present $[V(0) - V] / V(0)$

$S(0)/V(0)$ means the ratio of the initial surface area to the initial volume of the powder,

$\varphi(z)$ means the shape function of the powder, which takes into account the geometrical conditions during burning rate (sphere, flake, cylinder, n hole powder) ($\varphi(z)$ = current surface area/initial surface area)

$\dot{e}(p_{ref})$ means the linear burning velocity at the reference gas pressure p_{ref}

p_{ref} is the reference gas pressure and

α is the pressure exponent.

The equation for the burning rate rate dz/dt can also be written in the form

$$\frac{dz}{dt} = A \times \varphi(z) \times p^{\alpha}$$

and is then called Charbonnier's Equation.

The parameter $A = (S(0) \times V(0)) \times \varphi(z) \times \dot{e}(p_{ref})/p_{ref}^{\alpha}$ is called the "vivacity" or "quickness" factor".

The pressure exponent α typically has a value close to 1 for propellant charge powder (burning rate at high pressure level). At low pressure ranges (rocket burning rate) it can be brought close to zero ("plateau burning rate") or even less than zero ("mesa burning rate") by suitable additives to the propellant.

When the geometry of the propellant is known, the linear burning rate and the pressure exponent of a propellant can be determined experimentally in a → *ballistic bomb*.

If the gases flow continously out, as in the case of a rocket motor, the pressure remains almost constant throughout the combustion period. The linear burning rate and its variation with the temperature and pressure may be determined in a → *Crawford Bomb*. The temperature coefficient of the burning rate is the variation per degree of temperature increase at constant pressure. The dependance on pressure is characterized by the pressure exponent (see above).

For details on relevant theoretical and practical relationships see:
Barrère, Jaumotte, Fraeijs de Veubeke, Vandenkerckhove: "Raketenantriebe", Elsevier Publ. Co., Amsterdam 1961, p. 265ff.; *Dadieu, Damm, Schmidt*: "Raketentreibstoffe", Springer, Wien 1968.

Other relevant keywords are: → *Solid Propellant Rockets,* → *Specific Impulse,* → *Thermodynamic Calculation of Decomposition Reactions,* → *Thrust.*

Bus Wire

Antenne für Parallelschaltung; antenne pour le couplage en parallele

Two wires that form an extension of the lead line and connecting wire and common to all caps in parallel. In parallel firing, each of the two wires of each electric blasting cap is connected to a different bus wire. For series in parallel firing each side of the series is connected to a different bus wire (→ *Parallel Connection*).

Butanediol Dinitrate

1,3-Butylenglykoldinitrat; dinitrate de butyléneglycol

$$\begin{array}{l} CH_3 \\ | \\ CH-O-NO_2 \\ | \\ CH_2 \\ | \\ CH_2-O-NO_2 \end{array}$$

colorless liquid
empirical formula: $C_4N_8N_2O_6$
molecular weight: 180.1
oxygen balance: −53.3%
nitrogen content: 15.56%
density: 1.32 g/cm^3
lead block test: 370 cm^3/10 g

Butanetriol dinitrate is insoluble in water, but is soluble in solvents for nitroglycerine; it is more volatile than nitroglycerine. Soluble guncotton is readily gelatinized. The nitrate is formed by reaction of butylene glycol with a nitric acid-sulfuric acid mixture as in the nitroglycerine synthesis, but the product is very easily destroyed by oxidation; the reaction mixture decomposes generating heat and nitrous gases. The product cannot be obtained under industrial conditions and has not found practical application for this reason.

Butanetriol Trinitrate

1,2,4-Butantrioltrinitrat; trinitrate de butanetriol

$$\begin{array}{l} CH_2-O-NO_2 \\ | \\ CH_2 \\ | \\ CH-O-NO_2 \\ | \\ CH_2-O-NO_2 \end{array}$$

pale yellow liquid
empirical formula: $C_4H_7N_3O_9$
molecular weight: 241.1
energy of formation: −379.2 kcal/kg = −1586.4 kJ/kg
enthalpy of formation: −402.5 kcal/kg = −1683.9 kJ/kg
oxygen balance: −16.6%
nitrogen content: 17.43%
refractive index: n_D^{20} = 1.4738
volume of explosion gases: 836 l/kg

heat of explosion
(H$_2$O liq.): 1439 kcal/kg = 6022 kJ/kg
(H$_2$O gas): 1327 kcal/kg = 5551 kJ/kg
density: 1.52 g/cm^3 (20/4)
solidification point: -27 °C = -17°F
impact sensitivity: 0.1 kp m = 1 N m

1,2,4-Butanetriol is nitrated with a mixture of nitric and sulfuric acids. The nitrated product is very stable. It is, like nitroglycerine, gelatinized by nitrocellulose.

Butanetriol trinitrate was used in the manufacture of tropic-proof double base powders. Isomers of butantriol trinitrate were also studied and utilized in practical work; these include methyl glycerol trinitrate and 1,2,3-butanetriol trinitrate, which have similar properties.

N-Butyl-N-(2-nitroxyethyl)nitramine

N-Butyl-nitratoethyl-nitramin, BuNENA

$$O_2N-N\begin{matrix}CH_2-CH_2-O-NO_2\\ \\ CH_2-CH_2-CH_2-CH_3\end{matrix}$$

colorless liquid
empirical formula: C$_6$H$_{13}$N$_3$O$_5$
molecular weight: 207,19
energy of formation: -803.34 kJ/kg
enthalpy of formation: -928.94 kJ/kg
oxygen balance: -104.25%
nitrogen content: 20.28%
density: 1.22 g/cm^3
melting point: -9 °C

This compound is prepared from N-butylethanolamine and nitric acid with acetic anhydride and a chloride source for catalysis. N-BuNENA is an energetic plasticizer for propellant formulations

Calcium Nitrate

Calciumnitrat; Kalksalpeter; nitrate de calcium

hydrated: Ca(NO$_3$)$_2 \cdot$ 4H$_2$O
colorless crystals
anhydrous product: Ca(NO$_3$)$_2$
white powder

The following data refer to the anhydrous product:

> molecular weight: 164.1
> energy of formation: −1352.1 kcal/kg = −5657.3 kJ/kg
> enthalpy of formation: −1366.6 kcal/kg = −5717.7 kJ/kg
> oxygen balance: +48.8 %
> nitrogen content: 17.07 %
> melting point: 561 °C = 1042°F
> very hygroscopic

Calcium nitrate can be used as an oxidizer component of → *Slurries.*

Camphor

Campher, Kampfer; camphre

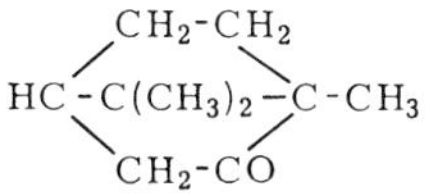

$$HC \underset{CH_2-CO}{\overset{CH_2-CH_2}{-C(CH_3)_2-C-CH_3}}$$

> empirical formula: $C_{10}H_{16}O$
> molecular weight: 152.3
> energy of formation: −480 kcal/kg = −2008 kJ/kg
> enthalpy of formation: −513 kcal/kg = −2146 kJ/kg
> oxygen balance: −283.8 %
> density: 0.98−0.99 g/cm^3
> melting point: 177−178 °C = 351−353°F
> boiling point: 209 °C = 408°F

This compound is utilized in celluloid industry, and also as gelatinizer in nitrocellulose gunpowders.

Specifications

> net content: not less than   99 %
> (analysis by titration with
> hydroxylamine)
> melting point: not less than   176 °C = 350°F
> insoluble in alcohol and ether: 0.1 %
> not more than
> chlorides: not more than   traces

Cap Sensitivity

Sprengkapsel-Empfindlichkeit; sensibilité au choc détonateur

Tests are carried out to determine the reaction of an explosive to a detonating cap. The results are used to determine the classification of the explosive as a transport hazard. The U.S. Department of Transportation has placed → *Blasting Agents* into a hazard category subject to regulations similar to those applicable to the former N.C.N. classification, i.e. much reduced in stringency. Explosives classified as blasting agents are those which can not be initiated by means of an explosive cap.

In Germany the following test for sensitivity to explosive caps has been developed:

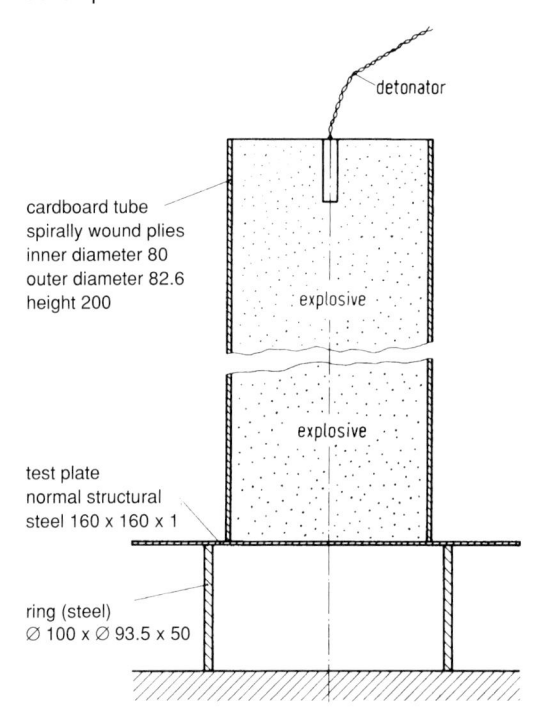

Fig. 7. Cap test (dimensions in mm)

The explosive is placed into a cardboard tube, 200 mm long, inside diameter of 80 mm, wall thickness between 1.3 and 1.4 mm. One end of the tube is sealed by a thin cardboard disk, which is glued into position. The density of the filling charge is determined by weighing (increase in weight after filling volume 1005 cm^3). The cap sensitivity can be influenced by the density of the charge. The test sample is placed upright onto a steel plate of 1 mm thickness, which is placed on a steel ring 50 mm in height, inside diameter of 100 mm, and wall thickness of 3.5 mm. A European test fuse (0.6 g PETN secondary charge) is inserted from the top throughout the full length of the tube, and initiated. No change in the condition of the plate or denting with or without fissure is classed as non-detonation. A circular hole indicates detonation.

Table 2. Cap test results.

Composition	Loading Density g/cm³	Test Result: Detonation
ANFO, porous prills	0.79–0.93	none
AN cryst. with TNT and fuels	0.82–1.07	always +
AN cryst. with TNT, DNT and fuels	0.82–1.07	always +
AN cryst. with DNT and fuels (N.C.N.'s)	0.75–1.10	always +*)
AN porous prills coated with DNT	0.82–0.84	none
AN cryst. with fuel	0.62–1.10	not at higher densities
AN cryst. with Al, earth alkaline nitrates, fuels and water (slurries)	1.13–1.26	+
(also with TNT)	1.37–1.60	none

AN = ammonium nitrate

The results are unchanged when explosives are tested at increased temperatures (30 °C). An exception to this are the AN prills coated with DNT.

DNT diffuses into the pores, the explosive becomes more homogeneous and therefore more sensitive. No change occurs when the European test fuse is replaced by a No. 8 detonator (0.75 g Tetryl).

A similar test has been developed in the USA (according the deformation of a lead block, using commercial caps with 0.4–0.45 g PETN) It is advisable to classify according to test results and not, as was the custom in the USA, by the classification of N.C.N. according to the explosive composition. As indicated above, ANFO's are not cap sensitive; mixtures of finely ground ammonium nitrate containing only 2% instead of 6% of oil or wax can, however, be cap sensitive.

---

* N.C.N.-explosives can be non-cap-sensitive at somewhat higher densities.

Carbamite

Denomination frequently used in English for → *Centralit I.*

Cardox

A physical explosion process which, like the Armstrong process and Airdox process, operates on the principle of a sudden release of compressed gas by means of a bursting disc. In the Cardox process, condensed CO_2 is brought to a high vapor pressure by means of a heating cartridge.

Cartridge

Patrone; cartouche

This term denotes any quantity of an explosive material or functional formulations thereof, which has been sheathed in order to improve handling, loading or dosing; for ammunition, "cartridge" most often means an assembly of an → *Igniter*, a → *Propellant* charge and a projectile, which may itself contain a high explosive charge with ignition mechanism. As applied to industrial explosives, the term "cartridge" denotes the amount of the explosive – which may vary between 50 g and several kg – enclosed in an envelope which is usually cylindical-shaped, and is made of paper, cardboard or plastic.

Cartridge Density

Patronendichte; densité de cartouche

(→ *Loading Density*) In industrial explosives, the ratio between the weight of an explosive cartridge and its volume.

Some manufacturers indirectly give the cartridge density on the package by stating the number of standard 1 1//4×8″ cartridges contained in a 50-pound case. The relationship is given in the following table:

Table 3. Cartridge density, weight of 1 1//4×8″ cartridge and number of 1 1//4 by 8″ cartridges in 50-pound case

Density g/cm³	Weight of 1 1/4×8″ Cartridge g	Number of 1 1/4×8″ Cartridges in 50-pound Case	Density g/cm³	Weight of 1 1/4×8″ Cartridge g	Number of 1 1/4 by 8″ Cartridges in 50-pound Case
0.62	100	227	1.18	190	120
0.68	110	206	1.24	200	114
0.75	120	189	1.31	210	108
0.81	130	175	1.37	220	103
0.87	140	162	1.43	230	99
0.93	150	151	1.49	240	95
0.99	160	142	1.55	250	91
1.06	170	134	1.62	260	87
1.12	180	126	1.68	270	84

Cartridge Strength

Synonymous with → *Bulk Strength*.
Also → *Strength;* → Weight Strength.

Case

Brennkammer; chambre de combustion; also: Kiste; caisse

Pressure vessel designed to contain propellant charge before and during burning.

Also: a large shipping container for explosive materials.

Case Bonding

This expression denotes a modern processing technique in the field of rockets driven by solid propellants. The pourable → *Composite Propellant* is cast directly into the combustion chamber, which has been pre-treated to produce a bonding and insulating layer and is allowed to harden (cure) in the chamber. Since temperature variations may be expected to produce major stresses, owing to the different values of thermal expansion coefficients, the success of the method depends to a large extent on the bonding forces acting between the bonding and insulating layer and the metal wall on the one hand, and the hardened propellant on the other, as well as on their elastomeric stress relaxation capability.

Caseless Ammunition

Hülsenlose Munition; munition sans douille

The requirement to improve portable firearms resulted in a reduction of the caliber (dimension 4–5 mm); and to reduce the ammunition weight led to the caseless ammunition project. Moreover, in the event of a crisis the problem of a worldwide shortage of nonferrous metals for cartridge cases will arise.

For a considerable time the caseless ammunition consisted of a compressed NC propellant body into which the bullet was inserted. However, this propellant tends to self-ignition even at relatively low temperatures (ca. 170 °C). Thus a "cook-off" may result, i.e. a premature ignition in a hot cartridge chamber which may occur with all automatic guns. In addition, with caseless ammunition the heat which is otherwise transferred to the cartridge case remains in the cartridge chamber. Therefore, in order to avoid the "cook-off"; HITPs (High Ignition Temperature Propellant) have been developed worldwide. DNAG used such a propellant for the first time with caseless cartridges for a newly developed gun (G 11) from Messrs. Heckler & Koch. The essential innovations with regard to previous developments are the use of a high-temperature-resistant, non-crystalline explosive as binding material, a special granular shape for the main energy component and the possibility to adjust the interior ballistics by porosity and stability of the propellant body. Further new developments are the combustible primer and the booster.

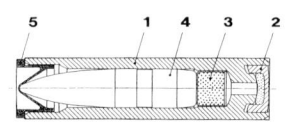

1 Cartridge body
2 Primer cap
3 Booster
4 Projectile
5 End cap

Fig. 8. Sectional view of the caseless cartridge body (Calibre 4.73 mm) for the G 11 weapon system.

Casting of Explosives

Giessen von Sprengladungen; coulée de charge de projectiles

Since the brisance of an explosive largely depends on its loading density, the highest possible loading densities are employed, in particular for military explosives. This density is attained by casting or pressing. The pressing operation requires a technical device. A cast charge is easier to fit into shells, mines and bombs, which have rather complex-shaped internal profiles.

Since → *TNT* is pourable at 80 °C (176°F), it is highly important in military technology. Since a considerable contraction takes place when the liquid explosive solidifies, good care must be taken during casting to ensure free access to all parts of the cast which have not yet solidified, in order to ensure proper replenishment of liquid material. Formerly, this was done by simple manual poking, but many automatic devices have been developed which do not involve any manual labor and which yield cavity-free casts.

Pure TNT tends to form very long, needle-shaped friable crystals, with a loose texture which does not correspond to the maximum density. Cast TNT charges must be fine-crystalline, mechanically firm and dense, with numerous crystallization nuclei; i.e., solid TNT must be finely dispersed in the cast. According to BOFORS, the texture of the cast can be improved by the addition of → *Hexanitrostilbene*.

Casting of Propellants

Giessen von Triebsätzen; coulée de propergols

Casting processes are needed especially in rocketry for the shaping of large propellant grains. Unlike in the casting of explosives, processes which cause shrinking and yield friable crystals cannot be applied.

There are two solutions to this problem:

a) hardening of polycondensates (e.g. polyurethanes or epoxys) with mechanically incorporated oxidizers, e.g. ammonium perchlorate (→ *Composite Propellants*); the hardened plastic material acts as fuel for the oxidizer;

b) converting of pourable nitrocellulose granules by treatment with liquid nitrate esters (e.g. with nitroglycerine). The granules can be poured dispersed in the liquid ("slurry casting"), or filled in the rocket motor shell and gelled in situ with the added nitrate ester.

CDB Propellants

→ *Composite Propellants.*

Centralite I

diethyldiphenylurea; symm. Diethyldiphenylharnstoff; diéthyldiphénylurée; Ethyl Centralite; Carbamite

$$\begin{array}{c} \diagup{C_2H_5} \\ N \\ \diagup\diagdown C_6H_5 \\ O=C \\ \diagdown\diagup C_6H_5 \\ N \\ \diagdown C_2H_5 \end{array}$$

colorless crystals
empirical formula: $C_{17}H_{20}N_2O$
molecular weight: 268.4
energy of formation: -68.2 kcal/kg = -285.6 kJ/kg
enthalpy of formation: -93.5 kcal/kg = -391.5 kJ/kg
oxygen balance: -256.4%
nitrogen content: 10.44%
density: 1.112 g/cm^3
melting point: 71.5–72 °C = 161–162°F
boiling point: 326–330 °C = 618–625°F

Centralite I, II and III are used as → *Stabilizers* in gunpowders, especially in nitroglycerine powders (→ *Double Base Propellants*). It is insoluble in water, but soluble in organic solvents.

Centralites are not only stabilizers, but gelatinizers as well. The latter property is taken advantage of in solvent-free manufacture of double base propellants.

Specifications

solidification point: not less than	71 °C = 160°F
molten material: bright clear pale liquid	
ashes: not more than	0.1%
volatiles: not more than	0.1%
acetone solution:	clear, no residue
secondary and tertiary amines: not more than	0.1%
chlorides as HCl: not more than	0.001%
reaction:	neutral
acidity: not more than	0.04%

Centralite II

dimethyldiphenylurea; Dimethyldiphenylharnstoff;
diméthyldiphénylurée

$$\begin{array}{c} \diagup^{CH_3} \\ N \\ O=C\diagup^{C_6H_5}_{C_6H_5} \\ N \\ \diagdown_{CH_3} \end{array}$$

colorless crystals
empirical formula: $C_{15}H_{16}N_2O$
molecular weight: 240.3
energy of formation: −37.3 kcal/kg = −156 kJ/kg
enthalpy of formation: −60.8 kcal/kg = −254 kJ/kg
oxygen balance: −246.3%
nitrogen content: 11.66%
melting point: 121−121.5 °C = 250−251°F
boiling point: 350 °C = 662°F

Specifications

same as for Centralite I, except
solidification point: not less than 119 °C = 246°F

Centralite III

methylethyldiphenylurea; Methylethyldiphenylharnstoff;
méthyléthyldiphénylurée

$$\begin{array}{c} \diagup^{CH_3} \\ N \\ O=C\diagup^{C_6H_5}_{C_6H_5} \\ N \\ \diagdown_{C_2H_5} \end{array}$$

colorless crystals
empirical formula: $C_{16}H_{18}N_2O$
molecular weight: 254.3
energy of formation: −94.7 kcal/kg = −396.1 kJ/kg
enthalpy of formation: −119.1 kcal/kg = −498.5 kJ/kg
oxygen balance: −251.7%
melting point: 57−58 °C = 135−138°F

Specifications

same as for Centralite I, except
solidification point: 57 °C = 135°F

Channel Effect

Kanaleffekt

Interruption in the detonation of an explosive column because of the compaction of the cartridges which have not yet exploded due to the gas shock wave front in the borehole. This happens very often if the borehole cross-section is large as compared to that of the cartridges.

Chlorate Explosives

Chloratsprengstoffe; explosifs chloratés

Explosive mixtures of alkali metal chlorates with carbon-rich organic compounds such as wood dust, petroleum, oils, fats and nitro derivatives of benzene and toluene; they may also contain nitrate esters.

Their strength is lower than that of ammonium nitrate explosives in powder form. Chlorate explosives must not be stored together with ammonium nitrate explosives, since ammonium chlorate, which is formed when these two substances are brought into contact, decomposes and explodes.

"Miedziankit" is the name of absorbent potassium chlorate particles, which are impregnated with a liquid fuel just before use, and then acquire explosive properties. Following the development of → *ANFO*, this explosive is no longer of interest.

Cigarette-Burning

Stirnabbrand; combustion en cigarette

→ *Face Burning*

Circuit Tester (Ohmmeter)

Zündkreisprüfer; éprouveur; blasting galvanometer

Instrument for electrical testing of misconnected circuits. The current intensity used in the testing must be well below the → *No Fire* condition of the electric detonator; the circuit tester is accordingly equipped with resistances at both poles. Only officially approved testers should be employed. The testers are of two kinds: conduction testers which show, by means of a visual indicator, whether or not current is flowing in the circuit, and ohmmeters which measure the resistance of the priming circuit.

Class A, Class B and Class C Explosives

Classification defined by the U.S. Department of Transportation:

Class A Explosives:

Explosives, which possess detonating or otherwise maximum hazard; such as dynamite, nitroglycerine, lead azide, TNT, Composition B, PBX, Octol, blasting caps and detonating primers.

Class B Explosives:

Explosives, which possess flammable hazard; such as, but not limited to, propellant explosives, photographic flash powders, and some special fireworks.

Class C Explosives:

Explosives, which contain class A or class B explosives, or both, as components but in restricted quantities.

Coal Dust

Kohlenstaub; poussiére

Mixtures of coal dust with air are explosive and their explosion by blasting must be prevented (→ *Permitted Explosives*).

Column Charge

Gestreckte Ladung; file de cartouches

A charge of explosives in a blast hole in the form of a long continuous unbroken column.

Combustibility

Feuergefährlichkeit; danger d'inflammation

Capability of burning. Flammable. The relative combustibility of materials in storage is defined as: hazardous – materials that by themselves or in combination with their packaging, are easily ignited and will contribute to the intensity and rapid spread of a fire; moderate – materials and their packaging both of which will contribute fuel to a fire; noncombustible – materials and their packaging that will neither ignite nor support combustion.

Combustible Cartridge Cases

Verbrennbare Kartuschhülsen; douilles combustibles

The propellant charge used for the shot from a weapon is introduced into cases or bags ("cartouche bags"); for metallic cartouche cases, the projectile is combined with the propellant charge and the propellant Charge igniter to form a "cartridge".

Now, combustible cartridge cases serve the purpose of making the case material contribute to the ballistic performance and to render unnecessary the removal of inert material from the weapon after the shot. Such case material has to be adapted to the combustion process of the powder. It consists of high-energy material, e.g. nitrocellulose, a structure-reinforcing additive, e.g. kraft-paper pulp, binders of plastic material, and further additives, e.g. stabilizers such as contained in the powder itself. The cases are made by filtration from a pulp, pressing, molding and drying.

Caseless ammunition is also available for infantry weapons; as the ejector mechanism can be dispensed with, it is possible to raise the number of shots in machine guns.

→ *"Caseless Ammunition"*

Combustion

Verbrennung; combustion

Any oxidation reaction, including those produced by introduction of atmospheric oxygen; many explosives are capable of burning without detonation if unconfined. Moreover, the oxidation reaction taking place in propellants without introduction of oxygen is also designated as combustion: it is preferable to denote this process as burning (→ *Burning Rate;* → *Deflagration*).

Combustion Chamber

Brennkammer; chambre de combustion; case

In rocket technology, the chamber in which the reaction of the propellants takes place.

In solid fuel rockets, the propellant container also serves as the combustion chamber; in liquid fuel rockets it is the chamber in which the injected liquid components of the propellant to react with one another. The combustion chamber must withstand the predetermined working pressure and the temperatures developing at the chamber walls. In liquid fuel rockets the chamber wall is externally cooled in most cases; in solid fuel rockets, in which internal charges bonded to the chamber walls are often employed, the required protection is afforded by the propellant itself. These conditions determine the choice of a suitable chamber material. Since the weight of the combustion chamber has a decisive effect on the range of the rocket, the walls should be as thin as possible. The use of thermally insulating and reinforced (e.g., with fiberglass) inserts made of plastic materials has already proved successful.

Standard combustion chambers and laboratory combustion chambers*) have been developed for testing the behavior of solid rocket fuels and for the determination of their characteristic properties.

Commercial Explosives

Gewerbliche Sprengstoffe; explosifs pour usage industriel

Explosives designed, produced, and used for commercial or industrial applications other than military.

Compatibility

Verträglichkeit; compatibilité

Ability of materials to be stored intimately without chemical reaction occurring.

Incompatibility may result in a loss of effectiveness or may be very hazardous. For example, → *Chlorate Explosives* and → *Ammonium Nitrate Explosives* are not compatible (formation of self-decomposing ammonium chloride). For compatibility testing → *Vacuum Test*.

Compatibility Group

Dangerous goods of Class 1 are to be assigned to a compatibility group (A to H, J, K, L, N, S) characterizing the kind of good (A to N) or the hazardous effects (S). A table in the transport regulations containing the compatibility groups is the basis of the provisions for mixed loading of Class 1 goods.

Composite Propellants

Verbundtreibsätze; poudres composites

Composite propellants are solid rocket fuels, consisting of oxygendonating inorganic salts and a binder made of plastic.

---

* *E. Haeuseler* and *W. Diehl*, Explosivstoffe, Vol. 15, p. 217 (1967).

The high-polymeric binders in use today include polysulfides (PS), polybutadieneacrylic acid (PBAA), polybutadiene-acrylonitrile (PBAN), polyurethane (PU) and carboxyl- and hydroxyl-terminated polybutadiene (CTPB and HTPB).

Nitrates and perchlorates, → *Ammonium Perchlorate* in particular, are used as oxidizers.

These propellants can be manufactured by casting or by pressing. The grain fineness of the salt employed affects the combustion properties to a significant extent. The mechanical (preferably rubber-elastic) properties of the plastic binders must satisfy special requirements.

CDB Propellants are combinations of composites with → *Double Base Propellants*, which achieve "plateaus" (→ *Burning Rate*) otherwise difficult to attain.

For details about composite propellants see:

Zähringer, A. F.: Solid Propellant Rockets, Wyandotte, New York 1958

Barrère, Jaumotte, Fraeijs de Veubeke, "Vandekerckhove": Rocket Propulsions, Elsevier Publ. Amsterdam 1961

Dadieu, Damm, Schmidt: Raketentreibstoffe, Springer, Wien 1968

Compositions A; A-2; A-3

Pressed charges made of phlegmatized → *Hexogen* differing from each other only by the various kinds of wax they contain.

detonation velocity, confined: 8100 m/s
at $\rho = 1.71$ g/cm^3

Compositions B; B-2

Hexolite; Hexotol

Castable mixtures of Hexogen (RDX) and TNT in the proportion of 60:40; some of them contain wax as an additive. They are used as fillings for bombs, mines and → *Hollow (Shaped) Charges*.

density: about 1.65 g/cm^3*)
detonation velocity, confined: 7800 m/s
at $\rho = 1.65$ g/cm^3

---

* Can be raised to >1.7 g/cm^3 by application of special casting techniques.

Composition C; C-2; C-3; C-4

Military plastic explosives, consisting of → *Hexogen* and a plasticizer, which itself may or may not be explosive. The respective formulations are:
Table 4.

Composition	RDX %	Plasticizer %	Type
C	88.3	11.7	non-explosive
C-2	80.0	20.0	explosive
C-3	78	22.0	explosive
C-4	90 (selected grain fractions)	10.0	polyisobutylene

(→ also *Plastic Explosives*)

Composition I; II

Eutectic mixtures of ammonium nitrate, sodium nitrate, dicyanodiamide and guanidine nitrate.

Table 5.

	Composition I	II
ammonium nitrate	65.5	60
sodium nitrate	10.0	24
dicyanodiamide	14.5	8
guanidine nitrate	10.0	8

Confined Detonation Velocity

Detonationsgeschwindigkeit unter Einschluss; vitesse de détonation sous confinement

The detonation velocity of an explosive or blasting agent in a container such as a borehole in contrast to detonating in the open (→ *Detonation Velocity*).

Confinement

Einschluss

Confinement is understood to mean an inert material of some strength and having a given wall thickness, situated in the immediate vicinity of an explosive. Priming or heating the explosive materials produces different results, according to whether they are located in a stronger or a weaker confinement. If confined by thick steel, almost any explosive will explode or detonate on being heated; on the other hand, they burn on contact with an open flame if unconfined (→ *Combustion;* → *Mass Explosion Risk*), except → *Initiating Explosives.*

The destructive (fragmentation) effect of an explosion becomes stronger if the explosive is confined (stemmed) in an enclosure such as a borehole. In the absence of natural confinement, the explosive charge is often embedded in an inert material such as clay. See also → *Mud Cap* and → *Stemming.*

Contained Detonating Fuze

Sprengschnur mit Schutzmantel; cordeau détonant gainé

Mild detonating fuze completely contained within a shock-absorbing sheath to prevent damage to the surroundings when the fuze is detonated.

Contour Blasting

Profilsprengen; saulage en profil

The purpose of controlled blasting is to produce an excavation contour, while leaving behind an intact, fissure-free formation ("prenotching", "pre-splitting off", "notching", "contour blasting"). This is done by the application of diminished-strength explosive charges, using numerous boreholes driven exactly in parallel (vacant boreholes; firing in a cavity; charge diameters small as compared to the total diameter of the borehole; fissurefree roof firing in salt mines).

For further details see: *Rune Gustavson:* Swedish Blasting Technique. SPI, Gothenburg, Sweden (1972)

Copper chromite

Kupferchromit; chromite de cuivre

$$(CuO)_x(Cr_2O_3)_y$$

dark brown to black powder

Copper chromite is the reaction product of copper oxide and chromium oxide. It is an important catalyst for the burning of rocket propellants and pyrotechnical compositions.

Specifications

sieve analysis:	
through mesh width 0.07 mm: at least	98 %
through mesh width 0.04 mm: at least	90 %
net content	
CuO: at least	79 %
not more than	85 %
Cr_2O_3: at least	13 %
not more than	19 %
Fe_2O_3: not more than	0.35 %
water-soluble matter: not more than	0.5 %

Cordite

Designation for double base (nitroglycerine-nitrocellulose) gun propellants in the United Kingdom.

Coruscatives

This is the name given by the American worker *Zwicky* to pairs of materials (other than the well-known thermites, → *delay compositions*) which react with each other without formation of gas.

The exothermal nature of certain components may be surprisingly high; the mixture Ti:Sb:Pb = 48:23:29 is primed at 570 °C (1060°F), and the reaction temperature attains 1000 °C (1830°F). Other combinations include magnesium-silicon, magnesium-tellurium, magnesium-tin and magnesium-phosphorus.

Coyote Blasting

Kammerminensprengungen; sautage par grands fourneaux de mines

In coyote blasting, which is practiced in open-pit mining and in stone quarries, tunnels are driven into the mined face and chambers are drilled which can accommodate large quantities (up to several tons) of explosives. The chambers – usually several chambers at once – are charged, stemmed and detonated. They must be primed with the aid of a → *Detonating Cord*.

Coyote blasting has now been almost completely displaced by → *Large Hole Blasting*, because the spaces accommodating the explosive can be produced more rationally in this way.

Crawford Bomb

A bomb used to determine the → *Burning Rate* of solid rocket propellants.

The propellant grains are in the form of thin rods ("strands") which may have been cut or extruded and protected against surface burning by mantle insulation. The strand is placed in a bomb and electrically initiated at one end, after which its combustion rate is recorded with the aid of wire probes. Using compressed nitrogen, the pressure at which the combustion take place is adjusted in the bomb; standard values are 20, 40, 70, 100, 130, 180, 250 bar at a temperature between −40 °C and 60 °C.

Crimping

Anwürgen; sertir

The act of securing a blasting cap to a section of safety fuse by compressing the metal shell of the cap against the fuse by means of a cap crimper.

Critical Diameter

Kritischer Durchmesser; diamètre critique

The critical diameter is the minimum diameter of an explosive charge at which detonation can still take place. It is strongly texture-dependent, and is larger in cast than in pressed charges. Finely dispersed gas inclusions considerably reduce the critical diameter.

In the case of very insensitive materials – ammonium nitrate, for example – the critical diameter may be very large.

Cumulative Priming

Kumulative Zündung

Counter-current priming, in which the explosive charge is simultaneously primed at two or more places, so that the detonation waves travel to meet one another, and their effect becomes additive.

Curing

Härten; aushärten; maturer

Polymerization of prepolymer or monomer component of mixed propellants to increase mechanical strength.

Cushion Blasting

Hohlraumsprengen; fir avec chambres d'expansion

A method of blasting in which an air space is left between the explosive charge and the stemming, or in which the blast hole is purposely drilled larger than the diameter of the explosive cartridge to be loaded; → *Contour Blasting*.

Cut Off

Abschlagen einer Sprengladung; decapitation

Separation of a part of a borehole charge by the blast effect of another shot in electrical delay firing circuits. Cut off can also occur to the whole burden of the borehole charge by previous shots; → *Permitted Explosives*.

Cutting Charges

Schneidladungen; charge creuse pour découpage

Cutting charges serve to cut through iron plates, cables, bridge trusses etc. They are constructed on the principle of → *Shaped Charges*, but are not rotationally symmetrical; their shape is that of long channels (grooves).

The cutting depth of these charges depends to a considerable extent on the thickness and lining material of the angular or semi-circular groove; in addition, the optimum distance from the target must be determined in advance.

As in rotationally symmetrical hollow charges, a jet of highly accelerated gases and metal fragments is produced.

Cyanuric Triazide

Cyanurtriazid; triazide cyanurique

$$\begin{array}{c} N_3 \\ | \\ N{\diagup}^C{\diagdown}N \\ \| \quad \| \\ N_3-C{\diagdown}_N{\diagup}C-N_3 \end{array}$$

colorless crystals
empirical formula: C_3N_{12}
molecular weight: 204.1
energy of formation: +1090.3 kcal/kg = +4561.9 kJ/kg
enthalpy of formation: +1072.9 kcal/kg = +4489.2 kJ/kg
oxygen balance: −47%
nitrogen content: 82.36% N
melting point (under decomposition): 94 °C = 201°F
lead block test: 415 cm^3/10 g
detonation velocity, unconfined:
  5500 m/s at ρ = 1.02 g/cm^3
deflagration point (explosion):
  200–205 °C = 390–400°F
friction sensitivity: 0.01 kp = 0.1 N pistil load

This compound is prepared by slowly introducing powdered cyanogen chloride into an aqueous solution of sodium azide with efficient cooling.

Cyanuric triazide is an effective initiating explosive. It is not employed in practice owing to its high vapor pressure.

Cyclotol

The name given to RDX – TNT mixtures with compositions varying between 50:50 and 75:25 (→ *Compositions B*).

Cyclotrimethylene Trinitrosamine

trinitrosotrimethylenetriamine; Cyclotrimethylentrinitrosamin; cyclotrimethylène trinitrosamine

$$\begin{array}{c} H_2 \\ \quad C \\ ON-N{\diagup}{\diagdown}N-NO \\ | \qquad | \\ H_2C{\diagdown}_N{\diagup}CH_2 \\ | \\ NO \end{array}$$

pale yellow crystals
empirical formula: $C_3H_6N_6O_3$
molecular weight: 174.1

energy of formation: +417.9 kcal/kg = +1748.4 kJ/kg
enthalpy of formation: +392.4 kcal/kg = +1641.7 kJ/kg
oxygen balance: −55.1 %
nitrogen content: 48.28 %
volume of explosion gases: 996 l/kg
heat of explosion
  (H_2O liq.): 1081 kcal/kg = 4525 kJ/kg
  (H_2O gas): 1051 kcal/kg = 4397 kJ/kg
density: 1.508 g/cm^3
melting point: 102 °C = 216°F
heat of fusion: 5.2 kcal/kg = 22 kJ/kg
detonation velocity, confined:
  7300 m/s = 24 000 ft/s at ρ = 1.49 g/cm^3

Cyclotrimethylene trinitrosamine is soluble in acetone, alcohol chloroform and benzene, and is sparingly soluble in water.

This nitroso compound, which is related to Hexogen, is prepared by treating hexamethylenetetramine with alkali metal nitrites in a dilute acid solution.

Since concentrated acid is not required in the preparation, large-scale manufacture of the product, under the name of R-salt, was under active consideration at one time during the Second World War. However, even though easily prepared and powerful, the explosive has not yet been used in practice owing to its limited chemical and thermal stability.

Cylinder Expansion Test

Experimental method to measure the effectiveness of an explosive. The radial expansion on detonation of a metallic cylinder (usually copper) filled with a high explosive is observed. A streak camera or a laser method might be used. The detonation velocity is determined simultaneously, using for example time-of-arrival pins. The → *Equation of State* (EOS) which is often the Jones-Wilkins-Lee (JWL) EOS of the detonation products is derived using Gurney theory.

Dangerous Goods Regulations

Gefahrgutverordnungen

Dangerous Goods Regulations, Rail (GGVE)
Dangerous Goods Regulations, Road (GGVS}
Dangerous Goods Regulations, Sea (GGVSea)
Dangerous Goods Regulations, Inland Waterways (GGVBinsch)

The Dangerous Goods Regulations are internationally harmonised regulations (→ *ADR*, → *RID*, → *IMDG Code*, → *ADNR*, → *ICAO TI*) for the transport of dangerous goods. All substances and articles that have defined explosive properties are assigned to Class 1 "Explosives and Articles with Explosive Substance". To classify into one of the 6 Risk Classes (sub-classes of Class 1), the hazardous property of the substance or article is studied, including in its dispatch packing. This examination takes place in accordance with the test methods described in the "Recommendations on the Transport of Dangerous Goods; Manual of Tests and Criteria, United Nations". The → *BAM* (Federal German Materials Testing Laboratory, → *BICT* for the military area) is the competent authority in Germany for classifying explosives, detonators, propellants, pyrotechnical mixtures and articles.

The purpose of the sub-classes 1.1, 1.2, 1.3, 1.4, 1.5 and 1.6 is to characterise the explosive properties of the substances and articles in Class 1 with regard to their activity and to some extent their sensitivity as well. The 13 Compatibility Groups A, B, C, D, E, F, G, H, J, K, L, N and S reflect mainly the specific type of explosives. The Classification Code, consisting of the Sub-Class and Compatibility Group (e.g. 1.1D for a mass-explodable detonating explosive or an article with such a substance), characterises goods in Class 1.

Classification into a sub-class and a compatibility group lead to particular rules specified in the Dangerous Goods Regulations for transporting these goods.

Dangerous Goods Regulations

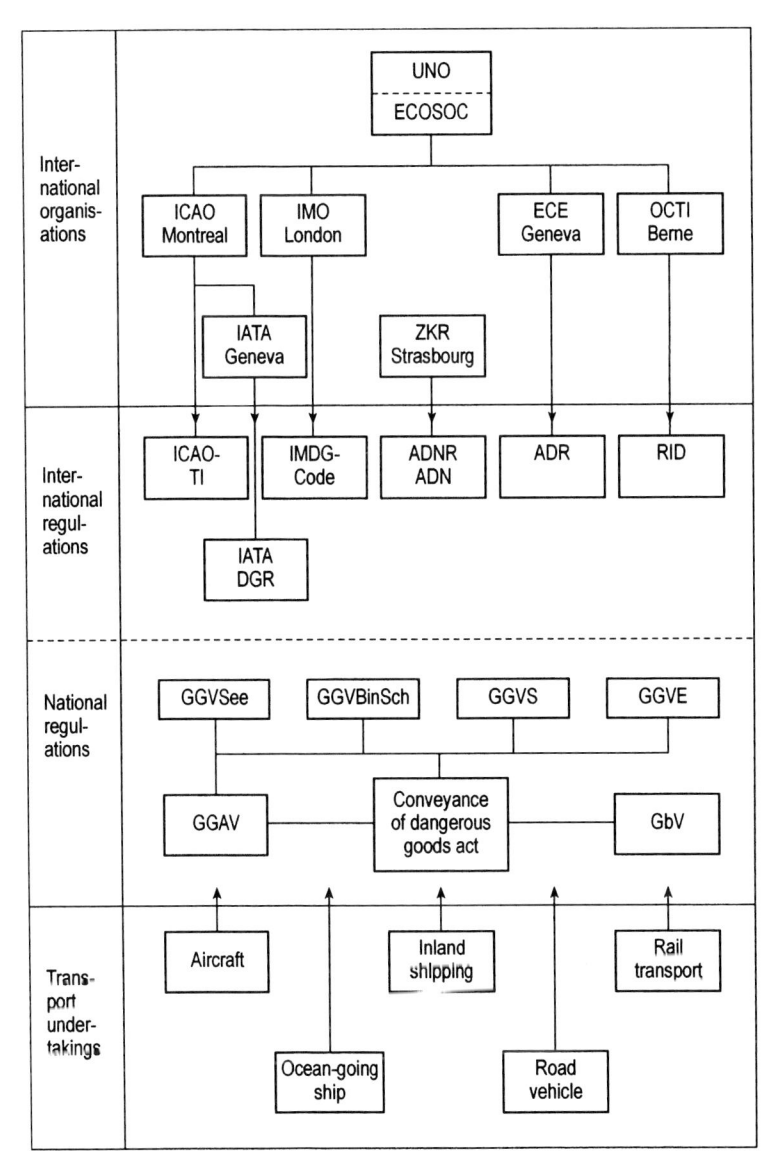

International organisations	UNO / ECOSOC — ICAO Montreal, IMO London, ECE Geneva, OCTI Berne; IATA Geneva; ZKR Strasbourg
International regulations	ICAO-TI, IMDG-Code, ADNR/ADN, ADR, RID; IATA DGR
National regulations	GGVSee, GGVBinSch, GGVS, GGVE; GGAV, Conveyance of dangerous goods act, GbV
Transport undertakings	Aircraft, Ocean-going ship, Inland shipping, Road vehicle, Rail transport

Fig. 9. Organisation of Dangerous Goods Transport.

Dautriche Method

A method for the determination of the detonation rate. The test sample of the explosive is accommodated in a column, which may or may not be enclosed in an iron tube; the length of the detonating column to be measured is marked out by means of two blasting caps, one at each end. A loop made of a detonating cord with a known detonation rate is connected to the caps and is passed over a lead sheet in its middle part. The cord is successively ignited at both ends, and the meeting point of the two detonation waves advancing towards each other makes a notch on the lead sheet. The distance between this meeting point and the geometric center of the cord is a measure of the reciprocal detonation rate to be determined:

$$D_x = D \times \frac{m}{2a}$$

where D_x is the detonation rate of the sample, D is the detonation rate of the detonator cord, m is the length of the distance to be measured, and a is the distance between the notch and the center of the cord length.

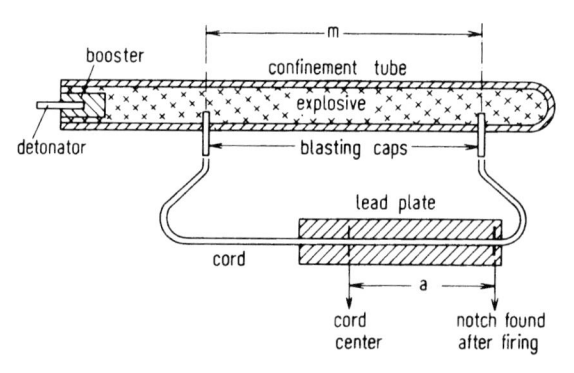

Fig. 10. Dautriche method.

The method is easy to carry out and no special chronometer is required.

DBX

A cast explosive charge, containing RDX, ammonium nitrate, TNT and aluminum powder in the proportions 21:21:40:18

Deckmaster

Trade name for primer charges with special delay inserts consisting of a sensor on one end and an aluminum shell delay cap on the other. Delay times: 0–500 milliseconds in 25 ms intervals. The Deckmaster-unit has to be connected with detonating cord with no more than 30 grains per ft (→ *Miniaturized Detonating Cord*). For varied delay steps in the hole, only one downline detonating cord is needed.

Deflagration

Explosive materials often decompose at a rate below the sonic velocity of the material. This type of reaction is known as deflagration. It is propagated by the liberated heat of reaction, and the direction of flow of the reaction products is opposite to that of decomposition propagation (unlike in → *Detonation*). The burning of a powder or of a rocket charge is a deflagration process (→ *Burning Rate*). The mode of reaction of an explosive material – deflagration or detonation – depends on its mode of actuation (→ *To Inflame,* → *Initiation*).

For transitions from deflagration to detonation (DDT) and vice versa see → *Detonation*.

It is important to prevent any deflagration of permitted explosives. Since the deflagration of an explosive proceeds at a much slower rate than its detonation, it may ignite methane-air and coal dust-air mixtures. This must be prevented by using suitable compositions (→ *Permitted Explosives*) and application techniques.

Deflagration Point

Verpuffungspunkt; température de décomposition

The deflagration point is defined as the temperature at which a small sample of the explosive, placed in a test tube and externally heated, bursts into flame, decomposes rapidly or detonates violently.

A 0.5-g sample (a 0.01-g sample in the case of → *Initiating Explosives*) is placed in a test tube and immersed in a liquid metal (preferably Wood's metal) bath at 100 °C (212°F), and the temperature is raised at the rate of 20 °C per minute until deflagration or decomposition takes place.

This method is identical with the official method laid down in *RID*. Nitrocellulose and nitrocellulose powder are tested in a stirred paraffin bath, heated at the rate of 5 °C per minute.

Delay

Verzögerung; retard

A pyrotechnic, mechanical, electronic, or explosive train component that introduces a controlled time delay in some element of the arming or functioning of a fuze mechanism.

delay, arming

The time or distance interval between the instant a device carrying the fuze is launched and the instant the fuze becomes armed.

delay compositions

Verzögerungssätze; compositions retardatrices

Delay compositions are mixtures of materials which, when pressed into delay tubes, react without evolution of gaseous products and thus ensure the minimum variation in the delay period. Examples of such mixtures are potassium permanganate with antimony; lead dioxide or minium with silicium; redox reactions with fluorides and other halides (→ also *Coruscatives* and → *delay, gasless*).

delay element

An explosive train component normally consisting of a primer, a delay column, and a relay detonator or transfer charge assembled in that order in a single housing to provide a controlled time interval.

delay function

The time or distance interval between the initiation of the fuze and the detonation.

delay fuze

Verzögerungszünder; fusée retardatrice

In the military, delay fuses are complete shell fuses which set off the explosive charge a definite time after impact.

delay, gasless

Verzögerung, gaslos; retard sans formation de gaz

Delay elements consisting of a pyrotechnic mixture that burns without production of gases.

delayed initiation; delayed inflammation

Zündverzug; Anzündverzug

In hypergolic pairs of rocket propellants (→ *Hypergolic*), a "delay" in inflammation is understood to mean the time which elapses from the moment of contact between the reaction partners up to the initiation; this delay is of the order of a few milliseconds, and must not exceed a certain limiting value; thus, e.g. the inflammation delay of the reagent pair furfuryl alcohol – nitric acid is about 20 milliseconds.

In the case of solid fuel rockets, the delay in inflammation, which is determined on a test stand, is understood to mean the time which elapsed between the moment of application of the initiation voltage to the electric inflammation element and the moment when about 10 % of the maximum pressure has been attained. Clearly, the magnitude of this parameter depends both on the nature of the firing charge employed and on the ease with which the solid propellant can be initiated. The permitted initiation delay will depend on the objective of the firing.

Density

Dichte; densité

Density is an important characteristic of explosives. Raising the density (e.g. by pressing or casting) improves → *Brisance* and *Detonation Velocity* (→ *Detonation, Hydrodynamic Theory of Detonation*). Low-density explosives, in contrast, produce a milder thrust effect (→ also *Loading Density;* → *Cartridge Density*).

Destruction of Explosive Materials

Vernichten von Explosivstoffen; dèstruchon de matières explosives

Destruction of explosives includes destruction of explosive materials and their waste which present a danger of explosion, removal of explosive residues on machines, instruments, pipes etc., and handling objects with adhering explosives (for the evacuation and handling of ammunition → *Dismantling of Explosive Objects, Especially Ammunition*). The destruction of explosives must be carried out under the supervision of an expert, who must be in charge of the entire operation.

The following techniques may be used in the destruction of explosive materials:

1. Combustion: this technique is applicable to most explosives apart from initiating explosives. However, this destruction technique, while important per se, can only be carried out by the manufacturer. Burning of explosives by the user can be dangerous.

2. The explosive is poured into a large volume of water and is mixed with it. This technique can be applied to materials which are soluble totally in water (black powder, ANFO).

3. Treatment with chemicals (acids, alkalis, boiling with water): lead azide is destroyed by treatment with nitric acid in the presence of sodium nitrite; lead trinitroresorcinate by treatment with nitric acid; mercury fulminate by prolonged treatment with boiling nitric acid.

4. Exploding the material: blasting operations must be carried out in a barricaded area licensed for the purpose, located at least 1000 ft away from any bulding which may be endangered by the explosion. A reinforced shelter is needed for protection of personnel; suitable protection from flying fragments (e.g. by walls; palisades) must be provided.

Destressing Blasting*)

Entspannungssprengung; sautage de détente

Destressing blasting serves to loosen up the rock mass in order to distribute high compressive loads more uniformly and to counteract the hazard of rockbursts. Rockbursts are particulary violent fracture processes, accompanied by considerable earth tremors. They mainly consist of a sudden thrust or ejection of the rock involved (coal; salts; massive rocks) and abrupt closure of the excavation. In coal seams, the risk manifests itself by abnormally great amounts of debris when drilling small holes (so-called test drilling). Destressing blasting is performed by contained detonations.

Detonating Cord, Detonation Cord

detonating fuse; Primacord; Sprengschnur; cordeau détonant; Cordtex

Detonating cords consist of a → *PETN* core (about 12 g/m) with wound hemp or jute threads and a plastic coating around it. The cord is initiated by a cap and its detonation velocity is about 7000 m/s. Special

---

* The article was made available by Dr. Bräuner, Bergbauverein Essen.
Publications:
Bräuner, G.: Gebirgsdruck und Gebirgsschläge. Verlag Glückauf. Essen (1981).
Bräuner, G.: Möglichkeiten der Gebirgsschlagbekämpfung im Ruhrbergbau unter besonderer Berücksichtigung des Entspannungssprengens.
NOBEL-Hefte July – September (1978), p. 91–97.
Bräuner, G.: Gebirgsdruck und Gebirgsschläge, Verlag Glückauf, Essen, 2nd Edition (1991).

fuses for the safe initiation of → *ANFO* contain 40 and 100 g/m PETN.

Detonating fuses serve to initiate blasting charges; the initiation is safe if the cord is coiled serveral times around the cartridge. To initiate several charges, branch cords are attached to a "main cord". In Germany, priming by detonating cords is mandatory in → *Large Hole Blasting* and in → *Coyote Blasting*.

Detonating cords are also employed for seismic shots in the desert and at sea. They are also used for clearing blasts in oil and gas wells, which restore the flow from blocked boreholes; special cords with a wire reinforced sheat are used for this purpose.

For the use of detonating cords in the determination of the detonation rate of explosives → *Dautriche Method*.

Transfer fuses which have no priming effect are manufactured in the USA. Those containing only a fraction of one gram of PETN per meter and a lead sheathing are known as "mild detonating fuses". Cords containing about 2 g of the explosive per meter inside a plastic-impregnated network are manufactured as "Primadet".

Detonating Cord Downline; Zündabzweigung

The section of detonating cord that extends within the blast hole from the ground surface down to the explosive charge.

Detonating Cord MS Connectors; Millisekunden-Verzögerer

Non-electric, short-interval (millisecond) delay devices for use in delaying blasts which are initiated by detonating cord.

Detonating Cord Trunirline; Leit-Sprengschnur; ligne de cordeau dètonant

The line of detonating cord that is used to connect and initiate other lines of detonating cord.

Detonation

Detonation; détonation

Detonation is a chemical reaction given by an explosive substance which produces a shock wave. High temperature and pressure gradients are generated in the wave front, so that the chemical reaction is initiated instantaneously. Detonation velocities lie in the approximate range of 1500 to 9000 m/s = 5000 to 30 000 ft/s; slower explosive reactions, which are propagated by thermal conduction and radiation, are known as → *Deflagration*.

Detonation

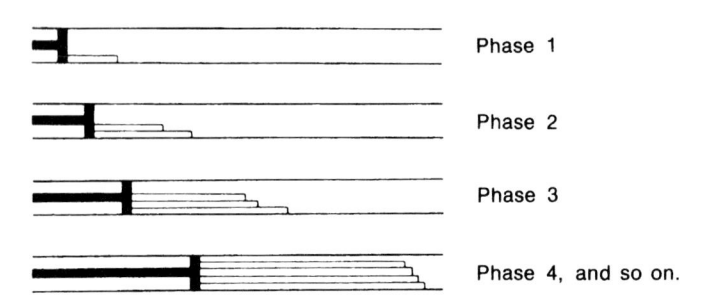

	Phase 1
	Phase 2
	Phase 3
	Phase 4, and so on.

Fig. 11. Generation of a plane shock wave.

1. Shock Wave Theory

Shock waves are also generated in non-explosive media by a sudden change in pressure. The generation of a shock wave in air (as a non-explosive gas) is illustrated by Fig. 11, which has been taken from *R. Becker:**).

Let a movable piston in a tube be suddenly accelerated from rest and then continue its motion at a constant rate (phase 1). The air in front of the piston must be compressed somewhat and warms up a little; the compression range is determined by the velocity of sound in the air.

The increase in pressure and the range of the increase after a short time are symbolized by the line drawn in front of the piston. Now let the piston accelerate again and continue its motion at the new, higher rate. The new compression is imparted to the medium, some of which is already in motion, as shown in phase 2 of Fig. 11; it is moving at a faster rate, the motion of the matter is superposed and, in addition, the sonic velocity has increased in the somewhat warmer medium. Phases 3, 4, etc. show that a steep pressure front is thus generated. A mathematical derivation of the relationships governing such a process would be beyond the scope of this book**).

---

* *R. Becker*, Zeitschrift für Physik 8, p. 321–362, (1922).
** For a detailed presentation see the reference list on page 89 ff.

The state variables will be denoted as follows:
Table 6.

	Undisturbed Medium	Medium in Shock Compression
pressure	p_0	p_1
temperature	T_0	T_1
density	ρ_0	ρ_1
specific volume ($v = 1/\rho$)	v_0	v_1
internal energy	e_0	e_1
sound velocity	c_0	c_1

If we limit our consideration to nearly ideal gases such as air, the following values for the rise in temperature, the speed of propagation of the shock wave D, and the rate of motion of matter behind the wave front W can be calculated as a function of the compression ratio p_1/p_0:

Table 7.

$\dfrac{p_1}{p_0}$	T_1 °C	D m/s	W m/s
2	63	452	175
5	209	698	452
8	345	875	627
10	432	978	725
20	853	1369	1095

and further increasing values.

It is seen from the Table that even if the extent of compression is relatively small, the propagation rate becomes distinctly higher than the velocity of sound (330 m/s); at higher compression ratios the resulting temperatures are so high that glow phenomena occur even in the absence of an energy-supplying reaction. If the medium is an explosive gas mixture rather than air, it is obvious that an explosive reaction will be instantly initiated in front of the shock wave.

Owing to the sudden pressure effect, all explosions produce a shock wave in the surrounding air; this compression shock is the principle of the long-distance effect of explosions. If the propagation of the shock wave is nearly spherical, the compression ratio $\dfrac{p_1}{p_0}$ decreases ses rapidly, and so does the po velocity of matter W; it becomes zero when the shock wave becomes an ordinary sound wave. If the explosion-generated shock wave is propagated in three-dimensional space, its effect decreases with the third power of the distance; this is the guideline adopted in the German accident prevention regulations, in which the safety distance (in meters) is calculated from the expression $f \cdot \sqrt[3]{M}$ where M is the maximum amount of explosives in kg which are present in the building at any time, whereas f is a factor which varies, according to the required degree of safety, from 1.5 (distance between two barricaded store houses) to 8 (distance from the non-dangerous part of the plant). The f-value stipulated by the regulations may be as high as 20 for residential areas in the vicinity of the plant.

The shock wave theory is easier to understand, if we consider a planar shock wave, such as the one shown in Fig. 11, on the assumption that the tube is indestructible (such shock wave tubes are utilized as research instruments in gas dynamics and in solid state physics; the shock sources are explosions or membranes bursting under pressure).

Comparative treatment of the behavior of the gas in the tube yields the following relationships.

From the law of conservation of mass:

$$\rho_0 D = \rho_1 (D-W) \text{ or } v_1 D = v_0 (D-W) \tag{1}$$

From the law of conservation of momentum:

$$p_1 - p_0 = \rho_0 DW \text{ or } v_0(p_1 - p_0) = DW \tag{2}$$

From the law of conservation of energy:

$$p_1 W = \varrho_0 D \left(e_1 - e_2 + \frac{W^2}{2}\right); \tag{3}$$

Rearrangements yield the so-called *Hugoniot* equation:

$$e_1 - e_0 = \frac{1}{2}(p_1 + p_0)(v_0 - v_1) \tag{4}$$

Equation (4) represents a curve in the $p - v$ diagram, the *Hugoniot* curve.

The following expression is obtained for the velocity D of the shock wave and for the velocity of matter W:

$$D = v_0 \sqrt{\frac{p_1 - p_0}{v_0 - v_1}} \tag{5}$$

and

$$W = \sqrt{(p_1 - p_0)(v_0 - v_1)} \tag{6}$$

These relationships are valid irrespective of the state of aggregation.

2. Detonation Wave Theory

If the medium is explosive, an explosive chemical reaction must be produced immediately in the wave front because of the drastic temperature and pressure conditions. The propagation of the shock wave is maintained by the energy of the reaction.

The equations developed above are still valid, but the meaning of the equation parameters are:

p_1 – detonation pressure;
ρ_1 – density of gaseous products in the front of the shock wave; this density is thus higher than the density of the explosive ρ_0;
D – detonation rate;
W – velocity of gaseous products (fumes).

Equation (1) remains unchanged.

Since ρ_0 is negligibly small as compared to the detonation pressure p1, we can write equation (2) as

$$p_1 = \rho_0 DW \qquad (2\,d)^*$$

The detonation pressure in the wave front is proportional to the product of the density, the detonation rate, and the fume velocity, or – since the fume velocity is proportional to the detonation rate – to the square of the detonation rate. For a given explosive, the detonation velocity rises with increasing density. It is clearly seen from equation (2 d) that the detonation pressure increases very considerably if the initial density of the explosive can be raised to its maximum value – e.g., by casting or pressing – or if the density of the explosive is intrinsically high (TNT 1.64; RDX 1.82; Octogen 1.96). High density of the explosive is important if high → Brisance is needed, whereas the blasting performance (→ Strength) is less affected by it. The importance of the maximum possible compaction of explosives is demonstrated by the → Hollow Charge technique.

Conversely, the detonation pressure and detonation rate may be reduced by reducing ρ_0, i.e., by employing a more loosely textured explosive. This is done if the blasting has to act on softer rocks and if a milder thrust effect is required (see below: explanation of the concept of impedance).

The determination of the maximum detonation pressure p_1, in equation (2 d) has been studied by X ray measurements. While the detonation velocity can be measured directly by electronic recorders or by the → Dautriche Method, there is no direct measurement possibility for the fume velocity W, but it can be estimated by the flow off angle of the fumes behind the wave front; this angle can be taken from X ray flash photographs. The relation between D and W is

---

* Equations of the detonation wave theory are denoted by numbers corresponding to the respective equations of the shock wave theory, with a suffix "d" (for "detonation").
The pressure maximum p_1 in the wave front is also called "Neumann spike" p_N.

$W = \dfrac{D}{\gamma + 1}$; γ is denoted as the "polytrop exponent" in the modified state equation

$p = C\rho^\gamma$ C = const.*)

The value of y is about 3, so that equation (2 d) can be written

$$p_1 = \varrho_0 \dfrac{D^2}{4} \qquad (2\,d)$$

Equation (2) above can be recalculated to

$$p_1 - p_0 (v_0 - v^1) \, \rho_0^2 \, D^2 \qquad (7\,d)$$

represented in the pressure-volume diagram (Fig. 11) by a straight line with the slope $-\rho_0^2 \, D^2$, known as the *Rayleigh* line. The *Hugoniot* equation (4), applied to the detonation process involving the chemical energy of reaction q, becomes:

$$e_1 - e_0 = \dfrac{1}{2}(p_1 + p_0)(v_0 + v_1) + q \qquad (4\,d)$$

Equations (5) and (6) remain unchanged, but D now denotes the detonation rate, while W stands for fume velocity.

In a detonation process, the positions of the Hugoniot curve and the Rayleigh line on the *pv*-diagram are as shown in Fig. 12.

The dotted part of the Hugoniot curve shown in Fig. 12 does not describe real detonation states, because here the term under the square root in equation 5 becomes negative, and D contains the factor −1. The curve now consists of two separate segments: the one situated in the higher pressure area represents detonation, while the one located in the lower pressure area represents → *Deflagration*. The *Rayleigh* line is tangent to the Hugoniot curve at the *Chapman-Jouguet* (CJ) point**) (all state parameters assigned to the "CJ state" are indexed CJ). These parameters describe a "stable" detonation, i. e., a detonation which, unlike a shock wave, can pass through the medium in a stationary manner, that is, at constant intensity and constant velocity. The following equation is then also valid

$$D_{CJ} = W_{CJ} + C_{CJ} \qquad (8\,d)$$

i. e., the detonation rate is the sum of fume velocity and sound velocity.

All the equations given above involve no assumption as to the → *Equation of state* of the medium; they are thus valid irrespective of its state

---

* A detailed report is given by *H. Hornberg*. The State of the Detonation Products of Solid Explosives, Propellants and Explosives 3, 97–106 (1978).
** *Chapman* and *Jouguet* are pioneers of the shock wave theory development; also *Riemann*, *Hugoniot* and *Rayleigh*.

of aggregation. They yield no information as to the thickness of the reaction zone; as a matter of fact, the transitions from v_0 and p_0 to v_1 and Pl are mathematically discontinous. In reality, the thickness of the reaction zone is about 1 mm, and may be deduced from the effects of friction and thermal radiation, which were ignored in the treatment given above. The physical meaning of the imaginary part of the *Hugoniot* curve is that there is no continuous transition between detonation and deflagration. In practice, however, transition between these two phenomena may take place in either direction. Roth*) compared both these types of reactions on → *Nitroglycol*. Table 8 is a comparison of the reaction performance of nitroglycol (ρ_0 = 1.5×10^3 kg/m³**) during detonation and deflagration respectively.

Table 8.

	Deflagration	Detonation
propagation rate D, m/s	3×10^{-4}	7.3×10^3
mass reacted $m = \rho_0 D$, kg/m²s	4.5×10^{-1}	11×10^6
reaction energy q per kg	460 kcal = 1.93×10^{-3} kJ	1600 kcal = 6.7×10^3 kJ
output, kcal/m²s	2.1×10^2	1.8×10^{10}
output ratio deflagration: detonation	about 1:	10^8
width b of reaction zone	1×10^{-2} m	1×10^{-3} m
energetic load of reaction zone $m \cdot q/b$, kcal/m³h	7.5×10^7	6.6×10^{16}

The value of 6.6×10^{16} kcal/m³h for the energetic load may be compared with the maximum value of "only" 10^9 kcal/m³ which can be attained in chemical reactor technology.

The physical treatment of the detonation process involves yet another magnitude known as "impedance"***); this is the product of the density and the detonation rate and represents the material throughput. It has the dimension of a resistance, and reflects the fact that the progress of the detonation through the explosive medium becomes the more difficult, with increasing density of the explosive (i. e., if the density of the explosive has been increased by casting or pressing).

---

* J. F. Roth. Article "Sprengstoffe" in Ullmanns Encyklopädie der technischen Chemie, 3rd ed., Vol. 16, p. 58 (1965).
** The unconventional dimension of kg/m³ is the result of our consistent application of the SI rather than the older CGS system of units. The fundamental SI units are meter, kilogram (mass), second, ampere, Kelvin (K) an Candela, while force, weight, pressure etc. are derived magnitudes. For conversion tables see the back flyleaf of this volume.
*** Sprengtechnik – Begriffe, Einheiten, Formelzeichen. DIN 20, 163 (1973), Beuth-Vertrieb GmbH. *Roth*, Explosivstoffe, Vol. 6, p. 26 (1958)

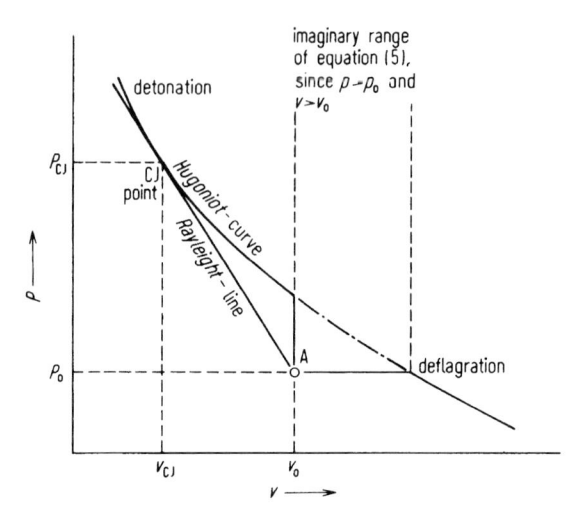

Fig. 12. The *Hugoniot* curve and the Rayleigh line in the p–v diagram.

3. Selective Detonation

Selectivity in the course of a detonation process, as described by *Ahrens*, is noted when processes with very different sensitivities, and thus also with very different induction periods, participate in the intensive chemical reaction (→ 2. *Detonation Wave Theory*) produced by the shock wave. If the intensity of the shock wave is very low owing to external conditions – explosion in an unconfined space, for example – the induction periods of less sensitive reactions may become infinite, i. e., the reaction may fail to take place.

This selectivity is important for ion-exchanged → *Permitted Explosives*. The proportion of the nitroglycerine-nitroglycol mixture in these types of permitted explosives is chosen so that it would just produce a detonation as if it were dispersed in an inert salt bed. The decomposition reaction of the ion exchanged salt pairs $NaNO_3$ (or KNO_3) + NH_4Cl NaCl (or KCl) + N_2 + 2 H_2O + 1/2 O_2 is insensitive and only takes place if the detonation process is favored by confinement; otherwise, the mixture will behave as an inert salt. Thus, if the explosive is detonated while unconfined (e.g. in angle-shot mortar test or because the confinement was destroyed in the previous blast), the only reaction which takes place is that of the nitroglycerine-nitroglycol mixture which is fast and is limited by its relative proportion and is thus firedamp safe. If the explosive is detonated in an undamaged borehole, double decomposition will take place, and the explosive can develop its full strength.

4. Sympathetic Detonation

gap test; flash over; Übertragung; coefficient de self-excitation

These terms denote the initiation of an explosive charge without a priming device by the detonation of another charge in the neighborhood. The maximum distance between two cartridges in line is determined by flash-over tests, by which the detonation is transmitted. The transmission mechanism is complex: by shock wave, by hot reaction products, by flying metallic parts of the casing (if the donor charge is enclosed) and even by the → *Hollow Charge* effect.

In the EU a method for determining the transmission of detonation is standardized as EN 13631-11. Two cartridges are coaxially fixed to a wooden rod with an air gap between them. Depending on the type of explosive the test is done with or without confinement (e. g. steel tube). One cartridge (donor) is initiated and it is noted whether the second cartridge (acceptor) detonates. The complete detonation of the acceptor is verified by measuring the velocity of detonation in it. The result of the test is the largest air gap in cm for which the detonation of the acceptor was proved. For cartridged blasing explosives which shall be used in the EU a minimum transmission distance of 2 cm is required.

In Germany, the ion-exchanged → *Permitted Explosives* are also gap tested in a coal-cement pipe; these are cylinders made of a bonded mixture of cement with coal dust in the ratios of 1:2 and 1:20 and provided with an axial bore.

In the studies so far reported, donor and receiver cartridges consisted of the same explosive. The transmission of a standard donor cartridge through varying thicknesses of a stopping medium can also be employed to determine the sensitivities of different explosives. Recent practice in the United States is to insert cards (playing cards, perspex sheets etc.) between the donor cartridge and the receiver cartridge. Tests of this kind are named gap tests. In a more sophisticated method, the gap medium (e.g. a plexiglas plate, see Fig. 13 below) stops flying particles and directs heat transmission completely (shock-pass heat-filter). The shock wave is the only energy transmission to the acceptor charge.

For a 5 cm long and 5 cm diameter Tetryl donor charge with a density of 1.51 g/cm³, the pressure p in the plexiglas as a function of the plexiglas length d according to *M. Held**) is given by

---

* *M. Held*, Initiierung von Sprengstoffen, ein vielschichtiges Problem der Detonationsphysik, Explosivstoffe **16**, 2–17, (1968) and *J. Jaffe, R. Beaugard* and *Amster*: Determination of the Shock Pressure Required to Initiate Detonation of an Acceptor in the Shock Sensitivity Test – ARS Journal **32**, 22–25, (1962).

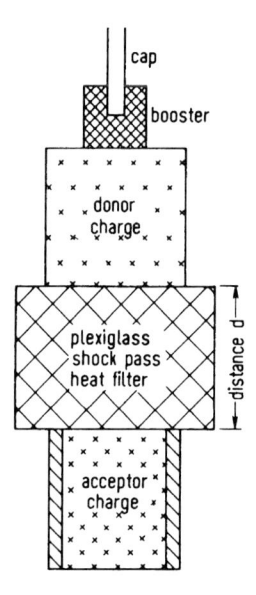

Fig. 13. Gap test

$p = 105\ e^{0.0358d}$

p in kbar, d in mm.

The result of the gap test is recorded as the minimum pressure at which the acceptor charge detonates.

F. *Trimborn* (Explosivstoffe vol. **15**, pp. 169–175 (1967) described a simple method in which water is used as the heat blocking medium; the method can also be used to classify explosives which are hard to detonate and are insensitive to blasting caps.

The gap test explosive train is directed from bottom to top. The donor charge (Hexogen with 5% wax) is placed into a plexiglas tube and covered with water. The acceptor charge to be tested is introduced into the water column from above. The distance between the two charges can be easily varied.

A detonating cord, terminating on a lead plate, serves as evidence for detonation.

Some results: see Table 9.

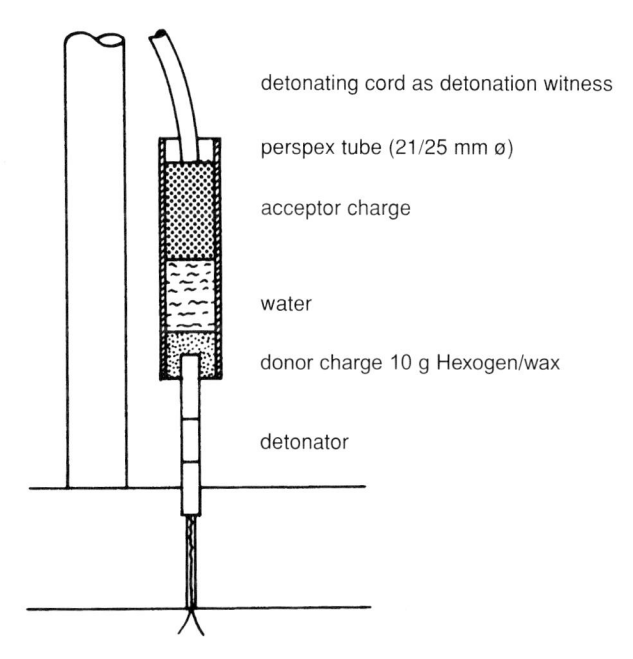

Fig. 14. Gap test according to *Trimborn*

5. Detonation Velocity

Detonationsgeschwindigkeit; vitesse de détonation

The detonation velocity is the rate of propagation of a detonation in an explosive; if the density of the explosive is at its maximum value, and if the explosive is charged into columns which are considerably wider than the critical diameter, the detonation velocity is a characteristic of each individual explosive and is not influenced by external factors. It decreases with decreasing density of packing in the column. It is measured by ionisation probes or fibre optical sensors.

The detonation velocities of confined and unconfined nitroglycerine and nitroglycol explosives have very different values; these values are known as upper and lower detonation velocities respectively. The velocity measured in a steel pipe confinement is not attained in a borehole. Special seismic explosives (e.g. → *Geosit*) detonate at the same high detonation rate as measured in the steel pipe, whether confined or not.

Table 9.

Explosive	State	Density g/cm³	No mm	Detonations at Distance in Water		Initiating Pressure for Detonations kbar
				50% mm	100% mm	
composition B	cast	1.68	18	17	16	17
Hexogen, 5% wax	pressed	1.63	22	21	20	12
PETN, 7% wax	pressed	1.60	29	28	27	7
Pentolite 50/50	cast	1.65	23	22	20	12
picric acid	pressed	1.58	17	16	15	18
Tetryl	pressed	1.53	24	–	23	10
TNT	pressed	1.53	22	21	20	12
TNT	cast	1.58	7	6	5	38
TNT	cast	1.61	6	5;4	2	43

Pressure values comply well with those published in other literature.

6. Detonation Development Distance

Anlaufstrecke; distance d'évolution de détonation

A term denoting the distance required for the full detonation rate to be attained. In initiating explosives, this distance is particularly short.

The detonation development distance, especially that of less sensitive explosives, is strongly affected by the consistency, density and the cross-section of the charge.

References:

Riemann, B.: Abh. Ges. Wiss. Göttingen, Math. Phys. Kl 8, 43 (1860)
Rankine, W. J.: Trans. Roy. Soc. (London) 160, 277–288 (1870)
Hugoniot, H.: Journal de l'ecole polytechnique (Paris) 58, 1–125 (1889)
Becker, R.: Z. Phys. 8, 321–362 (1922)
Jouguet, E.: Proc. Int. Congr. Appl. Mech. 1926, 12–22
Bolle, E.: Explosion und Explosionswellen in: Auerbach und Hort: Handbuch der physikalischen und technischen Mechanik, Leipzig 1928
Schmidt, A.: Z. ges. Schieß- und Sprengstoffw. 27, 145–149; 184–188; 225–228; 264–267; 299–302 (1932) and 33, 280–283; 312–315 (1938)
Bechert, K.: Ann. Phys. (5) 37, 89–123 (1940); (5) 38, 1–25 (1940); (5) 39, 169–202 (1941); (5) 39, 357–372 (1941)
Courant, R. and *Friedrich K. O.*: Supersonic Flow and Shock Waves, Interscience Publ. Inc., New York 1948
Wecken, F. and *Mücke, L.*: Rapport 8/50, Deutsch-Franz. Forschungsinstitut St. Louis 1950
Bowden, F. P. and *Yoffe, A. D.*: Initiation and Growth of Explosions in Liquids and Solids, Cambridge University Press, Cambridge 1952
Taylor, J.: Detonation in Condensed Explosives, Clarendon Press, Oxford 1952
Cook, M. A.: The Science of High Explosives, Reinhold, New York 1958
Roth, J. F.: Explosivstoffe, 23–31; 45–54 (1958)
Zeldovich, J. B. and Kompaneets, *A. S.*: Theory of Detonation, Academic Press, New York and London 1960
Cachia, G. P. and *Whitebread, e. g.*: The initiation of Explosives by Shock, Proc. Ray. Soc. A 246 (1958) 268–273. Card Gap Test for Shock Sensitivity of Liquid Monapropellant, Test Nr. 1, Recommended by the JANAF Panel an Liquid Monopropellants Toot Methods, March 1960
Amster, A. B., Noonan, E. C. and *Bryan, G. J.*: Solid Propellant Detonability, ARS-Journal 30, 960–964 (1960)
Price, D. and *Jaffe, J.*: Large Scale Gap Test: Interpretation of Results for Propellants, ARS-Journal 31, 595–599 (1961)
Wagner, H. G.: Gaseous Detonations and the Structure of a Detonation Zone (in: Fundamental Data obtained from Shock Tube Experiments, Editor: Ferri, A.). Pergamon Press, Oxford 1961
Cook, M. A., Keyes, R. T. and *Ursenbach, W. O.*: Measurements of Detonation Pressure, J. Appl. Phys. 33, 3413–3421 (1962)
Berger, J. and *Viard, J.*: Physique des explosifs solides, Dunod, Paris 1962
Dinegar, R. H., Rochester, R. H. and *Millican, M. S.*: The Effect of Specific

Surface on Explosion Times of Shock Initiated PETN, Am. Chem. Soc., Div. Fuel Chem. 7 (Nr. 3), 17–27 (1963)

Andrejev, K. K. and *Beljajev, A. F.*: Theorie der Explosivstoffe, Svenska National Kommittee for Mechanik, Stockholm 1964 (Translation into German)

Rempel, G. G.: Determination of Speeds of Shock Waves Necessary to Trigger Detonation of Explosives, in: Andrejev, K. K. et. al.: Theory of Explosives (Original Russian, Moscow 1963), Translation into English: Foreign Techn. Div., Wright Patter, Ohio (Clearinghouse) 1964, p. 746–815 (N 65–13494)

Roth, J. F.: Torpedierungsprengungen in großen Tiefen. Prüfung der Sprengstoffe und Zündmittel unter entsprechenden Bedingungen. Nobel Hefte 31, 77–101 (1965)

Mills, E. J.: Hugoniot Equations of State for Plastics: a Comparsion, AIAA-Journal 3, 742–743 (1965)

Zeldovich, J. B. and *Raizer, J.*: Physics of Shock Waves and High Temperature, Hydrodynamic Phenomena, Academic Press, New York, London (1966)

Price, D., Jaffe, J. and *Robertson, G. E.*: Shock Sensitivity of Solid Explosives and Propellants, XXXVI. Int. Kongress f. Industrielle Chemie, Brüssel 1966

Lee, J. H., Knystautas, R. and *Bach, G. G.*: Theory of Explosion, McGill University Press, Montreal 1969

Kamlet, M. J. and *Jacobs, S. J.*: Chemistry of Detonations, a Simple Method for Calculation Detonating Properties of CHNO-Explosives, Journal of Chem. Phys. 48, 23–50 (1968)

Tiggelen, A. van.: Oxydations et Combustion, Tome II, Publications de l'Institut Francais du Petrole, Paris 1968

Johannson, C. H. and *Persson, P. A.*: Detonics of High Explosives, Academic Press, London and New York 1970

Hornberg, H.: The State of the Detonation Products of Solid Explosives, Propellants Explos. 3, 97–106 (1978)

Fickett, W. and *Davis, W. C.*: Detonation, University of California Press, Berkeley 1979

Mader, Ch.: Numerical Modeling of Detonation, University of California Press, Berkeley 1979

LASL Explosive Property Data. Editor: *Gibbs, T. R., Popolato, A.*, University of California Press, Berkeley, California 1980

LASL Phermex Data, Va. 1–3. Editor: Mader, CH. L., University of California Press, Berkeley, California 1980

LASL Shock Hugoniot Data. Editor: *March, St. P.*, University of California Press, Berkeley, California 1980

Explosives Performance Data. Editor: *Mader, Ch. L., Johnson, J. N., Crane, Sh. L.*, University of California Press, Berkeley, California 1982

Shock Wave Prof Data. Editor: *Morris, Ch. E.*, University of California Press, Berkeley, Los Angeles, London 1982

Shock Waves, Explosions and Detonations, Editor: *Bowen, J. R., Manson, N., Oppenheim, A. K.* and *Soloukhin, R. I.*, AIAA, New York 1983 (Progress in Astronautics and Aeronautics, Vol. 87)

Dynamics of Shock Waves, Explosions and Detonations, Editor: *Bowen, J. R., Manson, N., Oppenheim, A. K.* and *Soloukhin, R. I.*, AIAA, New York 1984 (Progress in Astronautics and Aeronautics, Vol. 94)

Kinney, G. F. and *Kenneth, J. G.*: Explosive Shocks in Air, 2nd ed., Springer, Berlin, Heidelberg, New York 1985

Dynamics of Explosions, Editor: *Bowen, J. R., Leyer, J. C.* and *Soloukhin, R. I.*, AIAA, New York 1986 (Progress in Astronautics and Aeronautics, Vol. 106)
Dynamics of Explosions, Editor: *Kuhl, A. L., Bowen, J. R., Leyer, J. C.* and *Borisov, A.*, AIAA, New York 1988 (Progress in Astronautics and Aeronautics, Vol. 114)
Cheret, R.: La detonation des explosifs condenses, part 1 and 2, Masson, Paris 1988/89 Medard, L.: Accidental Explosions, Vol. 1: Physical and Chemical Properties, Vol. 2: Types of Explosive Substances, Ellis Horwood Ltd. Chichester 1989 (English translation)

Detonator

détonateur

Part of an explosive train which initiates the → *detonation* of high explosives, especially of insensitive ones. It may itself be triggert by a seperate → *primer* of primary explosives or an integrated primer. Detonators are classified by the method of initiation: percussion, stab, electrical impulse, or flash. Laser initiation is also used. Depending on the application detonators can include a delay mechanism. Explosive charge placed in certain equipment and set to destroy the equipment under certain conditions (→ *Initiator*).

1,1-Diamino-2,2-dinitroethylene

DADNE, DADE, FOX-7

$$\begin{array}{c} H_2N \\ H_2N \end{array} \!\!\! \diagdown \!\!\! = \!\!\! \diagup \!\!\! \begin{array}{c} NO_2 \\ NO_2 \end{array}$$

Yellow crystals
sum formula: $C_2H_4N_4O_4$
molecular weight: 148.08 g
energy of formation: −119 kJ/mole
enthalpy of formation: −133.9 kJ/mole
oxygen balance: −21.61 %
volume of explosion gases 779 l/kg
heat of explosion (calculated): 4091 J/g (H2O gas);
    4442 J/g (H2O liq.)
density:
    α polymorph 1.89 g/cm^3
    β polymorph 1.80 g/cm^3
specific energy: 1156 J/g
detonation velocity 8869 m/s
deflagration point 215 °C
mech. sensitivity 20−40 Nm (impact); > 550 N (friction)

DADNE is a relatively new, low sensitive high explosive developed by the Swedish Defence Research Agency FOI. It is insoluble in cold water, slightly soluble in acetonitrile and cyclohexanone, soluble in DMSO, dimethylformamide and N-methylpyrrolidinone. Besides, it can cause allergic skin reactions.

DADNE is of interest for the development of IM-propellants and explosives. Three polymorphs α, β and γ were found by means of X-ray diffraction, but only α-polymorph is stable at room temperature. Transitions occur on heating from α to β and β to γ at 113 and 173 °C, respectively.

Diamyl Phthalate

Diamylphthalat; phthalate diamylique

$$\text{C}_6\text{H}_4(\text{CO-O-C}_5\text{H}_{11})_2$$

colorless liquid
empirical formula: $C_{18}H_{26}O_4$
molecular weight: 306.4
energy of formation: -692.0 kcal/kg = -2895.2 kJ/kg
enthalpy of formation: -721.0 kcal/kg = -3016.5 kJ/kg
oxygen balance: -235.0%

Diamyl phthalate is used as an additive to gunpowders, both for the purpose of gelatinization and to effect → *Surface Treatment.*

Diazodinitrophenol

diazodinitrophénol; Dinol, Diazol; D.D.N.P.

$$O_2N\text{-}C_6H_2(NO_2)(O^-)\text{-}N\equiv N^+$$

red yellow amorphous powder
empirical formula: $C_6H_2N_4O_5$
molecular weight: 210.1
energy of formation: $+236.4$ kcal/kg = $+988.9$ kJ/kg
enthalpy of formation: $+220.8$ kcal/kg = $+924.0$ kJ/kg
oxygen balance: -60.9%
nitrogen content: 26.67%
density: 1.63 g/cm^3
lead block test: 326 cm^3/10 g
detonation velocity, confined: 6600 m/s = 21 700 ft/s

at $\rho = 1.5$ g/cm^3
deflagration point: 180 °C = 356°F
impact sensitivity: 0.15 kp m = 1.5 N m

The compound is sparingly soluble in water, soluble in methanol and ethanol, and readily soluble in acetone, nitroglycerine, nitrobenzene, aniline, pyridine, and acetic acid. It rapidly darkerns in sunlight. It is of interest for → *Lead-free Priming Compositions*.

It is prepared by diazotization of → *Picramic Acid* with sodium nitrite in a hydrochloric acid solution with efficient cooling. The dark brown reaction product is purified by dissolution in hot acetone and reprecipitation with iced water.

In the USA, this diazo compound is used as an initiating explosive. It is more powerful than mercury fulminate and slightly less so than lead azide.

For more information on Diazophenols see: *Lowe-Ma, Ch., Robin, A. N.* and *William, S. W.*: Diazophenols – Their Structure and Explosive Properties, Naval Weapons Center, China Lake, CA 9355–6001; Rept.-Nr.: WC TP 6810 (1987)

Dibutyl Phthalate

Dibutylphthalat; phthalate dibutylique

    CO-O-C$_4$H$_9$
    CO-O-C$_4$H$_9$

colorless liquid
empirical formula: $C_{16}H_{22}O_4$
molecular weight: 278.4
energy of formation: −696 kcal/kg = −2913 kJ/kg
enthalpy of formation: −723 kcal/kg = −3027 kJ/kg
oxygen balance: −224.2 %
density: 1.045 g/cm^3
boiling point at 20 mm Hg: 205 °C = 401°F

Dibutyl phthalate is insoluble in water, but is readily soluble in common organic solvents. It is used a as gelatinizer and to effect → *Surface Treatment* in gunpowder manufacture.

Specifications

net content: no less than (analysis by saponification)	99 %
ashes: not more than	0.02 %
density:	1.044–1.054 g/cm^3
reaction in alcoholic solution:	neutral to phenolphthaleine

Diethyleneglycol Dinitrate

Diglykoldinitrat; Dinitrodiglykol; dinitrate de diéthylèneglycol

$$\begin{array}{l} CH_2\text{-}O\text{-}NO_2 \\ | \\ CH_2 \\ \diagdown \\ O \\ \diagup \\ CH_2 \\ | \\ CH_2\text{-}O\text{-}NO_2 \end{array}$$

colorless oil
empirical formula: $C_4H_8N_2O_7$
molecular weight: 196.1
energy of formation: -506.7 kcal/kg = -2120.0 kJ/kg
enthalpy of formation: -532.3 kcal/kg = -2227.3 kJ/kg
oxygen balance: -40.8%
nitrogen content: 14.29%
volume of explosion gases: 991 l/kg
heat of explosion
   (H_2O liq.): 1091 kcal/kg = 4566 kJ/kg
   (H_2O gas): 990 kcal/kg = 4141 kJ/kg
specific energy: 120.2 mt/kg = 1178 kJ/kg
density: 1.38 g/cm³
refractive index: n_{25}^D = 1.4498
melting point: 2 °C = 35.6°F (stable modification)
-10.9 °C = $+12.4$°F (unstable modification)
vapor pressure:

Pressure millibar	Temperature °C	°F
0.0048	20	68
0.17	60	140

lead block test: 410 cm³/10 g
detonation velocity, confined:
   6600 m/s = 21 700 ft/s at ρ = 1.38 g/cm³
deflagration point: 190 °C = 374°F
impact sensitivity: 0.01 kpm = 0.1 Nm

This compound is miscible at ordinary temperatures with nitroglycerine, nitroglycol, ether, acetone, methanol, chloroform and benzene, and even with its precursor compound – the diglycol prior to nitration. It is not miscible with ethanol and is sparingly soluble in carbon tetrachloride. It has a low hygroscopicity and is sparingly soluble in water, but more soluble than nitroglycerin. Its vapors produce headaches, though these are not as strong as those produced by nitroglycol vapors.

Diethyleneglycol dinitrate, like nitroglycerine, is prepared by nitrating diethylene glycol with mixed acid in batches or continuously. The diglycol is produced by synthesis. Since the waste acid is unstable, special formulations of mixed acid must be employed, and the mixed acid must be denitrated at the end of the nitration stage.

Diglycol dinitrate was used extensively in the Second World War by the German side as one of the main components of → *Double Base Propellants*. The explosion heat of diglycol in powder form can be kept lower than the heats of the corresponding nitroglycerine powders; they represented the first step towards the so-called cold powders. Diglycol dinitrate and triglycol dinitrate are also employed in double base rocket propellants.

Specifications for diethyleneglycol as a nitration raw material

clear, colorless liquid	
density (20/4):	1.1157–1.1165 g/cm^3
reaction:	neutral
boiling analysis	241 °C = 466°F
beginning, fifth drop: not below	
distillation at:	246.5 °C = 475.5°F
the end: not above	250 °C = 482°F
moisture: not more than	0.5 %
glow residue: not more than	0.02 %
acidity as H_2SO_4: not more than	0.01 %
chlorides:	traces only
saponification number: not above	0.02 %
reducing substance (test with ammoniacal solution of $AgNO_3$):	none
viscosity at 20 °C = 68°F	35.7 cP

An additional specification was required in Germany:
content of (mono-) ethyleneglycol:
not more than   2 %

determination: 4 cm^3 diethyleneglycol and 4 cm^3 NaOH solution (which contains 370 g NaOH per liter) are mixed and cooled, and 2 cm^3 of copper sulfate solution (which contains 200 g $CuSO_4 \cdot 5\, H_2O$ per liter) is added and shaken. The color is compared with the color obtained by standard mixtures of pure diethyleneglycol with 0.5, 1.5 and 2 % ethyleneglycol after the same reaction.

Differential Thermal Analysis

Thermoanalyse; analyse thermique différentielle

All methods in which the sample to be analyzed is gradually heated and its calorimetric behavior studied. The method includes thermogravimetry (TG) and differential thermal analysis (DTA).

In thermogravimetry, the sample is placed in an oven and heated the desired rate; the loss in weight of the sample is then recorded. Such changes in weight can be due, for example, to the evaporation of hygroscopic moisture, evolution of gases, or chemical decomposition reactions. The thermal balance can also be applied in this manner to the study of thermal stability of explosive materials.

Thermal balance can also be combined with differential thermal analysis. DTA registers small temperature differences, which appear during simultaneous heating of the sample and a standard. In this way all physical and chemical processes, which are accompanied by an additional absorption or evolution of heat by the substance, are recorded. Examples of such processes are changes taking place in the crystal lattice, melting, evaporation, chemical reactions, and decompositions. Thus, the application of DTA gives more selective information about the behavior of explosive materials as a function of the temperature than does the determination of the → *Deflagration Point*.

See also: *Krien*, Explosivstoffe, Vol. 13, p. 205 (1965). An extensive report is given by Krien in an internal paper of the Bundesinstitut für Chemisch-Technische Untersuchungen: Thermoanalytische Ergebnisse der Untersuchung von Sprengstoffen, Az.: 3.0–3/3960/76 (1976).

Diglycerol Tetranitrate

Tetranitrodiglycerol; Tetranitrodiglycerin; tetranitrate de diglycérine

CH_2——O——CH_2
CH-O-NO_2 CH-O-NO_2
CH_2-O-NO_2 CH_2-O-NO_2

yellow oil
empirical formula: $C_6H_{10}N_4O_{13}$
molecular weight: 346.2
oxygen balance: −18.5%
nitrogen content: 16.18%
density: 1.52 g/cm^3
lead block test: 470 cm^3/10 g
impact sensitivity: 0.15 kp m = 1.5 N m

Pure tetranitrodiglycerol is a very viscous oil, which is non-hygroscopic, insoluble in water, and readily soluble in alcohol and ether. It has a lower explosive power than nitroglycerine, is less sensitive to impact, and its gelatinizing effect on nitrocellulose is not as satisfactory.

Prolonged heating of glycerol yields diglycerol and a small amount of other polyglycerols. If such mixtures of glycerol and diglycerol are nitrated, mixtures of nitroglycerol and tetranitroglycerol are obtained; they have a lower solidification temperature than pure nitroglycerine.

Tetranitrodiglycerol was used in the manufacture of non-freezing dynamites when sufficient quantities of glycol from largescale industrial syntheses were not available.

Diluent

An additive, usually inert, used to regulate the burning rate or temperature.

Dimethylhydrazine, unsymmetrical

Dimethylhydrazin; diméhylhydrazine; UDMH

$$H_2N-N\begin{smallmatrix}CH_3\\CH_3\end{smallmatrix}$$

colorless liquid
empirical formula: $C_2H_8N_2$
molecular weight: 60.1
energy of formation: +247 kcal/kg = +1035 kJ/kg
enthalpy of formation: +198 kcal/kg = +828 kJ/kg

UDMH is used in liquid-fuel rockets both as fuel and as – *Monergol* by catalytic decomposition. Precision pulses in U.S. space technique are givon by UDMH.

Dingu and Sorguyl*)

dinitroglycolurile and tetranitroglycolurile; glycolurile dinitramine et glycolurile tétranitramine

The reaction between glyoxal O=CH–CH=O and urea H_2N–C=O–NH_2 yields glycolurile with the structural formula

$$\begin{array}{c} \quad\quad H \\ \quad\quad | \\ HN-C-NH \\ | \quad\quad | \\ O=C \quad\quad C=O \\ | \quad\quad | \\ HN-C-NH \\ \quad\quad | \\ \quad\quad H \end{array}$$

The dinitration of the compound yields "Dingu":

$$\begin{array}{ccc}
\begin{array}{c} H \\ | \\ HN-C-NNO_2 \\ |\quad\quad | \\ O=C \quad C=O \\ |\quad\quad | \\ O_2NN-C-NH \\ | \\ H \end{array} \text{or} &
\begin{array}{c} H \\ | \\ O_2NN-C-NNO_2 \\ |\quad\quad | \\ O=C \quad C=O \\ |\quad\quad | \\ HN-C-NH \\ | \\ H \end{array} \text{or} &
\begin{array}{c} H \\ | \\ HN-C-NNO_2 \\ |\quad\quad | \\ O=C \quad C=O \\ |\quad\quad | \\ HN-C-NNO_2 \\ | \\ H \end{array}
\end{array}$$

colorless crystals
empirical formula: $C_4H_4N_6O_6$
molecular weight: 232.1
oxygen balance: −27.6 %
nitrogen content: 36.21 %
density: 1.94 g/cm^3
detonation velocity, confined: 7580 m/s = 24 900 ft/s at ρ = 1.75 g/cm^3
misfire at maximum density
deflagration point: 225–250 °C = 437–482 °F
decomposition begins at 130 °C = 266 °F
impact sensitivity: 0.5–0.6 kp m = 5–6 N m
friction sensitivity: 20–30 kp = 20–300 N pistil load

The product is easily decomposed by alkaline hydrolysis. It is stable in contact with neutral or acid water. It is insoluble in most solvents and in molten TNT; it is soluble in dimethylsulfoxide (DMSO).

Nitration with a $HNO_3 - N_2O_5$ mixture yields the tetranitramine "Sorguyl":

$$\begin{array}{c} H \\ | \\ O_2NN-C-NNO_2 \\ |\quad\quad | \\ O=C \quad C=O \\ |\quad\quad | \\ O_2NN-C-NNO_2 \\ | \\ H \end{array}$$

---

* Dingu and Sorguyl were developed by SOCIÉTÉ NATIONALE DES POUDRES ET EXPOLOSIVS, Sorgues, France.

colorless crystals
empirical formula: $C_4H_2N_8O_{10}$
molecular weight: 322.1
oxygen balance: +5.0%
nitrogen content: 34.79%
density: 2.01 g/cm^3
detonation velocity, confined:
9150 m/s = 30000 ft/s at ρ = 1.95 g/cm^3
deflagration point: 237 °C = 459°F
impact sensitivity: 0.15−0.2 kp m = 1.5−2 N m

The product is interesting because of its high density and also high detonation velocity.

Sorguyl is not hygroscopic, but it decomposes easily by hydrolysis. It is insoluble in hydrocarbons and chlorinated hydrocarbons, but soluble in numerous solvents.

It decomposes when mixed with molten → *TNT.*

Dinitrobenzene

→ Metadinitrobenzene

4,6-Dinitrobenzofuroxan

4,6-Dinitrobenzfuroxan; 4,6-dinitrobenzofurazan-1 -oxide; dinitro-dinitrosobenzene

yellow-gold needles
empirical formula: $C_6H_2N_4O_6$
molecular weight: 226.1
oxygen balance: −49,5%
nitrogen content: 24,78%
melting point: 172 °C

Dinitrobenzofuroxan is practically insoluble in water, alcohol and benzine. It is readily soluble in aromatic hydrocarbons and boiling acetic acid.

The compound is obtained by means of direct nitrating of benzofurazan-1-oxide with concentrated nitric and sulfuric acid, or by heating → *Trinitrochlorbenzene (Picrylchloride)* with sodium azide in acetic acid in a water bath.

Dinitrobenzofuroxan has a somewhat more explosive power than *Picric Acid*, but due to its slightly acidic properties and its relatively high production cost it has yet to become widely-used.

Of particular interest are the potassium and barium salts, both of which are thermally very stable and low → *Initiating Explosive* materials. In the categories of impact and friction sensitivity, the potassium-dinitrobenzofuroxan (KDNBF) falls between → *Mercury Fulminate* and → *Lead Azide*. It has been used mainly in the USA in explosive-initiating compositions for both military and commercial applications since the early 1950s.

Dinitrochlorobenzene

1,2,4-Chlordinitrobenzol; dinitrochlorbenzène

pale yellow crystals
empirical formula: $C_6H_3N_2O_4Cl$
molecular weight: 202.6
energy of formation: -13.8 kcal/kg = -57.8 kJ/kg
enthalpy of formation: -28.6 kcal/kg = -120 kJ/kg
oxygen balance: -71.1%
nitrogen content: 13.83%
density: 1.697 g/cm^3
boiling point: $315\,°C = 599°F$
solidification point: $43\,°C = 109°F$ (isomere mixture)
lead block test: 225 cm^3/10 g
deflagration point: evaporation without deflagration
impact sensitivity: up to 5 kp m = 50 N m no reaction
friction sensitivity: up to 36 kp = 353 N pistil load no reaction
critical diameter of steel sleeve test: at 1 mm⌀ no reaction

Dinitrochlorobenzene is insoluble in water, but is soluble in hot ethanol, ether and benzene.

It is prepared by nitration of chlorobenzene, which yields a mixture of the 2,4- and the 2,6-isomers, with melting points of 53.4 °C (127.5°F) and 87–88 °C (190–192°F) respectively.

Dinitrochlorobenzene is not an explosive. It serves as an intermediate in many syntheses (→ *Hexanitrodiphenylamine; Trinitrochlorobenzene; Trinitroaniline;* etc.).

Dinitrodimethyloxamide

Dinitrodimethyloxamid; dinitrodiméthyloxamide

$$\begin{array}{l} CO-N(NO_2)-CH_3 \\ | \\ CO-N(NO_2)-CH_3 \end{array}$$

colorless needles
empirical formula: $C_4H_6N_4O_6$
molecular weight: 206.1
energy of formation: -331.2 kcal/kg = -1385.8 kJ/kg
enthalpy of formation: -354.2 kcal/kg = -1482.0 kJ/kg
oxygen balance: -38.8%
nitrogen content: 27.19%
density: 1.523 g/cm^3
lead block test: 360 cm^3/10 g
detonation velocity, confined: 7100 m/s = 23 300 ft/s
  at $\rho = 1.48$ g/cm^3
impact sensitivity: 0.6 kp m = 6 N m

The compound is insoluble in water, sparingly soluble in ether and chloroform and soluble in acetone. It is chemically stable.

It is prepared by nitration of dimethyloxamide with a sulfuric acidnitric acid mixture.

Dinitrodioxyethyloxamide Dinitrate

Dinitrodioxiethyloxamiddinitrat, Dinitrodiethanoloxamiddinitrat, Neno; dinitrate de dioxyéthyl-dinitroxamide; N, N'-dinitro-N, N' bis (2-hydroxyethyl)-oxamide dinitrate

$$\begin{array}{l} NO_2 \\ | \\ CO-N-CH_2-CH_2-O-NO_2 \\ | \\ CO-N-CH_2-CH_2-O-NO_2 \\ | \\ NO_2 \end{array}$$

colorless flakes
empirical formula: $C_6H_8N_6O_{12}$
molecular weight: 356.2
energy of formation: -355.5 kcal/kg = -1487.2 kJ/kg
enthalpy of formation: -377.1 kcal/kg = -1577.7 kJ/kg
oxygen balance: -18.0%
nitrogen content: 23.60%
melting point: 88 °C = 190°F

This compound is readily soluble in acetone and in hot alcohol, and is insoluble in cold water. It is prepared by nitration of diethanoloxamide, the latter being prepared by condensation of monoethanolamine with oxalic acid.

Dinitrodiphenylamine

Dinitrodiphenylamin; dinitrodiphénylamine

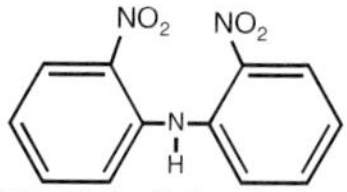

2,2'-Dinitrodiphenylamine

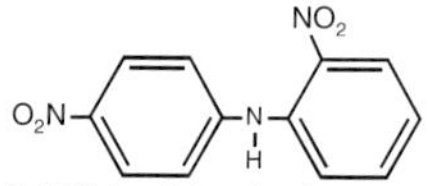

2,4'-Dinitrodiphenylamine

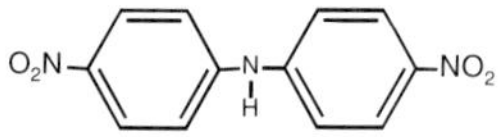

4,4'-Dinitrodiphenylamine

red crystals
empirical formula: $C_{12}H_9N_3O_4$
molecular weight: 259.2
energy of formation: +39.4 kcal/kg = +165 kJ/kg
enthalpy of formation: +21.1 kcal/kg = +88.3 kJ/kg
oxygen balance: −151.2 %
nitrogen content: 16.22 %
density: 1.42 g/cm^3
melting point
  2,2'-isomer: 173 °C
  2,4–isomer: 220 °C
  2,4'-isomer: 156–167 °C
  2,6–isomer: 107 °C
  4,4'-isomer: 217–218 °C

Dinitrondiphenylamine is formed in nitrocellulose propellants stabilized by diphenylamine (→ *Stability*).

Dinitroaphthalene

1,5-; 1,8-Dinitronaphthalin; dinitronaphthalène; Dinal

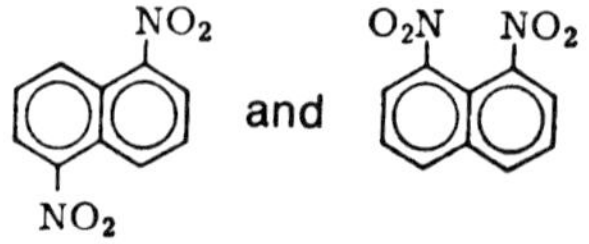

1,5-   and   1,8-

grey yellow powder
empirical formula: $C_{10}H_6N_2O_4$

molecular weight: 218.2
energy of formation:
1,5–isomer: +49.7 kcal/kg = +208.1 kJ/kg
1,8–isomer: +57.5 kcal/kg = +240.7 kJ/kg
enthalpy of formation:
1,5–isomer: +33.5 kcal/kg = +140.0 kJ/kg
1,8–isomer: +41.25 kcal/kg = +172.6 kJ/kg
oxygen balance: −139.4%
nitrogen content: 12.84%
volume of explosion gases: 488 l/kg
heat of explosion (H_2O liq.):
1,5–isomer: 725 kcal/kg = 3031 kJ/kg
1,8–isomer: 732 kcal/kg = 3064 kJ/kg
specific energy: 58 mt/kg = 569 kJ/kg
melting point: 1,5–isomer: 216 °C = 421°F
1,8–isomer: 170 °C = 338°F
deflagration point: 318 °C = 605°F

This material is prepared by a two-step nitration of naphthalene with nitric acid. The commercial product, which is a mixture of isomers, melts above 140 °C = 276°F. It is readily soluble in benzene, xylene, and acetone and is sparingly soluble in alcohol and ether. It has been used in French explosive mixtures (schneiderites) as fuel mixed with ammonium nitrate.

Dinitroorthocresol

Dinitro-o-kresol; dinitroorthocrésol

$$O_2N-C_6H_2(OH)(CH_3)-NO_2$$

yellow crystals
empirical formula: $C_7H_6N_2O_5$
molecular weight: 198.1
energy of formation: −221.8 kcal/kg = −928.1 kJ/kg
enthalpy of formation: −241.3 kcal/kg = −1009.4 kJ/kg
oxygen balance: −96.9%
nitrogen content: 14.51%
volume of explosion gases: 832 l/kg
heat of explosion (H_2O liq.): 724 kcal/kg = 3027 kJ/kg
specific energy: 70.5 mt/kg = 691 kJ/kg
melting point: 86 °C = 187°F
impact sensitivity: up to 5 kp m = 50 N m no reaction

friction sensitivity:
up to 36 kp = 353 N pistil load no reaction

o-Dinitrocresol is prepared by introducing o-nitrophenyl glyceryl ether into mixed acid at 25–30 °C = 77–85°F. It is insoluble in water and readily soluble in acetone; it is a poor gelatinizer of nitrocellulose.

Dinitrophenoxyethylnitrate

Dinitrophenylglykolethernitrat; nitrate de 2,4-dinitrophénoxyéthyle

$$\text{CH}_2\text{-O-C}_6\text{H}_3(\text{NO}_2)_2$$
$$|$$
$$\text{CH}_2\text{-O-NO}_2$$

pale yellow crystals
empirical formula: $C_8H_7N_3O_8$
molecular weight: 273.2
energy of formation: -236.8 kcal/kg = -990.6 kJ/kg
enthalpy of formation: -256.3 kcal/kg = -1072.2 kJ/kg
oxygen balance: -67.4%
nitrogen content: 15.38%
density: 1.60 g/cm^3
solidification point: 64 °C = 147°F
lead block test: 280 cm^3/10 g
detonation velocity, confined:
6800 m/s = 22 300 ft/s at ρ = 1.58 g/cm^3
deflagration point: over 300 °C = 570°F
impact sensitivity: 2 kp m = 20 N m

The compound is insoluble in water, but soluble in acetone and toluene. It is prepared by dissolving phenyl glycol ether in sulfuric acid and pouring the reaction mixture into mixed acid at 10–20 °C (50–68°F).

It is a nitrocellulose gelatinizer.

Dinitrophenylhydrazine

Dinitrophenylhydrazin

$$\text{C}_6\text{H}_3(\text{NH-NH}_2)(\text{NO}_2)_2$$

empirical formula: $C_6H_6N_4O_4$
molecular weight: 198.1

energy of formation: +81.2 kcal/kg = +339.6 kJ/kg
enthalpy of formation: +60.3 kcal/kg = +252.1 kJ/kg
oxygen balance: −88.8 %
nitrogen content: 28.28 %

According to the studies performed by the Bundesanstalt für Materialprüfung, Germany (BAM), this compound may explode when dry, but in the presence of 20 % water there is no longer any danger of explosion. It is widely used in analytical organic chemistry for the preparation of dinitrophenylhydrazon and its derivates from ketones and aldehydes.

Dinitrosobenzene

Dinitrosobenzol; dinitrosobenzène

empirical formula: $C_6H_4N_2O_2$
molecular weight: 136.1
oxygen balance: −141 %
nitrogen content: 20.58 %
melting point: decomposition
lead block test: 138 cm^3/10 g
deflagration point: 178−180 °C = 352−355°F
impact sensitivity: 1.5 kp m = 15 N m
friction sensitivity:
up to 36 kp = 353 N pistil load no reaction
critical diameter of steel sleeve test: 2 mm

This substance is explosive despite its low oxygen content. It will explode in a 1-in steel pipe if actuated by a primer.

Dinitrotoluene

Dinitrotoluol; dinitrotoluène; DNT

2,4-isomer   2,6-isomer

yellow needles
empirical formula: $C_7H_6N_2O_4$
molecular weight: 182.1
energy of formation:
   2,4-isomer: −70.0 kcal/kg = −292.8 kJ/kg
   2,6-isomer: −38.1 kcal/kg = 159.5 kJ/kg
enthalpy of formation:
   2,4-isomer: −89.5 kcal/kg = −374.7 kJ/kg
   2,6-isomer: −57.6 kcal/kg = −241.2 kJ/kg
oxygen balance: −114.4 %
nitrogen content: 15.38 %
volume of explosion gases: 807 l/kg
heat of explosion:
   2,4-isomer, (H_2O liq.): 763 kcal/kg = 3192 kJ/kg
              (H_2O gas): 729 kcal/kg = 3050 kJ/kg
   2,6-isomer, (H_2O liq.): 795 kcal/kg = 3325 kJ/kg
              (H_2O gas): 761 kcal/kg = 3183 kJ/kg
specific energy: 70 mt/kg = 687 kJ/kg
density: 2,4-isomer 1.521 g/cm^3
          2,6-isomer 1.538 g/cm^3
melting point, pure 2,4-isomer: 70.5 °C = 159°F
   natural isomer mixture: about 35 °C = 95°F
vapor pressure of the 2,4-isomer:

Pressure millibar	Temperature °C	°F
0.014	35	95
0.11	70	158 (melting point)
0.83	100	212
8.5	150	302
50.5	200	392
223	250	482
300	300	572

heat of fusion:
  2,4-isomer: 26.1 kcal/kg = 109 kJ/kg
  2,6-isomer: 22.5 kcal/kg = 94 kJ/kg
lead block test: 240 cm^3/10 g
deflagration point: ignition at 360 °C = 680°F
impact sensitivity: up to 5 kp m = 50 N m no reaction
friction sensitivity:
  up to 36 kp = 353 N pistil load no reaction
critical diameter steel sleeve fest: 1 mm

Dinitrotoluene is sparingly soluble in water, alcohol and ether, but readily soluble in acetone and benzene. It is formed as an intermediate in → *TNT* Synthesis.

The product, which is obtained as a low-melting mixture of six isomers, is an important component in the manufacture of both gelatinous and powdery commercial explosives; owing to its negative oxygen balance, it also serves as a carbon carrier. It is readily miscible with nitroglycerine and gelatinizes soluble guncotton.

A purer product, consisting mainly of the 2,4-isomer, is also employed as a component of gunpowder.

The MAK value is 1.5 mg/m^3.

Specifications

moisture: not more than	0.25%
benzene insolubles: not more than	0.10%
acidity as H_2SO_4: not more than	0.02%
tetranitromethane:	none
solidification point,	
gunpowder grade	68.0 ± 2.5 °C (154°F)
for industrial explosives:	as low as possible

Table 10. Data of the other DNT isomers.

Dinitrotoluene Isomer	Density g/cm^3	Melting Point °C	Melting Point °F	Energy of Formation kcal/kg	Energy of Formation kJ/kg	Enthalpy of Formation kcal/kg	Enthalpy of Formation kJ/kg
2,3-	1.2625	59.5	139	−1.1	−4.6	−20.9	− 87.5
2,5-	1.2820	50.5	123	−25.3	−106	−45.0	−188
3,4-	1.2594	59.5	139	0	0	−19.2	− 80.4
3,5-	1.2772	93	199.5	−37.3	−156	−57.1	−239

Dioxyethylnitramine Dinitrate

Nitrodiethanolamindinitrat;
dinitrate de dioxydthylnitramine; DINA

$$\text{N-NO}_2 \begin{cases} \text{CH}_2\text{-CH}_2\text{-O-NO}_2 \\ \text{CH}_2\text{-CH}_2\text{-O-NO}_2 \end{cases}$$

colorless crystals
empirical formula: $C_4H_8N_4O_8$
molecular weight: 240.1
energy of formation: -249.8 kcal/kg = -1045.1 kJ/kg
enthalpy of formation: -274.4 kcal/kg = -1148.2 kJ/kg
oxygen balance: -26.6%
nitrogen content: 23.34%
volume of explosion gases: 924 l/kg
heat of explosion
 (H_2O liq.): 1304 kcal/kg = 5458 kJ/kg
 (H_2O gas): 1201 kcal/kg = 5025 kJ/kg
specific energy: 133 mt/kg = 1306 kJ/kg
density: 1.488 g/cm^3
melting point: 51.3 °C 124.3 °F
detonation velocity, confined:
 7580 m/s = 25 000 ft/s at ρ = 1.47 g/cm^3
impact sensitivity: 0.6 kp m = 6 N m

This compound is prepared from diethanolamine and nitric acid with acetic anhydride as a dehydrating agent and in the presence of hydrochloric acid as a catalyst. The nitration product is stabilized by boiling in water, followed by dissolution in acetone and reprecipitation with water.

It is a satisfactory gelatinizer for nitrocellulose and is a powerful explosive, comparable to Hexogen and PETN. Double base propellants based on DINA instead of nitroglycerine are named "Albanite".

Dipentaerythritol Hexanitrate

Hexanitrodipentaerythrit; hexanitrate de dipentaérythrite; DIPEHN

$$\begin{matrix} O_2N\text{-O-H}_2C \\ O_2N\text{-O-H}_2C\text{-C-CH}_2\text{-O-CH}_2\text{-C} \\ O_2N\text{-O-H}_2C \end{matrix} \begin{matrix} CH_2\text{-O-NO}_2 \\ CH_2\text{-O-NO}_2 \\ CH_2\text{-O-NO}_2 \end{matrix}$$

colorless crystals
empirical formula: $C_{10}H_{16}N_6O_{19}$
molecular weight: 524.2
energy of formation: -424.2 kcal/kg = -1771 kJ/kg
enthalpy of formation: -446 kcal/kg = -1867 kJ/kg

oxygen balance: -27.5%
nitrogen content: 16.03%
volume of explosion gases: 878 l/kg
heat of explosion
   (H_2O liq.): 1229 kcal/kg = 5143 kJ/kg
   (H_2O gas): 1133 kcal/kg = 4740 kJ/kg
specific energy: 125 mt/kg = 1223 kJ/kg
density: 1.63 g/cm^3
melting point: 72 °C = 162°F
detonation velocity, confined:
   7400 m/s = 24300 ft/s at $\rho = 1.6$ g/cm^3
deflagration point: 200 °C = 392°F
impact sensitivity: 0.4 kp m = 4 N m

The compound is soluble in acetone, but insoluble in water. When technical grade pentaerythritol is nitrated, a certain amount of di-pentaerythritol hexanitrate is formed as a by-product.

Diphenylamine

Diphenylamin; diphénylamine

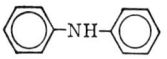

$\bigcirc$–NH–$\bigcirc$

colorless crystals
empirical formula: $C_{12}H_{11}N$
molecular weight: 169.2
energy of formation: +204.6 kcal/kg = +856.0 kJ/kg
enthalpy of formation: +183.6 kcal/kg = +768.2 kJ/kg
oxygen balance: -278.9%
nitrogen content: 8.28%
density: 1.16 g/cm^3
melting point: 54 °C = 129°F
boiling point: 302 °C = 576°F

Diphenylamine is sparingly soluble in water, but is readily soluble in alcohol and acids. It may be used as reagent for nitric acid and nitrates. Its use as a → *Stabilizer* is particularly important.

Specifications

solidification point:	51.7–53 °C = 125–127.4°F
insolubles in benzene: not more than	0.02%
moisture: not more than	0.2%
solution in ether-alcohol:	clear
ashes: not more than	0.05%
aniline: not more than	0.1%

acidity; as HCl: not more than 0.005 %
alkalinity, as NaOH: not more than 0.005 %

Diphenylurethane

Diphenylurethan; diphénylurethane

$$O=C\begin{smallmatrix}N-C_6H_5\\ C_6H_5\\ O-C_2H_5\end{smallmatrix}$$

empirical formula: $C_{15}H_{15}NO_2$
molecular weight: 241.3
energy of formation: -256.0 kcal/kg = -1071.1 kJ/kg
enthalpy of formation: -278.1 kcal/kg = -1163.5 kJ/kg
oxygen balance: -235.4 %
nitrogen content: 5.81 %

Diphenylurethane is used as a gunpowder stabilizer and gelatinizer.

Specifications

snow-white powder
solidification point:
not less than   70 °C = 158°F
melt: clear, colorless
volatiles: not more than   0.1 %
ashes: not more than   0.1 %
insolubles in ether   none
chlorides, as NaCl: not more than   0.02 %
reaction: neutral
acidity, n/10 NaOH/100 g:
not more than   0.1 cm^3

Dipicrylurea

Hexanitrocarbanilid; dipicrylurée

$$O_2N-\underset{NO_2}{\overset{NO_2}{\bigcirc}}-NH-CO-NH-\underset{NO_2}{\overset{NO_2}{\bigcirc}}-NO_2$$

pale yellow crystals
empirical formula: $C_{13}H_6N_8O_{13}$
molecular weight: 482.2
oxygen balance: -53.2 %
nitrogen content: 23.24 %
melting point: 208–209 °C = 406–408°F
deflagration point: 345 °C = 655°F

Dipicrylurea is prepared by nitration of carbanilide in one or more stages.

Dismantling of Explosive Objects, Especially Ammunition

A principal distinction must be made between two kinds of ammunition: ammunition of known origin, which has been properly stored and which has to be separated out for routine reasons (aging; replacement by other types of ammunition), and ammunition found lying around or acquired as booty. The latter kind of ammunition may well have been exposed to strong corrosive agents, and its delayed action fuses may no longer be controllable. Handling abandoned ammunition is one of the most dangerous tasks in explosive handling and must be left to top experts in this field (familiarity with the regulations concerning explosive substances is NOT enough); this includes the very first attempt to move the ammunition while still in situ. A detailed discussion of the deactivation of abandoned ammunition is beyond the scope of this book.

Explosive objects are classified according to the potential danger they present. The criteria of such a classification include the nature of the explosive object, whether or not they contain detonators or primers, and whether or not they present a → *Mass Explosion Risk.* Dangerous mechanical tasks, such as unscrewing the detonators or sawing them off, cutting, milling, or sawing, must in any case be performed under remote control.

Fusible explosives such as TNT and TNT mixtures may be melted out of their containers (grenades, bombs, mines) after removal of detonators and booster charges. The material thus obtained may be purified and re-used for non-military purposes.

Case bonded → *Composite Propellants* are unloaded from their casing by a remote controlled lathe or water gun; also → *Case Bonding*

Also → *Destruction of Explosive Materials*

Ditching Dynamite

A mixed dynamite, containing about 50 % non-gelatinized nitroglycerine, used for ditch blasting. This explosive displays a particularly strong tendency to flash over. Usually only the first charge is initiated by the cap. The following charges for the excavation of the "ditch" are exploded by the effect of the first detonation (shock wave) (→ *Detonation, Sympathetic Detonation*).

Dithekite

A U.S. trade name for an explosive liquid mixture of nitric acid, nitrobenzene and water.

Donarit

Trade names of ammonium-nitrate based nitroglycerine-sensitized powder-form explosives distributed in the Germany and exported by WASAGCHEMIE.

Donor Charge

Geberladung; charge excitatrice

An exploding charge producing an impulse that impinges upon an explosive *"Acceptor"* charge.

Double Base Propellants

nitroglycerine powders; POL-Pulver;
poudres a base de la nitroglycérine

This term denotes propellants containing two main components: nitrocellulose and nitroglycerine or other liquid nitrate esters. Double base powders are important solid rocket propellants.

Double base compositions can be manufactured without the application of organic solvents by heated rolling and pressing of → *Paste*.

Drop Test

→ *Bomb Drop Test*

Dutch Test

Holland Test

A method developed in 1927 by the Dutchman *Thomas* for the determination of the chemical stability of propellants. The parameter which is determined in the method is the weight loss which takes place after 72-hours heating at 105 °C (221°F) (multibase propellants) or at 110 °C (230°F) (single-base propellants). This loss, after subtracting the loss occurring after the first 8 hours of heating, must not exceed 2%.

An advantage of this test is that not only nitrogen oxides, but also all the other decomposition products of the propellants – in particular CO_2 and N_2 – are determined by it. In order to work under reproducible experimental conditions, precision-stoppered tubes of an identical type, equipped with calibrated capillaries, are employed.

Since the heating temperature is rather high, especially so for multi-base powders, it was proposed by *Siebert* to determine the weight loss at a lower temperature and not to limit the duration of heating, but to continue it until some auto-catalytic or other evident decomposition becomes apparent. This test, which should be carried out at 90, 75 and 65 °C (149°F), may also be employed to indicate the loss of stability on storage (shelf life) of a propellant.

Dwell Time

In press loading of powders into cavities, the dwell time is the interval of time that the powder is held at the full loading pressure.

Dyno Boost®

Dyno Boost® is the trade name of a booster charge made by Orica. The system, which consists of high-power explosive, can be detonated using a standard blasting cap.

Density:	1.6 g/ml
Weight:	1.7 kg
Detonation velocity:	7000 m/s

Dynacord®

Trade name of a detonating cord distributet in Germany and exportet by Orica Germany GmbH.

Dynaschoc®

Dynaschoc® is the trade name of a non-electrical detonating system (see also → *Nonel*) made by Orica. In this system the detonation pulse is propagated at about 2000 m/s via a thin plastic tube whose internal surface is dusted with about 16 mg of explosive per m. The tube is not destroyed by this detonation pulse.

Dynatronic®

Dynatronic® is the trade name of a series of programmable detonators together with the associated programming and control devices made by Orica.

Dynamite LVD; MVD

Compositions for defined detonation velocities:

LVD (low-velocity dynamite):
RDX	17.5%
TNT	67.8%
PETRIN	8.6%
binder (Vistac and DOS)	4.1%
acetylcellulose	2.0%

MVD (medium-velocity dynamite):
RDX	75%
TNT	15%
starch	5%
oil	4%
Vistanex oil gel	1%

Dynamites

Dynamite was the first trade name introduced for a commercial explosive by *Alfred Nobel;* it was nitroglycerine absorbed in kieselguhr (Guhr dynamite). Bonding of nitroglycerine by gelatinization with nitrocellulose was discovered by *Nobel at* a later date.

At first, active absorbents such as a mixture of sodium nitrate with wood dust were employed instead of the inert kieselguhr. The result was the development of nitroglycerine-sensitized powdery explosives, which are still known as "dynamites" in English-speaking countries; → also *Ditching Dynamite.*

Variations in the concentration of gelatinized nitroglycerine (the concentrated product is known as blasting gelatine) by the addition of sodium nitrate and wood dust or cereal meal yielded gel dynamites, which are known as "gelignites" in English-speaking countries; in Germany, the old designation of "dynamite" has been retained. In the meantime, they have been placed by ammonium-nitrate-based → *Ammongelit.* These products contain nitroglycol rather than nitroglycerine, with improvement in the safety of handling and transportation.

Dynamites are no longer manufactured in Germany.

Ednatol

A cast explosive charge employed in the USA. It consists of a mixture of → *Ethylendinitramine* and TNT in the ratio of 55:45.

    casting density: 1.62 g/cm^3
    detonation velocity at casting density,
    confined: 7300 m/s = 23 900 ft/s

EED

Electro-explosive device; elektrischer Zünder

Any detonator or initiator initiated by an electric current.

One-Ampere/One-Watt Initiator = EED that will not fire when one ampere of current at one watt of power is supplied to a bridgewire for a specified time.

Emulsion Slurries

Emulsion Explosives

Emulsion slurries (→ *Slurries*) are based on a "water-in-oil emulsion" which is formed from a saturated nitrate solution and a mineral oil phase. Additions controlling the density (formation of gas bubbles or → *Microballoons*) are used to achieve a density that can be adjusted within an range between primer sensitivity (booster charge) and cap sensitivity.

The density is slightly higher when compared with water gels and results in higher performance; explosion temperature and detonation pressure are higher. There is a positive oxygen balance. Due to the fact that the mixture is substantially more intimate, there are differences in detonation kinetics compared to water gels. The blasting efficiency is higher, particularly in hard rock.

Emulsion slurries can be applied using mobile pumping and mixing devices (in large hole blasting), as well as in the form of cartridges of varying length and diameter. In cartridged form, the emulsion slurries are replacing the "classic" nitroglycerine-nitroglycol-based gelatinous explosives.

End of Burning

Brennschluß; fin de combustion

The moment at which emission of the gas jet by a rocket ends. In solid propellant rockets this moment corresponds to a complete burnout of the propellant; in liquid fuel and hybrid rockets, reignition can take place.

End Burning Velocity

Brennschlußgeschwindigkeit; vitesse en fin de combustion

Velocity attained by a rocket at the moment at which combustion ceases. It is a function of the → *Gas Jet Velocity*, the → *Mass Ratio*, and the burning time.

Endothermal

Reaction that occurs with the absorption of heat (opposite of *exothermal*).

Energetic Binders

Energetische Binder; Aktive Binder; active Binders

In the realm of modern, nitric-ester-free → LOVA Gun Propellants, the widely used inert binders consume energy and to some extend have an undesirably high overall phlegmatizing effect on the explosivematerial. The high filler content of an explosive substance has a detrimental effect on the mechanical strength of this type of propellant.

An active binder would be a preferable alternative if it combined a high energy content with favorable mechanical properties, together with a thermal stability higher than that of → Nitrocellulose or of → Polyvinylnitrate while remaining relatively uncomplicated to process.

At present, the difficult task of developing such improved active binders has yielded only two usable compositions:

1. → Polynitropolyphenylene, a non-crystalline explosive material that withstands high temperatures. This polymer is a gelatinous type of binder and is combined with small amounts of softeners, inert binders and → Hexogen or → Octogen. By means of organic solvents, it is processed into the corresponding LOVA composition.

2. → Glycidyl Azide Polymer; a gas-producing glycerin derivate. The glycidyl azide polymer belongs to the group of reactivive polymers (thermoset materials) and is processed together with a main energy conductor, small amounts of softener, inert binders, curing agents and where necessary, accelerating agents. The type of acceleration and curing agents is determined not only by the final matrix structure, but also by the heat processing time of the respective composition and especially by the processing temperature.

Energy of Formation; Enthalpy of Formation

Bildungsenergie, Bildungsenthalpie; chaleur de formation

These thermodynamic concepts denote the energy which is bound during the formation of a given compound from its constituent elements at constant volume (energy of formation) or at constant pressure (enthalpy of formation, which includes the mechanical work performed at the standard state*) (25 °C = 77°F and a pressure of 1 bar). The data are tabulated in accordance with thermodynamic convention: if the formation of a compound from its elements is accompanied by a release of energy, the energy of formation is considered to be negative.

The knowledge of the energies of formation of an explosive or an inflammable mixture on one hand, and of the energies of formation of the presumed reaction products on the other, makes it possible to calculate the → *Heat of Explosion:* → also *Thermodynamic Calculation of Decomposition Reactions.* The values for the most important components of explosives and propellants are given in Table 31, p. 325–329. An extensive collection of tabulated data for energies and enthalpies of formation, including source references, was published by the Fraunhofer-Institut für Chemische Technologie (ICT), Berghausen, 1972*). These values were incorporated into the ICT-Database of Thermochemical Values, containing data of more than 6000 substances, which is available from ICT.

References

Médard, M. L.: Tables Thermochemiques. Mémorial de l'Artillerie Française, Vol. 23, pp. 415–492 (1954). (The given data are valid at 18 °C and for carbon as diamond.)

Ide, K. H., Haeuseler, E. and *Swart, K.-H.:* Sicherheitstechnische Kenndaten explosionsfähiger Stolle. II. Inform., Explosivstoffe, Vol. 9, pp. 195–197 (1961). *Urbanski, T.:* Chemistry and Technology of Explosives. Pergamon Press, London 1964–1967, Vol. 1–4.

Swart, K.-H., Wandrey, P.-A., Ide, K. H. and *Haeuseler, E.:* Sicherheitstechnische Kenndaten explosionsfähiger Stoffe. III Inform. Explosivstoffe, Vol. 12, pp. 339–342 (1965).

JANAF National Bureau of Standards, Midland, Michigan 1971; Supplements 1974–1982.

Shorr, M. and *Zaehringer, A. J.:* Solid Rocket Technology. John Wiley and Sons, Inc., New York 1967.

Sutton, E. S., Pacanowsky, E. J. and *Serner, S. F.:* ICRPG/AIAA. Second Soli Propulsion Conference, Anaheim, Calif., June 6–8, 1967.

---

* *F. Volk, H. Bathelt* and R. Kuthe, Thermodynamische Daten von Raketentreibstoffen, Treibladungspulvern und Sprengstoffen. Edited by the Fraunhofer Institut für Chemische Technologie, Berghausen 1972. Supplement Vol. 1981.

Dadieu, A., Damm, R. and *Schmidt, E. W.:* Raketentreibstoffe. Springer-Verlag, Wien – New York 1968.
Selected Values of Chemical Thermodynamic Properties. NBS, Technical Note 270, p. 3 (1968).
Stull, D. R., Westurm, E. F. and *Sinke, G. C.:* The Chemical Thermodynamics of Organic Compounds. John Wiley and Sons, Inc., New York 1969.
Tavernier, P., Boisson, J. and *Crampel, B.:* Propergols Hautement Energéliques Agardographie No. 141 (1970).
Cox, J. D. and *Pilcher, G.:* Thermochemistry of Organic and Organometallic Compounds. Acad. Press, London 1970.
Cook, M. A.: The Science of Industrial Explosives, Salt Lake City, Utah, 1974.
James, R. W.: Propellants and Explosives, Noyes Data Corporation, Park Ridge N. J. 1974.
Stull, D. A., Journ. of Chem. Educat. **48**, 3 (1971) A 173–182.
Volk, F., Bathelt, H. and *Kuthe, A.:* Thermochemische Daten von Raketentreibstoffen, Treibladungspulvern sowie deren Komponenten. Original print of the Institut für Chemie der Treib- und Explosivstoffe, Berghausen 1972.
Cook, M. A.: The Science of Industrial Explosives, Salt Lake City, Utah, USA, 1974.
James, A. W.: Propellants and Explosives, Noyes Data Corporation, Park Ridge N. J. 1974.
Fair, H. D. and *Walker, A. F.:* Energetic Materials, Vols. 1 and 2, Plenum Press, New York 1977.

Environmental Seal

Schutzmembran; diaphragme de protection

Diaphragm having very low moisture vapor transmission rate, used over generator outlets to provide a hermetic seal.

Eprouvette

This is an instrument to determine the performance of → *Black Powders*. It is a small mortar, positioned vertically upwards; a known amount (10 g) of the black powder sample is charged and set off with the aid of a fuse passing through a priming hole; the mortar is closed with a projectile guided upwards by two steel rods; the projectile gains its maximum height and is then locked. The height of the locked projectile is determined; it is a measure of the performance of the black powder sample.

Equation of State

Zustandsgleichung; l'équation d'état; EOS

The internal ballistic pressure resulting from an explosion of a powder propellant can reach up to 600 MPa and a temperature of up to 4000 K. Under such conditions of extreme pressure and temperature, the calculation of thermodynamic data is possible only using a suitable equation of state, whereby pressure P, temperature T, the density of Gas ρ and the specific number of moles n_s are associated.

For internal ballistics one ordinarily uses a truncated virial equation which breaks off after the third term and is in the form:

$$P = n_s \cdot R \cdot T \cdot \rho \, (1 + n_s \cdot \rho \cdot B + n_s^2 \cdot \rho^2 \cdot C)$$

P: Pressure [Pa]
n_s: Specific number of moles [kmol/kg]
R: Gas constant [J/(kmol·K)]
T: Explosion temperature [K]
ρ: Density of gas [kg/m^3]
B: Second viral coefficient [m^3/kmol]
C: Third virial coefficient [m^6/kmol^2]

The temperature-dependent second and third virial coefficient describe the increasing two- and three-particle collisions between the gas molecules and their accompanying increase in gas density. The virial coefficients are calculated using a suitable intermolecular potential model (usually a 12-6 Lennard-Jones Potential) from rudimentary statistical thermodynamics.

The detonation pressure behind the → *Shock Wave* of a liquid or solid explosive substance is between 2 GPa and 50 GPa whereby the temperature at the wave front can reach up to 5000 K.

Next to the → *Chapman-Jouget* theory, during the last 50 years, the principal methods of calculating detonation pressure and the velocity of flat detonation waves have been the Becker-Kistiakowsky-Wilson (BKW), the Lennard-Jones-Devonshire (LJD) and the Jacobs-Cowperthwaite-Zwisler (JCZ) equations of state.

All of these methods employ model equations which do not quite satisfactorily yield the condition of the highly dense and heated detonation products. This is shown in particular in the semiempirical BKW equation of state, which in addition to five parameters for the calibrating of experimental measurements values, requires two separate sets of data for the calculations involving explosives of either an extremely high or slightly negative oxygen balance or a positive oxygen balance.

The LJD and the JCZ equations of state represent methods, which, when used in conjunction with an intermolecular potential rudiment, employ lattice models.

With lattice models it is assumed that the molecules in the fluid phase repose on the lattice points of three dimensional lattice, while entering into an exchange effect with the adjacent molecules.

Among the more recent and theoretically-based equations of state in detonation physics are the perturbation-theoretical methods. First used by R. Chirat and G. Pittion-Rossillion, these methods were considerably improved later by F. Ree.

The perturbation theory is one of the processes that in the last fifteen years has achieved the most significant advances in the area of statistical thermodynamics.

R. Chirat and G. Pittion-Rossillion employ a simplified Weeks-Chandler-Andersen (WCA) perturbation theory while F. Ree uses the Mansoori-Canfield-Rasaiah-Stell (MCRS) hardsphere variational theory. Both methods build on the α-Exp-6 potential and yield the theoretical Chapman-Jouget detonation velocities and pressures, which for a large number of explosives lie within the measurement accuracy of practically obtained values.

Despite the advances made over the last several decades in the field of detonation physics, there still exist many phenomena that quantitatively are not understood. Among these in particular are the unstationary, multidimensional detonation processes of gaseous, liquid or condensed bodies.

References:

R. Becker: Z. Phys. 4, 393 (1921)
R. Becker: Z. techn. Phys. 3, 249 (1922)
M. Cowperthwaite and *W. H. Zwisler:* Proceedings of the Sixth International Symposium on Detonation, edited by D. J. Edwards, ACR-221 (Office of Naval Research, Department of the Navy), 162 (1976)
F. Volk and *H. Bathelt:* Propellants Explos. 1, 7 (1976)
H. Hornberg: Propellants Explos. 3, 97 (1978)
C. L. Mader: Numerical Modeling of Detonation, University of California Press, Berkeley (1979)
R. Chirat und *G. Pittion-Rossillion:* J. Chem Phys. 74, 4634 (1981)
R. Chirat und *G. Pittion-Rossillion:* Combust. Flame 45, 147 (1982)
F. H. Ree: J. Chem. Phys. 81, 1251 (1984)
F. H. Ree: J. Chem. Phys. 84, 5845 (1986)

ERLA

Abbreviation for an epoxy compound for the formation of binders in → *Composite Propellants*.

Structural formula:

$$(H_2C\overset{O}{-}\underset{H}{C}-CH_2)_2N-\bigcirc-O-CH_2-\underset{H}{C}\overset{O}{-}CH_2$$

empirical formula: $C_{15}H_{19}NO_4$
molecular weight: 277.16
density: 1.20–1.21 g/cm^3

Erosion

Wearing away of a material due to high gas velocities and entrained particles.

Erosive Burning

erosiver Abbrand; combustion érosive

Term used in solid fuel rocket technology to describe the anomalous increase in the burning rate. This increase is thought to originate from turbulent instead of laminar gas flow along the burning surface, which leads to a higher feed back of heat energy onto this surface, and thus a higher rate of burning. Mechanical erosion may also take place by gases enriched with solid particles, e.g., Al_2O_3.

Resonance combustion is defined as the generation of pressure maxima in the combustion chamber and the consequent irregularity of the burning rate; these maxima originate from the interaction between the gas stream and the flame and become apparent as a kind of vibration.

Star-shaped grooves in case-bonded charges tend to equalize the pressure and suppress the tendency to resonate. Other relevant keywords are: → *Burning Rate,* → Solid Propellant Rockets.

Erythritol Tetranitrate

Tetranitroerythrit; tétranitrate d'érythrite

$$\begin{array}{l} CH_2\text{-}O\text{-}NO_2 \\ CH\text{-}O\text{-}NO_2 \\ CH\text{-}O\text{-}NO_2 \\ CH_2\text{-}O\text{-}NO_2 \end{array}$$

colorless crystals
empirical formula: $C_4H_6N_4O_{12}$
molecular weight: 302.1
oxygen balance: +5.3 %
nitrogen content: 18.55 %
volume of explosion gases: 704 l/kg
heat of explosion
  (H_2O liq.): 1519 kcal/kg = 6356 kJ/kg
  (H_2O gas): 1421 kcal/kg = 5943 kJ/kg
specific energy: 111 mt/kg = 1091 kJ/kg
density: 1.6 g/cm^3
melting point: 61.5 °C = 143°F
deflagration point: 154–160 °C = 309–320°F violent explosion
impact sensitivity: 0.2 kp m = 2 N m

Erythrol tetranitrate is insoluble in cold water, but is soluble in alcohol and ether. It is prepared by dissolving erythrol in concentrated nitric acid with efficient cooling, and precipitating the product by concentrated sulfuric acid. It crystallizes out of alcohol as colorless plates.

Erythritol tetranitrate serves as an effective cardial medicine (in a low percentage mixture with milk sugar).

The pure substance is extremely sensitive to shock and friction.

Ethanolamine Dinitrate

Monoethanolamindinitrat; dinitrate d'éthanolamine

$$\begin{array}{c} NH_2 \cdot HNO_3 \\ | \\ CH_2-CH_2-O-NO_2 \end{array}$$

colorless crystals
empirical formula: $C_2H_7N_3O_6$
molecular weight: 169.1
oxygen balance: -14.2%
nitrogen content: 24.85%
volume of explosion gases: 927 l/kg
heat of explosion
   (H_2O liq.): 1254 kcal/kg = 5247 kJ/kg
   (H_2O gas): 1089 kcal/kg = 4557 kJ/kg
specific energy: 118.8 mt/kg = 1165 kJ/kg
density: 1.53 g/cm^3
melting point: 103 °C = 217°F
lead block test: 410 cm^3/10 g
deflagration point: 192 °C = 378°F

This compound is readily soluble in water, sparingly soluble in cold alcohol, and somewhat hygroscopic. It is prepared by dissolution of monoethanolamine in concentrated nitric acid and precipitation from alcohol or ether with cooling.

Ethriol Trinitrate

trimethylolethylmethane trinitrate; trimethylolpropane trinitrate; trinitrate de trimethylolethylméthane

$$C_2H_5-C\begin{array}{c} {\nearrow}CH_2-O-NO_2 \\ -CH_2-O-NO_2 \\ {\searrow}CH_2-O-NO_2 \end{array}$$

colorless crystals
empirical formula: $C_6H_{11}N_3O_9$
molecular weight: 269.4
energy of formation: -401 kcal/kg = -1678 kJ/kg
enthalpy of formation: -426 kcal/kg = -1783 kJ/kg

oxygen balance: −50.5 %
nitrogen content: 15.62 %
volume of explosion gases: 1009 l/kg
heat of explosion
   (H_2O liq.): 1014 kcal/kg = 4244 kJ/kg
   (H_2O gas): 936 kcal/kg = 3916 kJ/kg
density: 1.5 g/cm^3
melting point: 51 °C = 124°F
lead block test: 415 cm^3/10 g
detonation velocity, confined:
   6440 m/s = 21 100 ft/s at ρ = 1.48 g/cm^3

This compound is prepared by nitrating trimethylolpropane (obtained by condensing formaldehyde with butyraldehyde in the presence of lime) with a mixture of nitric acid and sulfuric acid.

Ethylenediamine Dinitrate

dinitrate d'éthylène diamine: PH-Satz; EDD

$$\begin{array}{l} CH_2 - NH_2 \cdot HNO_3 \\ | \\ CH_2 - NH_2 \cdot HNO_3 \end{array}$$

colorless crystals
empirical formula: $C_2H_{10}N_4O_6$
molecular weight: 186.1
energy of formation: −807.4 kcal/kg = −3378.2 kJ/kg
enthalpy of formation: −839.2 kcal/kg = −3511.3 kJ/kg
oxygen balance: −25.8 %
nitrogen content: 30.11 %
volume of explosion gases: 1071 l/kg
heat of explosion
   (H_2O liq.): 912 kcal/kg = 3814 kJ/kg
   (H_2O gas): 739 kcal/kg = 3091 kJ/kg
density: 1.577 g/cm^3
melting point: 188 °C = 370°F
volume of detonation gases: 945.5 l/kg
lead block test: 350 cm^3/10 g
detonation velocity, confined:
   6800 m/s = 22 300 ft/s at ρ = 1.53 g/cm^3
deflagration point: 370−400 °C = 700−750°F
impact sensitivity: 1.0 kp m = 10 N m
friction sensitivity:
   at 36 kp = 353 N pistil load no reaction
critical diameter of steel sleeve test: 2 mm

Ethylenediamine dinitrate is somewhat hygroscopic and is readily soluble in water. It is prepared by saturating an aqueous solution of ethylenediamine with nitric acid.

It forms an eutectic mixture (melting point 100 °C = 212°F) when mixed with an equal amount of ammonium nitrate.

Ethylenedinitramine

N,N,-dinitroethylene diamine; éthylène dinitramine; Haleite; Halite; EDNA

$$\begin{array}{l} CH_2\text{-}NH\text{-}NO_2 \\ | \\ CH_2\text{-}NH\text{-}NO_2 \end{array}$$

empirical formula: $C_2H_6N_4O_4$
molecular weight: 150.1
energy of formation: -137.7 kcal/kg = -576.2 kJ/kg
enthalpy of formation: -165.3 kcal/kg = -691.6 kJ/kg
oxygen balance: -32.0%
nitrogen content: 37.33%
volume of explosion gases: 1017 l/kg
heat of explosion
 (H_2O liq.): 1123 kcal/kg = 4699 kJ/kg
 (H_2O gas): 1023 kcal/kg = 4278 kJ/kg
density: 1.71 g/cm^3
melting point: 176.2 °C = 349.2°F (decomposition)
lead block test: 410 cm^3/10 g
detonation velocity, confined:
 7570 m/s = 24 800 ft/s at $\rho = 1.65$ g/cm^3
deflagration point: 180 °C = 356°F
impact sensitivity: 0.8 kp m = 8 N m

This compound behaves as a dibasic acid and forms neutral salts. It is insoluble in ether, sparingly soluble in water and alcohol and soluble in dioxane and nitrobenzene; it is not hygroscopic.

It possesses considerable brisance, combined with a high chemical stability and relatively low mechanical sensitivity.

Ethylenedinitramine is prepared by nitration of ethylene urea with mixed acid, to yield dinitroethylene-urea; the latter compound liberates carbon dioxide and forms ethylenedinitramine.

Ethylene-urea is prepared by reacting ethylene diamine to ethyl carbonate under elevated pressure.

Pourable mixtures of ethylenedinitramine with TNT are known in the USA as Ednatol.

Ethyl Nitrate

Ethylnitrat; nitrate d'éthyle

C_2H_5—O—NO_2

empirical formula: $C_2H_5NO_3$
molecular weight: 91.0
energy of formation: −470.4 kcal/kg = −1968 kJ/kg
enthalpy of formation: −499.5 kcal/kg = −2091 kJ/kg
oxygen balance: −61.5 %
nitrogen content: 15.24 %
volume of explosion gases: 1101 l/kg
heat of explosion (H_2O liq.): 993 kcal/kg = 4154 kJ/kg
density: 1.10 g/cm^3 melting point: −102 °C = −152°F
lead block test: 420 cm^3/10 g
detonation velocity, confined:
  5800 m/s =19000 ft/s at ρ = 1.1 g/cm^3

This compound is a colorless, mobile liquid with a pleasant smell. It is practically insoluble in water, but is soluble in alcohol and in most organic solvents. Ethyl nitrate vapors readily form explosive mixtures with air even at room temperature; the lower explosion limit is at 3.8 % ethyl nitrate.

Ethyl nitrate explodes when brought into contact with alkali metals.

Ethylphenylurethane

Ethylphenylurethan; éthylphénylinréthane

$$O=C\begin{matrix}\diagup N\diagdown^{C_2H_5}_{C_6H_5}\\ \diagdown OC_2H_5\end{matrix}$$

colorless liquid
empirical formula: $C_{11}H_{15}NO_2$
molecular weight: 193.2
energy of formation: −492.5 kcal/kg = −2060.5 kJ/kg
enthalpy of formation: −520.1 kcal/kg = −2175.9 kJ/kg
oxygen balance: −227.7 %
nitrogen content: 7.25 %

Ethylphenylurethane is a gelatinizing → *Stabilizer* especially for → *Double Base Propellants.*

Specifications
  clear, colorless liquid
  density at 20 °C = 68°F: 1.042−1.044 g/cm^3

refractive index n_D^{20}: 1.504–1.507
boiling analysis at 760 Torr: 252–255 °C = 485–491°F
acidity, as HCl: not more than 0.004 %
reaction: neutral

Ethyl Picrate

2,4,6-trinitrophenetol; Ethylpikrat; picrate d'ethyle

$$\underset{\underset{NO_2}{}}{O_2N-\text{C}_6H_2(O-C_2H_5)-NO_2}$$

pale yellow needles
empirical formula: $C_8H_7N_3O_7$
molecular weight: 257.2
energy of formation: –167.1 kcal/kg = –699 kJ/kg
enthalpy of formation: –186.7 kcal/kg = –781 kJ/kg
oxygen balance: –77.8 %
nitrogen content: 16.34 %
volume of explosion gases: 859 l/kg
heat of explosion
  (H_2O liq.): 840 kcal/kg = 3515 kJ/kg
  (H_2O gas): 805 kcal/kg = 3369 kJ/kg
specific energy: 86 mt/kg = 847 kJ/kg
melting point: 78 °C = 172°F
detonation velocity, confined:
  6500 m/s =21 300 ft/s at ρ = 1.55 g/cm^3

The preparation of this compound resembles that of → *Trinitroanisol.*

Ethyltetryl

2,4,6-trinitrophenylethylnitramine; trinitrophényléthylnitramine

$$O_2N-\text{C}_6H_2(NO_2)_2-N(C_2H_5)-NO_2$$

green yellow crystals
empirical formula: $C_8H_7N_5O_8$
molecular weight: 301.2
energy of formation: +5.4 kcal/kg = +22.5 kJ/kg
enthalpy of formation: –14.3 kcal/kg = –59.8 kJ/kg

oxygen balance: −61.1 %
nitrogen content: 23.25 %
volume of explosion gases: 874 l/kg
heat of explosion
 (H_2O liq.): 970 kcal/kg = 4058 kJ/kg
 (H_2O gas): 939 kcal/kg = 3930 kJ/kg
specific energy: 109 mt/kg = 1069 kJ/kg
density: 1.63 g/cm^3
melting point: 95.8 °C = 204.4°F
heat of fusion: 18.7 kcal/kg = 78 kJ/kg
lead block test: 325 cm^3/10 g
impact sensitivity: 0.5 kp m = 5 N m
friction sensitivity:
 up to 36 kp = 353 N pistil load no reaction

The properties of this compound resemble those of Tetryl; it can be prepared from mono- or diethylaniline.

Since the melting point of ethyltetryl is lower than that of Tetryl, the former can be more readily employed in energy-rich pourable mixtures.

EURODYN 2000®

EURODYN 2000® is the trade name of a gelatinous rock explosive made by Eurodyn Sprengmittel GmbH. In contrast to the classical → *Ammongelites*, this explosive does not contain any nitro-aromatics harmful to health, such as → Dinitrotoluene and Trinitrotoluene.

EWALID W

EWALID W is the trade name of a new rock explosive made by the WASAG CHEMIE Sythen GmbH Company. EWALID W is manufactured without any nitro glycerine or nitro glycol as well as any nitrous aromatic compounds (DNT. TNT) that are rated carcinogenic. It's a safehandling, water resistant explosive and well suited to be used as booster.

Exothermal

Process characterized by the evolution of heat (opposite of endothermal).

Explode

Explodieren; exploser

To be changed in chemical or physical state, usually from a solid or liquid to a gas (as by chemical decomposition or sudden vaporization) so as to suddenly transform considerable energy into the kinetic form (→ *Explosion*).

Exploding Bridgewire

Detonator or initiator that is initiated by capacitor discharge that explodes (rather than merely heats) the bridgewire. Cannot be initiated by any normal shock or electrical energy.

Exploding BridgeWire Detonator (EBW)

An initiating device which utilizes the shock energy from the explosion of a fine metallic wire to directly initiate a secondary explosive train. Invented by Luis Alvarez for the Manhatten project in the early 1940's, the basic EBW consists of a fine wire (typically gold, 0.038 mm in diameter, 1 mm long), next to a secondary explosive such as → *PETN* or → *RDX*. A large, fast current pulse (>200 amps in approximately 1 microsecond) through the wire causes it to rapidly vaporize generating a shock wave of about 15 kilobars. This intense shock wave is sufficient to directly initiate the low density explosive next to the exploding wire. The low density explosive is than used to initiate a higher density explosive output pellet which in turn can initiate main charge explosives.

Exploding Foil Initiator (EFI, Slapper)

Similar in some respects to an Exploding BridgeWire Detonator, the Exploding Foil Initiator uses a high electrical current to vaporize a foil and accelerate a dielectric flyer down a short barrel (typically about 0.2 mm long). The kinetic energy of the flyer is sufficient to initiate high density secondary explosives such as HNS directly. Invented in 1965 by John Stroud of the Lawrence Livermore National Laboratory.

Explosion

An explosion is a sudden increase in volume and release of energy in a violent manner, usually with generation of high temperatures and release of gases. An explosion causes pressure waves in the local medium in which it occurs. Explosions are categorized as deflagrations if these waves are subsonic and detonations if they are supersonic (shock waves).

Explosion Heat

→ Heat of Explosion

Explosion Temperature

Explosionstemperatur; température d'explosion

Explosion temperature is the calculated temperature of the fumes of an explosive material which is supposed to have been detonated while confined in a shell assumed to be indestructible and impermeable to heat; the calculation is based on the → *Heat of Explosion* and on the decomposition reaction, with allowance for the dissociation equilibria and the relevant gas reaction (→ *Thermodynamic Calculation of Decomposition Reactions*). The real detonation temperature in the front of the shock wave of a detonating explosive can be estimated on the strength of the hydrodynamic shock wave theory, and is higher than the calculated explosion temperature.

Explosive Forming and Cladding

Metallbearbeitung durch Sprengstoffe; traitement des métaux par explosion

The applicability of explosive matrials for metal forming have been studied with three different objectives in view: sheet forming and matrix forming of flat items by pressure impact; metal plating; surface hardening of manganese hard steel.

The application of the pressure chock of an explosive to form very large workpieces is primarily intended, to achieve the shaping of a workpiece without using presses, which are very expensive. The transmission of the pressure impact takes place under water. Preliminary experiments gave encouraging results, but a large-scale industrial application has not yet been developed.

The development of explosive cladding is very much more advanced: the metal sheet to be cladded is exploded onto the base material, parallel to it or at a certain angle. In this way it is possible to effect cladding tasks which would be impossible to fulfil by manual welding, owing to the formation of brittle intermediate alloys between the plating material and the base material – as, for instance, in plating titanium onto a steel surface.

On the surface of manganese steel, the impact of the explosive layer onto the steel surface results in hardening; the only objective of this process is that it enables repair work to be carried out on railway tracks in remote regions, and there is no need to convey the defective parts over long distances. In densely populated areas, forming explosions are difficult to perform.

Explosive Bolt

Sprengriegel; verrou destructif

A bolt that is intended to be fractured by a contained or inserted explosive charge.

Explosive Loading Factor

Spezifischer Sprengstoffverbrauch; consommation specitique d'explosits

The amount of explosive used per unit of rock, usually expressed as pounds of explosives per cubic yard of rock or tons of rock per pound of explosives, or their reciprocals.

Explosive Materials

Sprengmittel; materiaux explosif (→ Table 11)

These include explosives, blasting agents and detonators. The term includes, but is not limited to, dynamite and other high explosives, slurries and water gels, blasting agents, black powder pellet powder, initiating explosives, detonators, safety fuses, squibs, detonating cord, igniter cord and igniters. A list of explosive materials determined to be within the coverage of "18 U.S.C. Chapter 40, Importation, Manufacture, Distribution and Storage of Explosive Materials" is issued at least annually by the Director of the Bureau of Alcohol, Tobacco and Firearms of the Department of the Treasury.

The United States Department of Transportation classifications of explosive materials used in commercial blasting operations are not identical with the statutory definitions of the Organized Crime Control Act of 1970, Title 18 U.S.C., Section 841. To achieve uniformity in transportation, the definitions of the United States Department of Transportation in Title 49 Transportation CFR, Parts I-999 subdivides these materials into:

→ Class A Explosives – Detonating, or otherwise maximum hazard.
→ Class S Explosives – Flammable hazard.
→ Class C Explosives – Minimum hazard.
Oxidizing Material – A substance that yields oxygen readily to stimulate the combustion of organic matter (→ *Oxidizer*)

A list of energetic materials is also available from the German BAM.

Explosive Train

A train of combustible and explosive elements arranged in order of decreasing sensitivity. The explosive train accomplishes the controlled augmentation of a small impulse into one of suitable energy to actuate a main charge. A fuze explosive train may consist of a primer, a detonator, a delay, a relay, a lead and booster charge, one or more of which may be either omitted or combined. If the bursting charge is added to the foregoing train it becomes a bursting charge explosive train. A propelling charge explosive train might consist of a primer, igniter or igniting charge, usually black powder, and finally, any of the various types of propellants (→ *Igniter Train*).

Explosives

Explosivstoffe; explosifs

1. Definition

Explosives are solid or liquid*) substances, alone or mixed with one another, which are in a metastable state and are capable, for this reason, of undergoing a rapid chemical reaction without the participation of external reactants such as atmospheric oxygen. The reaction can be initiated by mechanical means *(impact,* → Impact Sensitivity; friction, → Friction Sensitivity), by the action of heat (sparks, open flame, red-hot or white-hot objects), or by detonating shock (→ *Blasting Cap* with or without a → Booster charge). The resistance of the metastable state to heat is known as → *Stability*. The ease with which the chemical reaction can be initiated is known as → *Sensitivity*.

The reaction products are predominantly gaseous (→ *Fumes*). The propagation rate from the initiation site outwards through the explosive

---

* Of course, gases and gaseous mixtures can also be explosive. Explosive mixtures are often generated spontaneously (leaks in gas pipes, solvent tanks; firedamp in coal mining).

material may be much slower than the velocity of sound (→ *Deflagration;* → Gunpowder) or may be supersonic (→ *Detonation*). Explosives are solid, liquid, or gelatinous substances or mixtures of individual substances, which have been manufactured for blasting or propulsion purposes. For their effectiveness: → *Strength;* → Burning Rate; → Brisance.

Materials which are not intended to be used for blasting or shooting may also be explosive. They include, for example, organic peroxide catalysts, gas-liberating agents employed in the modern manufacture of plastic materials and plastic foams, certain kinds of insecticides etc. Table 11 gives a an overview of explosive materials.

2. Important Explosives

Of the many explosive chemicals discussed in this book, the following are, at present, of industrial or military importance:

Nitro compounds:

> → *TNT* in various degrees of purity, as defined by the solidification point of the material; pure 2,4- and 2,6-isomers of dinitrotoluene (as propellant components) and low-melting isomer mixtures (for commercial explosives);

Aromatic nitramines:

> → *Tetryl* (trinitrophenylnitramine) for booster charges and secondary blasting cap charges;

Aliphatic nitramines:

> → *Hexogen* (RDX) and → *Octogen* (HMX) as components for high-brisance compositions (→ *Compositions B;* → Hollow Charges); → *Nitroguanidine* as the main component in powders with low explosion heat and in rocket propellants.

Nitrate esters:

> → *Nitroglycerine,* which is still of primary importance in commercial explosives, smokeless powders and rocket propellants;
> → *Nitroglycol* in commercial explosives only;*)
> → *PETN* as a high-brisance component, which is phlegmatized and pressed for booster charges; it is also employed as a secondary charge of blasting caps and as a detonating cord charge;

---

* Nitroglycol-based gelatinous explosives being replaced by → *Emulsion Slurries.*

→ *Diethyleneglycol Dinitrate* for smokeless (cold) powders;
→ *Nitrocellulose*, which is the most important component of single-base and double-base powders and multibase rocket propellants. It is also used to gelatinize commercial explosives. Outside the explosives industry, it is also used in the manufacture of lacquers and varnishes.

Initiating explosives:

→ *Mercury Fulminate*, and other fulminates, which are now used to a much smaller extent;
→ *Lead Azide*, alone and in mixtures with *Lead Trinitroresorcinate*, as primary charges in blasting caps; also for firedamp-proof cooper caps in coal mining, and in military primers of all kinds;
→ *Lead Styphnate (Lead Trinitroresorcinate)* mixtures, which may contain → Tetrazene, for percussion caps.

Many nitro derivatives of benzene and naphthalene were of importance in the past, since toluene – the starting compound in the manufacture of TNT – could only be prepared by distillation of coal. Owing to the advances in petrochemistry, toluene is now available in practically unlimited amounts; the bulk of the toluene now produced is employed as the starting material for the preparation of toluene diisocyanate (TDI) used in the production of plastics.

3. Quality Requirements for Industrial and Military Explosives

The quality requirements for industrial explosives are quite different from those valid for military explosives. It follows that their compositions and the mode of their preparation must be different as well. Table 12 gives an overview.

The combustion behavior of propellants (→ *Burning Rate*), which is affected by the ignition and by the design of the grain configuration in the combustion chamber, must be exactly reproducible.

Primary explosives, when set off by a flame, must detonate immediately, and their detonation development distance must be as short as possible.

4. Acquisition, Handling, and Storage

Almost every country has its own laws and regulations governing the acquisition and utilization of explosives. These laws were passed in order to protect the public and to make the use of explosives for criminal purposes a heavily punishable offence. Generally speaking,

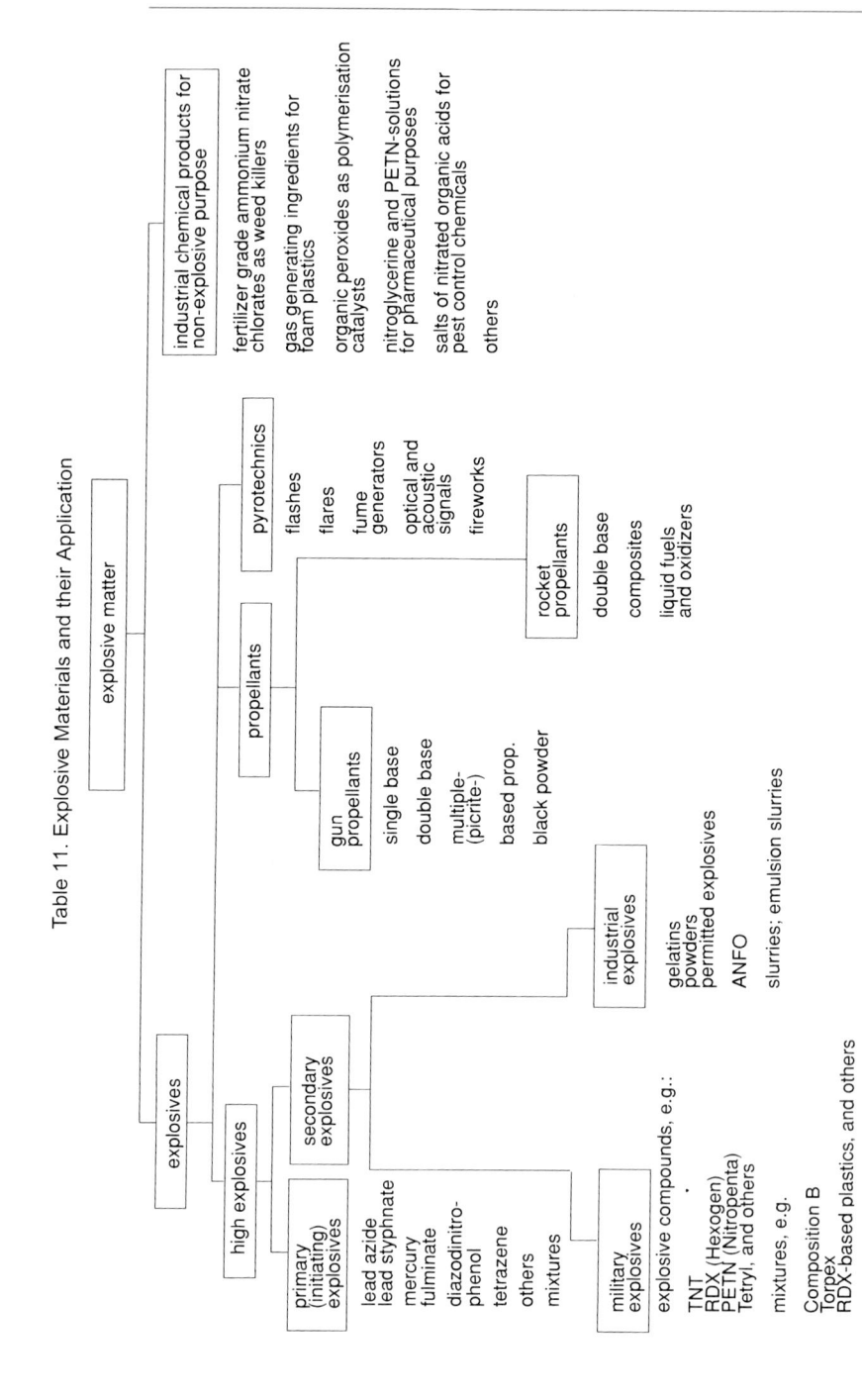

Explosives

Table 11. Explosive Materials and their Application

- explosive matter
  - explosives
    - high explosives
      - primary (initiating) explosives
        - lead azide
        - lead styphnate
        - mercury fulminate
        - diazodinitrophenol
        - tetrazene
        - others
        - mixtures
      - secondary explosives
        - military explosives
          - explosive compounds, e.g.:
            - TNT
            - RDX (Hexogen)
            - PETN (Nitropenta)
            - Tetryl, and others
          - mixtures, e.g.
            - Composition B
            - Torpex
            - RDX-based plastics, and others
        - industrial explosives
          - gelatins
          - powders
          - permitted explosives
          - ANFO
          - slurries; emulsion slurries
    - propellants
      - gun propellants
        - single base
        - double base
        - multiple- (picrite-) based prop.
        - black powder
      - rocket propellants
        - double base
        - composites
        - liquid fuels and oxidizers
    - pyrotechnics
      - flashes
      - flares
      - fume generators
      - optical and acoustic signals
      - fireworks
  - industrial chemical products for non-explosive purpose
    - fertilizer grade ammonium nitrate
    - chlorates as weed killers
    - gas generating ingredients for foam plastics
    - organic peroxides as polymerisation catalysts
    - nitroglycerine and PETN-solutions for pharmaceutical purposes
    - salts of nitrated organic acids for pest control chemicals
    - others

Table 12. Requirements on Industrial and Military Explosives

	Industrial Explosives	Military Explosives
performance	large gas volume and high heat of explosion = high strength high detonation velocity not needed, except: special gelatins for seismic prospecting	according to the purpose of the weapon: mines, bombs, mine projectiles, rocket war head charges: high gas impact large gas volume high heat of explosion (high detonation velocity not needed) grenades: high speed splinter formation high loading density high detonation velocity medium strength is sufficient shaped (hollow charge effect): extremely high values for density and detonation velocity (HMX best component) high strength + high brisance
sensitivity	safe in handling cap-sensitive (except: blasting agents and slurries) safe flash over capacity in long columns	as unsensitive as possible firing safety impact safety → projectile impact safety
stability and behavior on storage	storage life about 6 months or longer neutral (e.g. no nitric acid as component)	storage life 10 years or longer neutral no reaction with metals such as picrate formation
water resistance	when cartridged, should withstand 2 h in stagnant water (for seismic prospecting shots even longer)	completely waterproof, at least when loaded in the weapon
consistency	formable (gelatinous or powder form) to be able to introduce the cap	castable or pressible
thermal behavior	must not freeze above −25 °C (−13°F) must withstand about +60 °C (140°F) for several hours (e.g. in deep mining)	fully functional between −40 °C (−40°F) and 60 °C (140°F) or even higher for special purposes.

buyers and users must prove their competence to the authorities and are obliged by law to keep a storage record. The manner of construction of the buildings serving as storerooms, the permissible stacking height, and the minimum distance between residental buildings and buildings in which explosives are stored (or produced) are in most cases officially specified.

In order to estimate the safe distance required for an amount M of the explosive, it may be assumed that this distance increases with the cube root of M; we thus have $f \cdot \sqrt[3]{M}$. If M is given in kg, and the safe distance is to be obtained in meters, then f is about 16 for the distance to inhabited buildings, about 8 for the distance between dangerous and safe areas of the explosive-manufacturing plant, and about 1.5 for the distance between one storehouse and another.

Transport regulations for dangerous materials: → *RID;* → IATA; → IMO; → Mass Explosions Risk.

Exudation

Ausschwitzen; exsudation

The separation of oily ingredients out of explosives during prolonged storage, especially at elevated temperatures. It may be caused by low melting eutectics of isomers or primary products of the explosive material or by added ingredients. Exudation may particularly be anticipated in TNT shell charges; accordingly, the chemical purity standards of the product (→ *TNT,* specifications) are particularly important.

In propellant charges, exudation occurs if the percentage of e.g., nitroglycerine, aromatic compounds, gelatinizers, or vaseline is high. The propellant grains will agglomerate whereby ignition will suffer. The same disadvantage may be caused by crystalline separation of stabilizers. The ballistic performance can also be affected.

Prolonged storage, especially in wet climates, may cause exudation of gelatinous nitroglycerine explosives. Mostly, the exudated liquid consists of a watery ammonium nitrate solution; initiation sensitivity and performance may be affected. Highly dangerous is the exudation of unbonded nitroglycerine; it occurs when the gelatinization with nitrocellulose (blasting soluble) was faulty or the nitrocellulose of bad quality.

Face Burning

"Cigarette-Burning"; Stirnabbrand; combustion en cigarette

In rocket technology, a design of the propellant charge which results in the combustion process being restricted to the cross-section of the combustion chamber. This type of combustion is produced by coating all other surfaces with a non-flammable layer. In such rockets long, combustion times (10 minutes or more) at a nearly constant thrust can be achieved.

Fallhammer

mouton de choc

Fallhammer instruments are used to determine the → *Impact Sensitivity* of explosives.

Ferrocene

Ferrocen, Bis-cyclopentadienyl-Eisen; ferrocène

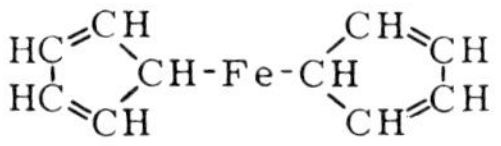

$$\mathrm{HC}\!\!\begin{smallmatrix}\nearrow CH \\ \searrow CH\end{smallmatrix}\!\!CH\text{-}Fe\text{-}CH\!\!\begin{smallmatrix}\nearrow CH \\ \searrow CH\end{smallmatrix}\!\!\begin{smallmatrix}CH \\ CH\end{smallmatrix}$$

empirical formula: $C_{10}H_{10}Fe$
molecular weight: 186.0
energy of formation: +214.9 kcal/kg = +899.2 kJ/kg
enthalpy of formation: +199.0 kcal/kg = +832.6 kJ/kg
oxygen balance: −223.6 %

Ferrocene is a combustion-modifying additive especially for → *Composite Propellants*.

Firedamp

Schlagwetter; grisou

Firedamp is an explosive mixture of marsh gas (methane, CH_4) with air. These mixtures are explosive at normal temperatures and pressures, and the explosion is propagated over large distances if the mixture contains 5–14 % methane. A methane-air mixture containing 8.5–9.5 % methane is prescribed for official tests of permissibles. The danger of explosion is greatest in this concentration range.

→ *Permitted Explosives*.

Firing Current

Zündstrom; courant de mise à feu

An electric current of recommended magnitude to sufficiently energize an electric blasting cap or a circuit of electric blasting caps.

Firing Line

Zündkabel; ligne de tir

The wire(s) connecting the electrical power source with the electric blasting cap circuit.

First Fire

Igniter composition used with pyrotechnic devices that is loaded in direct contact with main pyrotechnic charge. Pyrotechnic first fire composition compounded to produce high temperature. Composition must be readily ignitible, and be capable of igniting the underlying pyrotechnic charge.

Flame

Flamme; flamme

Chemical reaction or reaction product, partly or entirely gaseous, that yields heat and emits light. State of blazing combustion. The flame profile maybe represented by the temperature profile. Flame temperature is the calculated or determined temperature of the flame.

Flame Shield

Flammenschild; boinclier contre l'érosion

Thin metal shield adjacent to case insulation to prevent erosion of the insulation and to prevent objectionable insulation pyrolysis products from entering the gas stream.

Flare

Fackel; flambeau

A pyrotechnic device designed to produce a single source of intense light or radiation for relatively long durations for target or airfield illumination, signaling, decoy for guided missiles, or other purposes.

Flash Over

Übertragung; détonation par influence

→ *Detonation, Sympathetic Detonation*. "Flash over" means the transmission of detonation from a cartridge to another one in line. Explosives with extremely high flash over tendency can be initiated by the shock wave from one charged borehole to the next one, even at large distances (→ Ditching Dynamite).

Flash Point

Flammpunkt; point d'inflammation

The lowest temperature at which vapors above a volatile combustible substance ignite in air when exposed to flame.

Fly Rock

Steinflug; projections de roche

Rocks propelled from the blast area by the force of an explosion.

Fragmentation Test

Splittertest; epreuve de fracture

A USA standard test procedure for explosives of military interest.

The weight of each empty projectile and the weight of water displaced by the explosive charge is determined, from which the density of the charge is calculated. All 3-inch and 90-mm projectiles are initiated by M20 Booster pellets, and those used with 3-inch HE, M42AI, Lot KC-5 and 90-mm HE, I1I71, Lot WC-91 projectiles are controlled in weight and height as follows: 22.50 + 0.10 gm, and 0.480 to 0.485 inch.

The projectile assembled with fuze, actuated by a blasting cap, Special, Type II (Spec 49-20) and booster, is placed in boxes constructed of half-inch pine. The 90-mm projectiles are fragmented in boxes 21 × 10-1/2 × 10-1/2 inches and the 3-inch projectiles in boxes 15 × 9 × 9 inches external dimensions. The box with projectile is placed on about 4 feet of sand in a steel fragmentation tub, the detonator wires are connected, and the box is covered with approximately 4 feet more of sand. The projectile is fired and the sand runs onto a gyrating 4-mesh screen on which the fragments are recovered.

Fragment Velocity

Charges 10-1/8 inches long and 2 inches in diameter, containing a booster cavity, filled by a 72-gm Tetryl pellet (1-3/8 inches diameter, 2 inches long, average density 1.594) are fired in a model projectile of Shelby seamless tubing, 2 inches ID, 3 inches OD, SAE 1020 steel, with a welded-on cold-rolled steel base. The projectile is fired in a chamber, connected to a corridor containing velocity stations, so (protected sites for high-speed measuring equipment) that a desired wedge of projectile casing fragments can be observed. The fragment velocities are determined by shadow photographs, using flash bulbs, and rotating drum cameras, each behind three slits. The drum cameras have a writing speed of 30 meters per second.

Free-flowing Explosives

Rieselfähige Sprengstoffe; explosifs pulvérulents

Non-cartridged commercial explosives which can be poured into boreholes, mostly ammonium nitrate explosives containing anticaking agents. When ammonium nitrate became commercially available as → Prills (porous pellets), → ANFO blasting agents could also be utilized in the free-flowing form; → also Pellets.

Freezing of Nitroglycerine – based Explosives

Gefrieren von Nitroglycerin-Sprengstoffen; congélation d'explosifs à base de la nitroglycerine

Nitroglycerine may freeze at +10 °C. The frozen cartridges are unsafe to handle, and improvised thawing operations are risky. Freezing is prevented by adding nitroglycol to the nitroglycerine.

Friction Sensitivity

Reibempfindlichkeit; sensitiveness to friction; sensibilité au frottement

The sensitivity to friction can be determined by rubbing a small quantity of the explosive in an unglazed porcelain mortar. The sample being tested is compared with a standard specimen.

In the USA, the friction procedure is made by the friction pendulum test:

A 0.7-g sample of explosive, 5–100 mesh, is exposed to the action of a steel or fiber shoe swinging as a pendulum at the end of a long steel rod. The behavior of the sample is described

qualitatively, i.e., the most energetic reaction is explosion, and in decreasing order: snaps, cracks, and unaffected.

An improved method, developed by the Bundesanstalt für Materialforschung und -prüfung (→ BAM)*), Germany, yields reproducible numerical values.

---

* *Koenen* and *Ide*, Explosivstoffe, Vol. 9 pp. 4 and 10 (1961).

Sensitiveness of friction

This method is the recommended test method in the UN-recommendations for the transport of dangerous goods and it is standardized as EN 13631-3 as a so-called Harneonized European Standard.

Procedure

The sample is placed on a roughened 25×25×5 mm porcelain plate, which is rigidly attached to the sliding carriage of the friction apparatus. A cylindrical porcelain peg, 10 mm in diameter and 15 mm in height, with a roughened spherical end (radius of curvature 10 mm), is placed on top of the sample; the rod is tightly clamped and may be loaded with different weights with the aid of a loading arm. The load on the peg may vary

Table 13.

Explosive	Pistil Load kp	N
A. Initiating explosives, small machine		
lead azide	0.01	0.1
lead styphnate	0.15	1.5
mercury fulminate, gray	0.3	3
mercury fulminate, white	0.5	5
tetrazene	0.8	8
B. Secondary explosive materials, large machine		
PETN (Nitropenta)	6	60
RDX (Hexogen)	12	120
HMX (Octogen)	12	120
Tetryl	36	353
C. Industrial explosives, large machine		
blasting gelatin	8	80
Gelignite, 60% nitroglycerine	12	120
Ammongelit, 38% nitroglycol	24	240

(Gelatins with a low content of nitroglycerine or nitroglycol, powderform explosives, slurries and permitted explosives, ammonium nitrate; dinitrobenzene; nitroglycol; nitroglycerine, nitrocellulose up to 13.4% N, picric acid and TNT do not react up to a pistil load of 36 kp.)

between 0.01 and 1 kp in a small apparatus and between 0.5 and 36 kp in a large apparatus. The porcelain plate moves forward and back under the porcelain peg; the stroke length is 10 mm in each direction. The two ends of the peg will serve for two trials and the two friction surfaces of the plate will serve for three trials each.

Friction sensitivity of explosive materials

(Sensitiveness to explosive materials)

The figure reported is the smallest load of the peg which causes deflagration, crackling or explosion of the test sample at least once in six consecutive trials. The quantity of the test sample is 10 mm^3.

Fuel

Brennstoff; combustible

Most explosives and pyrotechnical compositions are prepared by a mixture of → *Oxidizers* and fuels. Fuel means any substance capable of reacting with oxygen and oxgen carriers (oxidizers) with the evolution of heat. Hence, the concept of fuel here has a wider significance than that of fuel in everyday language; thus, for instance, ammonium chloride in ion-exchanged → *Permitted Explosives* can act as a fuel.

Fuel Air Explosives

FAE; explosifs combustible-air; Brennstoff-Luft Sprengstoffe, Druckwellensprengstoffe

At the beginning of the seventies, the first useable FAE were developed at the U.S. Naval Air Warfare Centre Weapons Division NAWCWPNS, California. They are considered as the strongest non-nuclear chemical explosives. Primarily ethylene oxide (EO) or propylene oxide (PO) serve as fuels. These substances are atomised by explosive charges and ignited after mixing up with air. After intramoleculare decomposition the fuel reacts with atmospheric oxygen and starts a detonation with velocities about 2000 m/s. Peak pressure under the detonating cloud reaches up to 30 bar. The effectiveness of the blast wave exceeds TNT more than five times calculated for equivalent masses.

As a result from the rapid oxidation of surrounding oxygen a very strong suction phase is generated which is expressed by the term "vacuum bomb". EO and PO are toxic and carcinogenic which has led to the development of innocuous FAE in the last fifteen years (see also → *thermobaric explosives*).

For optimizing pressure wave propagation FAE are ignited similar to nuclear-weapons in a defined distance above ground zero. So they often produce an atomic-mushroom-like smoke signature and blast characteristics making them look like "mini Nukes".

Fields of deployment for FAE are the rapid removal of AP (antipersonal) mines and the production of highly effective blast waves. Being exposed to these long enduring pressure and suction phases may lead to heavy internal injuries up to lung rupture. Modern infantry troops have a much better fragment protection than in former days, but until now there is no effective protection against the effects of such strong blast waves.

Fumes

Schwaden; fumées de tir

The composition of the fumes produced by the detonation of an explosive can be determined by calculation (→ *Thermodynamic Calculation of Decomposition Reactions*) or by detonating a cartridge of the explosive in a closed vessel (→ *Bichel Bomb*) followed by gas analysis of the fumes.

In the case of industrial explosives containing an excess of oxygen (→ *Oxygen Balance*), it is conventionally assumed for the calculated values that only CO_2, but no CO, and also that only H_2O, N_2 and excess O_2 are contained in the fumes. In reality the reaction is much more complex, and the product may in fact include CO, NO, NO_2, CH_4 and many other substances, if the explosive contained sulfur and/or chlorine compounds.

It must always be assumed that explosive fumes and propellant fumes are to some extent toxic. Excess oxygen causes the formation of nitrogen oxides, deficiency carbon monoxide, both toxic. In the United States, the following classification of toxic fume components has been accepted: a 1–1/4 by 8″ cartridge in its cartridge paper is detonated in a → *Bichel Bomb*, and the fume composition is analysed. In the following Table "toxic gases" means the sum CO + H_2S (NO and NO_2 are not considered!) in ft^3/lb explosive:

A. Permitted explosives (as laid down by the Bureau of Mines, USA)

Table 14.

Fume Class	Toxic Gases ft^3/lb	Toxic Gases l/kg
A	less than 1.25	78
B	1.25–2.50	78–156
C	2.50–3.75	156–234

B. Rock-blasting explosives (as laid down by the → *IME: Institute of Makers of Explosives*, USA)

Table 15.

Fume Class	Toxic Gases ft^3/lb	Toxic Gases l/kg
1	less than 0.16	10
2	0.16–0.33	10–21
3	0.33–0.67	21–42

In the European Community the Standard EN 13631-16 "Detection and Measurement of toxic gases" specifies a method for quantification of nitrogen oxides and carbon oxides produced by the detonation of explosives for use in underground works.

The test is carried out in a blast chamber with a minimum volume of 15 m^3, which shall be designed to withstand the forces during the detonation of high explosives and to prevent a significant loss of blasting fumes. The chamber is equipped with a thick walled steel tube (inner diameter of 150 mm, length of 1400 mm) and with an effective mixing system to ensure a homogeneous gas phase. The chamber has ports for gas sampling and for measuring the ambient temperature and pressure.Cartridged explosives and bulk explosives filled in glass or aluminium tubes can be used. The explosive charge shall have the minimum diameter for application and a length of 700 mm or at least seven times the diameter. The explosive mass-to-chamber volume ratio shall be between 30 g/m^3 and 50 g/m^3.

The explosive charge is placed centrally in the bore of the steel tube and fired there. The initiation is done as recommended by the manufacturer. If booster charges are necessary their proportion in the fumes produced has to be considered in the calculation.

For sampling a gas extraction system has to be used, which prevents the condensation of water vapour and the subsequent dissolving of nitrous oxides.

The quantity of CO, CO_2, NO and NO_2 is measured simultaneously and continuously over a period 20 min. The concentration of CO and CO_2 are constant over the entire measuring period, provided the blast chamber is sufficiently gas tight. Since NO and NO_2 give subsequent secondary reactions, measured concentration is extrapolated to obtain the initial concentration. From the initial concentrations so determined, the volume of the chamber and the amount of the explosive fired, the amount of each toxic gas is calculated in litres per kilogram of explosive (at standard temperature and pressure.) The test is performed three times.

Limits for the toxic gases are not required in the European Standard. However, the measured amounts can be used by national authorities for regulations of the underground use of explosives.

Functioning Time

Ignition delay; Anzündverzugszeit; retard d'allumage

Lapsed time between application of firing current to start of pressure rise.

Fuse

An igniting or explosive device in form of a cord, consisting of a flexible fabric tube and a core of low or high explosive. Used in blasting and demolition work, and in certain munitions. A fuse with a black powder or other low explosive core is called a safety fuse or blasting fuse. A fuse with a → *PETN* or other high explosive core is called "detonating cord" or primacord.

Fuze

Zünder, Anzünder; fusée

A device with explosive or pyrotechnic components designed to initiate a train of fire or detonation.

Fuze, delay. Any fuze incorporating a means of delaying its action. Delay fuzes are classified according to the length of time of the delay.

Fuze, long delay. A type of delay fuze in which the fuze action is delayed for a relatively long period of time, depending upon the type, from minutes to days.

Fuze, medium delay. A type of delay fuze in which the fuze action is delayed for a period of time between that of short delay and long delay fuzes, normally four to fifteen seconds.

Fuze Head

Zündschraube, Anzündschraubo

A device for the ignition of a gun propellant. It consists of a percussion cap, containing a small amount of black powder booster in front, and a threaded armature part screwed into the base of the cartridge.

Gap Test

→ Detonation, Sympathetic Detonation; Flash Over.

Gas Generators

gaserzeugende Ladungen; charges génératrices de gaz

Pyrotechnic or propellant device in which propellant is burned to produce a sustained flow of gas at a given pressure on demand. Gas-generating units are employed in blasting operations conducted in mines without recourse to brisant explosives. The device consists of a non-detonating gas-generating material and a priming or a heating charge, which are confined together in a steel pipe. The heating charge evaporates the gas-generating substance such as liquid CO_2 (→ *Cardox*); another possibility is for the primer to initiate an exothermal chemical reaction (Chemecol process, Hydrox process). The gas-generating reaction may be the decomposition of nitrogen-rich compounds such as ammonium nitrate, or nitrate mixtures, or nitroguanidine in the presence of carbon carriers and sometimes in the presence of catalysts. When a given pressure has been reached, a bursting disc releases the gases in the pipe. The sudden gas expansion taking place in the borehole has an effect similar to that of an explosion.

Gas Jet Velocity

Nozzle Velocity; Ausströmgeschwindigkeit; vélocité à jet de gaz

In rocket technology, the velocity of the combustion gases discharged from the combustion chamber and passing the nozzle into the atmosphere. The jet velocity and the mass flow serve to calculate the → *Thrust*. The jet velocity will increase with the pressure in the combustion chamber, i.e., with the expansion ratio under passage through the → *Nozzle*. The pressure in the combustion chamber should not be adjusted too high, otherwise the wall thickness of the chamber (i.e. its weight) will become too great (→ *Mass Ratio*).

In accordance with the *Saint-Venant* and *Wantzel* formula

$$a = \sqrt{2 \frac{k}{k-1} \frac{RT}{M} \left[1 - \left(\frac{P_0}{P_1}\right)^{\frac{k-1}{k}}\right]}$$

where:
P_0 is the gas pressure at the nozzle exit (atmospheric pressure);
P_1 is the pressure in the combustion chamber;
k is the coefficient of specific heat;
R is the ideal gas constant measured in absolute units;
T is the flame temperature in Kelvin, and
M is the mean molecular weight of the combustion gases.

The jet velocity is proportional to the square root of the combustion temperature and inversely proportional to the square root of the mean molecular weight of the combustion gases.

Other details can be deduced from the formula.

Other keywords in this connection: → *Propellant Area Ratio* → *Solid Propellant Rockets.*

Gas Pressure

Gasdruck; pression de gaz

The pressure generated in the chamber of a weapon; its value depends to a large extent on the nature of the weapon and of the powder selected. Standard determinations of gas pressure are carried out with the aid of a crusher ("measuring egg") – a copper cylinder or a copper pyramid – the compression of which is a measure of the gas pressure.

A complete transient gas pressure curve can be plotted with the aid of piezo-quartz or other pressure transducer with an oscillograph (→ *Ballistic Bomb*).

GELANTINE DONARIT S

Gelatine Donarit S is a blasting cap sensitive gelatinous explosive. It contains specialsensitizers and can be used under high pressure.
seismic explosive. It will be delivered in screwableplastic tube which can be combined to build long loading units.

Gelatine Donarit S is manufactured by the Austin Powder GmbH, Austria (formerly Dynamit Nobel Wien).

density:	1,6 g/cm^3
gas volume:	775 l/kg
specific energy:	971 K.l/kg
velocity of detonation:	> 6000 m/s
(without confinement, diameter 32 mm)	

Gelatins; Gelatinous Explosives; Gelignites

They are dough-like Explosives based on nitroglycerine/nitroglycol gelatinized by nitrocellulose. Gelatine are being replaced by cartridged

→ *Emulsion Slurries.*

→ *Plastic Explosives.*

Geosit 3

Trade name of a sensitized gelatinous special explosive distributed in Germany and exported by WASAGCHEMIE. It is used for seismic prospecting and mud capping.
The explosive can be supplied as a cartridge in sealable plastic tubes.

density of cartridge:	1.6 g/cm^3
weight strength:	81%
detonation velocity at cartridge density, unconfined:	6100 m/s = 20000 ft/s

GGVE

Gefahrgutverordnung, Eisenbahn

German transport regulation; → *RID*

Glycerol Acetate Dinitrate

Acetyldinitroglycerin; acétate-dinitrate de glycérine

$$\begin{array}{l} CH_2\text{-}O\text{-}NO_2 \\ |\\ CH\text{-}O\text{-}CO\text{-}CH_3 \\ |\\ CH_2\text{-}O\text{-}NO_2 \end{array}$$

pale yellow oil
empirical formula: $C_5H_8N_2O_8$
molecular weight: 224.1
oxygen balance: −42.86%
nitrogen content: 12.50%
density: 1.412 g/cm^3
lead block test: 200 cm^3/10 g
deflagration point: 170–180 °C = 338–356°F

This compound is insoluble in water, but is readily soluble in alcohol, ether, acetone, and concentrated HNO_3.

It may be prepared by nitration of acetylglycerol with mixed acid containing a very large proportion of nitric acid.

Glycerol acetate dinitrate has been proposed as an additive to nitroglycerine in order to depress the solidification point of the latter. It has so far not been employed in practice.

Glycerol Dinitrate

Dinitroglycerin, Glycerindinitrat; dinitrate de glycérine

$$\begin{array}{ll} \text{CH}_2\text{-O-NO}_2 & \text{CH}_2\text{-O-NO}_2 \\ \text{CH-OH} & \text{CH-O-NO}_2 \\ \text{CH}_2\text{-O-NO}_2 & \text{CH}_2\text{-OH} \\ \alpha & \beta \end{array}$$

pale yellow oil
empirical formula: $C_3H_6N_2O_7$
molecular weight: 182.1
oxygen balance: -17.6%
nitrogen content: 15.38%
density: 1.51 g/cm^3
solidification point: $-30 \text{ °C} = -22\text{°F}$
lead block test: $450 \text{ cm}^3/10 \text{ g}$
deflagration point: $170 \text{ °C} = 338\text{°F}$
impact sensitivity: $0.15 \text{ kp m} = 1.5 \text{ N m}$

Glycerol dinitrate is a viscous liquid, but is more volatile and more soluble in water than nitroglycerine. It is hygroscopic and may be used as a gelatinizer of certain types of nitrocelluloses. It is more stable than glycerol trinitrate. Its vapors are toxic and cause headaches.

It is prepared by nitration of glycerol with nitric acid; such nitrations mostly yield mixtures of di- and trinitroglycerine.

Glycerol – 2,4-Dinitrophenyl Ether Dinitrate

Dinitrophenylglycerinetherdinitrat;
dinitrate de glycérine-dinitrophényléther, Dinitryl

$$\begin{array}{l} \text{CH}_2\text{-O-} \\ \text{CH-O-NO}_2 \\ \text{CH}_2\text{-O-NO}_2 \end{array} \!\!\!\!\bigcirc\!\!\begin{array}{l} \text{NO}_2 \\ \\ \text{NO}_2 \end{array}$$

pale yellow crystals
empirical formula: $C_9H_0N_4O_{11}$
molecular weight: 348.2
oxygen balance: -50.6%
nitrogen content: 16.09%
density: 1.60 g/cm^3
melting point: $124 \text{ °C} = 255\text{°F}$
lead block test: $320 \text{ cm}^3/10 \text{ g}$
deflagration point: $205 \text{ °C} = 400\text{°F}$
impact sensitivity: $0.8 \text{ kp m} = 8 \text{ N m}$

This compound is prepared by reacting glycerol nitrophenyl ether with a nitric acid – sulfuric acid mixture at 25 – 30 °C (~77 °F). It is insoluble in water, but is readily soluble in acetone. It is a poor gelatinizer of nitrocellulose.

Glycerol Nitrolactate Dinitrate

Dinitroglycerinnitrolactat; dinitrate-nitrolactate de glycérine

$$\begin{array}{l} CH_3 \\ | \\ CH\text{-}O\text{-}NO_2 \\ | \\ CO \\ \diagdown O \\ \diagdown CH_2 \\ | \\ CH\text{-}O\text{-}NO_2 \\ | \\ CH_2\text{-}O\text{-}NO_2 \end{array}$$

colorless liquid
empirical formula: $C_6H_9N_3O_{11}$
molecular weight: 299.2
oxygen balance: -29.7%
nitrogen content: 14.05%
density: 1.47 g/cm³
refractive index: $n_D^{25} = 1.464$
deflagration point: 190 °C = 374°F

Dinitroglycerol nitrolactate is practically insoluble in water, readily soluble in alcohol and ether, and is a good gelatinizer of nitrocellulose. It is more resistant to heat and less sensitive to impact than nitroglycerine.

Glycerol Trinitrophenyl Ether Dinitrate

*Trinitrophenylglycerinetherdinitrat;
dinitrate de trinitrophenyl-glycérineéther*

$$O_2N\text{-}\underset{NO_2}{\overset{NO_2}{\bigcirc}}\text{-}O\text{-}\underset{|}{CH_2}\text{-}\underset{|}{CH\text{-}O\text{-}NO_2}\text{-}CH_2\text{-}O\text{-}NO_2$$

the diole substituted prepolymer is forming yellowish,
    light-sensitive crystals
empirical formula: $C_9H_7N_5O_{13}$
molecular weight: 393.2
oxygen balance: -34.6%
nitrogen content: 17.81%
solidification point: 128.5 °C = 263.3°F

lead block test: 420 cm^3/10 g
deflagration point: 200–205 °C = 392–400°F
impact sensitivity: 0.4 kp m = 4 N m

Glycerol trinitrophenyl ether dinitrate is insoluble in water, but is readily soluble in acetone; it does not gelatinize nitrocellulose.

It is prepared by nitration of phenyl glycerol ether with a nitric acid-sulfuric acid mixture.

Glycidyl Azide Polymer

Glycidylazidpolymer; GAP

$$\left[-CH_2-CH-O- \atop \quad\quad | \atop \quad\quad CH_2-N_3 \right]_n$$

light-yellowish, viscous liquid
empirical formula of structural unit: $C_3H_5N_3O$
molecular weight of structural unit: 99.1
mean molecular weight: 2000
energy of formation: +1535.2 kJ/kg = +366.9 kcal/kg
enthalpy of formation: +1422.9 kJ/kg = +340.1 kcal/kg
oxygen value: –121.1 %
nitrogen content: 42.40 %
specific energy: 82.4 mt/kg = 808 kJ/kg
explosion heat (H_2O liq.): 3429 kJ/kg = 820 kcal/kg
normal volume of gases: 946 l/kg
viscosity: 4280 cP
density: 1.29 g/cm^3
deflagration temperature: 216 °C
impact sensitivity: 7.9 Nm = 0.8 kpm
sensitivity to friction: at 360 N = 37 kp pin load, no reaction

Glycidyl azide polymer is produced in a two-step process. First, epichlorohydrin in the presence of bortriflouride is polymerized into polyepichlorohydrin. Using dimethylformamide as a solvent, the polymer is then processed with sodium azide at high temperature. Nearly all the inorganic components as well as the solvent are removed, leaving the raw final product free of low molecular weight compounds.

Glycidyl azide polymer was originally developed in the USA as an → *Energetic Binder* for → *Composite Propellants.* Because this gas-producing component releases at the composition large amounts of nitrogen and thermal energy. It has been used in recent years as an energetic binder compound in → *LOVA* gun propellant and in gas generating propellants; potential for fast burning rocket propellants.

Grain

A single mass of solid propellant of the final geometric configuration as used in a gas generator or rocket motor.

Also used as a mass unit for gun propellants. 1 grain = 0.0648 g.

Granulation

Size and shape of grains of pyrotechnic or propellant ingredients (→ *Grist*)

Graphite

C
atomic weight: 12.01

serves for surface smoothing of flake-grained → *Gunpowder* and of → *Black Powder.*

Specifications

moisture: not more than	0.5%
reaction:	neutral
glow residue in natural graphite: not more than	25%
no scratching parts admitted silicic acid:	none

Grist

Particle size of pyrotechnic material (→ *Granulation*)

GSX

Cheap mixture of ammonium nitrate, water, aluminium powder and polystyrene adhesive as a bonding agent. First used in the 6.75 t free fall bomb BLU-82 ("Daisy cutter" or "aerosol bomb"). The peak pressure of the bomb ignited approximately 1 m above ground reaches 70 bar in a radius of 30 m. GSX has been used for mine field clearing and for "Instant helicopter landing zones" in the Viet Nam jungle-war.

Guanidine Nitrate

Guanidinnitrat; nitrate de guanidine

$$HN=C\begin{smallmatrix}NH_2 \cdot HNO_3\\ NH_2\end{smallmatrix}$$

colorless crystals
empirical formula: $CH_6N_4O_3$
molecular weight: 122.1
energy of formation: -726.1 kcal/kg = -3038 kJ/kg
enthalpy of formation: -757.7 kcal/kg = -3170.1 kJ/kg
oxygen balance: -26.2%
nitrogen content: 45.89%
volume of explosion gases: 1083 l/kg
heat of explosion
(H_2O liq.): 587 kcal/kg = 2455 kJ/kg
(H_2O gas): 447 kcal/kg = 1871 kJ/kg
specific energy: 72.6 mt/kg = 712 kJ/kg
melting point: 215 °C = 419°F
heat of fusion: 48 kcal/kg = 203 kJ/kg
lead block test: 240 cm³/10 g
deflagration point: decomposition at 270 °C = 518°F
impact sensitivity: up to 5 kp m = 50 N m no reaction
friction sensitivity:
 up to 36 kp = 353 N pistil load no reaction
critical diameter of steel sleeve test: 2.5 mm

Guanidine nitrate is soluble in alcohol and water. It is the precursor compound in the synthesis of → *Nitroguanidine*. It is prepared by fusing dicyanodiamide with ammonium nitrate.

Guanidine nitrate is employed in formulating fusible mixtures containing ammonium nitrate and other nitrates; such mixtures were extensively used during the war as substitutes for explosives, for which the raw materials were in short supply. However, a highbrisance explosive such as Hexogen or another explosive must usually be added to the mixtures. It was also proposed that guanidine nitrate be incorporated in → *Double Base Propellants* and gas generating propellants.

Guanidine Perchlorate

Guanidinperchlorat; perchlorate de guanidine

$$HN=C{\overset{NH_2 \cdot HClO_4}{\underset{NH_2}{\diagup}}}$$

empirical formula: $CH_6N_3O_4Cl$
molecular weight: 159.5
energy of formation: -440.1 kcal/kg = -1841.4 kJ/kg
enthalpy of formation: -466.1 kcal/kg = -1950.0 kJ/kg
oxygen balance: -5.0%
nitrogen content: 26.35%
melting point: 240 °C = 464°F
lead block test: 400 cm^3/10 g

This compound is prepared from guanidine hydrochloride and sodium perchlorate.

Guanidine Picrate

Guanidinpikrat; picrate de guanidine

$$HN=C{\overset{NH_2}{\underset{NH_2}{\diagup}}} \quad + \quad \underset{NO_2}{\underset{}{\overset{OH}{\underset{}{O_2N\diagdown\diagup NO_2}}}}$$

yellow crystals
empirical formula: $C_7H_8N_6O_7$
molecular weight: 288.1
oxygen balance: -61.1%
nitrogen content: 29.16%
melting point:
decomposition at 318.5–319.5 °C = 605–606°F
deflagration point: 325 °C = 617°F

Guanidine picrate is sparingly soluble in water and alcohol. It is prepared by mixing solutions of guanidine nitrate and ammonium picrate.

Guar Gum

Guarkernmehl; farine de guar

Guar gum is a water soluble paste made from the seeds of the guar plant Cyanopsis tetragonoloba. The product gels with water in the cold. It is added to commercial powder explosives so as to protect them from influx of water in wet boreholes. Guar gum gelled with water produces a barrier layer, which prevents any further penetration of water (→ *Water Resistance*; → *Slurries*).

Guarnylureadinitramide

GUDN, Guarnylureadinitramide, N-Guanylharnstoffdinitramid, FOX-12

$$\left[\begin{array}{c} \text{NH} \quad \text{O} \\ \| \quad \| \\ \text{H}_2\text{N}-\text{C}-\text{N}-\text{C}-\text{NH}_2 \\ | \\ \text{H} \end{array}\right] * \left[\begin{array}{c} \text{NO}_2 \\ | \\ \text{HN} \\ | \\ \text{NO}_2 \end{array}\right]$$

white crystals
sum formula: $C_2H_7N_7O_5$
molecular weight: 209.12 g
energy of formation: -332 kJ/mole
enthalpy of formation: -356 kJ/mole
oxygen balance: -19.13 %
volume of explosion gases 785 l/kg
heat of explosion (calculated): 2998 kJ/kg (H_2O gas);
  3441 kJ/kg (H_2O liq.)
density: 1.75 g/cm^3
specific energy: 950 kJ/kg
melting point: 215 °C

GUDN is a high explosive developed by the Swedish Defence Research Agency FOI. It provides good thermal stability, low water solubility and no hygroscopicty. It is used as a fuel in gas-generating compositions, and may be used for LOVA applications.

Gunpowder

propellant; Schiesspulver; poudre

The propellant which has exclusively been used for a long time in conventional military weapons is the smokeless (or, more accurately, low-smoke) powder. According to its composition, it can be classified as single-base powders (e.g., nitrocellulose powder), doublebase powders (e.g., nitroglycerine powder) and triple-base powders (e.g., nitrocellulose + nitroglycerine (or diglycol dinitrate) + nitroguanidine powders).

The main component of nitrocellulose powders is nitrocellulose, a mixture of guncotton (13.0–13.4 % nitrogen) and soluble guncotton (11–13 % nitrogen content). To manufacture the powder, the nitrocellulose mixture is gelatinized with the aid of solvents – mostly alcohol and ether. Additives – stabilizers in particular – can be incorporated at this stage. The plastic solvent-wet mass thus obtained is now shaped in extrusion presses to give strips or tubes and is cut to the desired length by a cutting machine. The residual solvents in the powder are removed by soaking the powder in water and drying. The dried powder is then polished in drums and is graphitized. A surface treatment is performed at the same time, using alcoholic solutions of Centralite, dibutyl phthalate, camphor, dinitrotoluene, or other phlegmatization agents.

To make nitroglycerine powder, nitrocellulose is suspended in water, the suspension is vigorously stirred, and nitroglycerine is slowly introduced into the suspension, when practically all of it is absorbed by the nitrocellulose. The bulk of the water (residual water content 25–35%) is then centrifuged off or squeezed out, and the powder paste is ground. It is then mixed by mechanical kneading with nitroglycerine-insoluble additives and is gelatinized on hot rollers, as a result of which the water evaporates, leaving behind a residual water content of about 1%.

This product, which is thermoplastic, can now be geometrically shaped as desired, in accordance with the type of the powder, using finishing rollers, cutting and punching machines, or hydraulic extrusion presses.

This solventless processing avoids variations in the characteristics of the products due to the presence of residual solvents. No prolonged drying operations are needed for ballistic stability of the gunpowder.

If the use of solvents is required in the production process of double and triple base propellants, the nitroglycerine can be introduced in the mixtures in the form of a "master mix", a gelatinized mixture consisting of 85% nitroglycerine and 15% alcohol-wet nitrocellulose of the same type as the prescribed powder component.

Depending on their intended use, nitroglycerine powders have a nitroglycerine content between 25 and 50%.

In the USA and in the United Kingdom, a large amount of nitroglycerine and nitroguanidine powders are still produced with the aid of solvents. Acetone is added to nitroglycerine in order to facilitate the kneading and pressing operations, but must be subsequently removed by drying.

A number of liquid nitrate esters other than nitrocellulose have been recently used, including diglycol dinitrate, metriol trinitrate, and butanetriol trinitrate, of which diglycol dinitrate has been the most extensively employed. Powders prepared with it or with triglycol dinitrate are lower in calories. This fact is relevant to the service life of the gunbarrels in which these powders are utilized. Such powders are known as "cold propellants".

Further research for gunbarrel-saving propellants led to the development of nitroguanidine powders, in which → *Nitroguanidine* (picrite) is the third energy-containing component, beside nitroglycerine (or diglycol dinitrate or triglycol dinitrate) and nitrocellulose. Powders containing more than 40% nitroguanidine can be made only with the aid of solvents.

Another special processing method is used for the manufacture of → *Ball Powder*. Floating spheres of concentrated nitrocellulose solutions are cautiously suspended in warm water; the solvent evaporates gradually and the floating spheres solidify. Finally, an intensive surface treatment is needed to reach the desired ballistic behavior. The ballistic properties of a powder are affected not only by its chemical composition, but also by its shape. Thus, in conventional weapons, it ought to bring about progressive burning, or at least ensure that the surface area of the grain remains constant during combustion.

The following geometric forms of powder grains are manufactured:

perforated long tubes	perforated tubes, cut short
multi perforated tubes	flakes
strips	ball powder
cubes	rods, cut short
rings	

Finer-grained powders are used for portable firearms; tubular powder is mostly employed for guns; powders in the form of flakes and short tubes are employed for mortars, howitzers, and other high-angle firearms.

Finer-grained powders can be improved in their ballistic behavior by → *Surface Treatment*. Phlegmatizers are infiltrated in the outer layer of the powder grains; the burning rate in the weapon chamber begins slowly and turns progressive.

Gurney energy

The energy E_G per unit mass available for the acceleration of fragments of detonating explosives. It consists of the kinetic energies of the moving accelerated fragments and fumes.

Gurney velocity

The velocity of fragments of explosives extrapolated to zero mass.
$v_G = \sqrt{2 E_G}$

Hangfire

Spätzündung; explosion tardive

The detonation of an explosive charge at some non-determined time after its normally designed firing time. This can be a dangerous phenomenon.

Hansen Test

In this stability test, which was proposed by Hansen in 1925, 8 samples of the material to be tested are heated up to 110 °C (230°F). Every hour one of the, samples is taken out of the oven, extracted with CO_2-free water, and the pH of the filtrate determined. Since the decomposition of propellants based on nitrates is usually accompanied by the liberation of CO_2, which interferes with the potentiometric determination, the results obtained are unsatisfactory, and the test is now hardly ever used.

HBX, HBX-1 etc.

These are pourable mixtures of TNT, Hexogen and aluminum (→ *Torpex*) containing phlegmatizing additives.

Heat of Combustion

Verbrennungswärme; chaleur de combustion

Unlike the heat of explosion, the heat of combustion represents the caloric equivalent of the total combustion energy of the given substance. It is determined in a calorimetric bomb under excess oxygen pressure. The heat of combustion is usually employed to determine the heat of formation.

The heat of combustion depends only on the composition of the material and not on any other factor, such as loading density or other factors.

Heat of Explosion

Explosionswärme; chaleur d'explosion

The heat of explosion of an explosive material, an explosive mixture, gunpowder or propellant is the heat liberated during its explosive decomposition. Its magnitude depends on the thermo-dynamic state of the decomposition products; the data used in practical calculations usually have water (which is a product of the explosion) in the form of vapor as the reference compound.

The heat of explosion may be both theoretically calculated and experimentally determined. The calculated value is the difference between the energies of formation of the explosive components (or of the explosive itself if chemically homogeneous) and the energies of formation of the explosion products (for more details → *Thermodynamic Calculation of Decomposition Reactions*). The advantage of the calculation method is that the results are reproducible if based on the same energies of formation and if the calculations are all conducted by the same method; this is often done with the aid of a computer.

The values of heats of explosion can also be more simply calculated from the *"partial heats of explosion"* of the components of the propellant (see below).

The calculated values do not exactly agree with those obtained by experiment; if the explosion takes place in a bomb, the true compositions of the explosion products are different and, moreover, vary with the loading density. In accurate calculations these factors must be taken into account. In difficult cases (strongly oxygen-deficient compounds and side reactions, such as the formation of CH_4, NH_3, HCN, or HCl), the only way is to analyze the explosion products. For standard values of heats of formation at constant volume or constant pressure → *Energy of Formation*.

The experimental determination takes place in a calorimetric bomb. The bomb volume is usually 20 cm^3, but can also be 300 cm^3. The sample quantity is usually so chosen as to obtain a loading density of 0.1 g/cm^3. If a powder refuses to explode – as is often the case if the heat of explosion is smaller than 800 cal/g – a "hot" powder with a known heat of explosion is added, and the heat of explosion of the sample powder is calculated from that of the mixture and that of the hot powder.

The heat of detonation under "CJ conditions" (→ *Detonation*) can differ from the explosion value, because the chemical reaction can be influenced by the conditions in the wave front (e.g., by the loading density of the explosive)*).

Moreover, the detonation energy is related to H_2O in the gaseous state. The calorimetric values as well as the calculated values given for the individual explosives in this book are based on H_2O in the liquid state as a reaction product.

---

* D. L. Ornellas, The Heat and Products of Detonation in a Calorimeter of CNO, HNO, CHNF, CHNO, CHNOF, and CHNOSi Explosives, Combustion and Flame 23, 37–46 (1974).

Partial Heat of Explosion

partielle Explosionswärme;
chaleur partielle d'explosion

A. Schmidt proposed a simplified way of estimating the probable heat of explosion of a propellant. In this method, a "partial heat of explosion" is assigned to each component of the powder. Materials with high negative oxygen balances (e.g., stabilizers and gelatinizers) are assigned negative values for the partial heat of explosion. The explosion heat of the propellant is calculated by the addition of the partial values weighted in proportion to the respective percentage of the individual components.

A number of such values have been tabulated. The value for trinitroglycerine is higher than its heat of explosion, since the excess oxygen reacts with the carbon of the other components.

Table 16. Values for the partial heat of explosion

Component	Partial Heat of Explosion	
	kcal/kg	kJ/kg
Akardite I	−2283	−9559
Akardite II	−2300	−9630
Akardite III	−2378	−9957
ammonium nitrate	+1450	+6071
barium nitrate	+1139	+4769
barium sulfate	+132	+553
butanetriol trinitrate (BTN)	+1400	+5862
camphor	−2673	−11192
Candelilla wax	−3000	−12561
carbon black	−3330	−13942
Centralite I	−2381	−9969
Centralite II	−2299	−9626
Centralite III	−2367	−9911
cupric salicylate	−1300	−5443
basic cupric salicylate	−900	−3768
diamyl phthalate (DAP)	−2187	−9157
dibutyl phthalate (DBP)	−2071	−8671
dibutyl tartrate (DBT)	−1523	−6377
dibutyl sebacate (DBS)	−2395	−10028
diethyleneglycol dinitrate (DGN, DEGN)	+1030	+4313
dioxyenitramine dinitrate (DINA)	+1340	+5610
diethyl phthalate (DEP)	−1760	−7369
diethyl sebacate (DES)	−2260	−9463
diisobutyl adipate (DIBA)	−2068	−8658

Partial Heat of Explosion

Component	Partial Heat of Explosion	
	kcal/kg	kJ/kg
dimethyl phthalate (DMP)	−1932	−8089
dinitrotoluene (DNT)	− 148	− 620
dioctyl phthalate (DOP)	−2372	−9931
diphenylamine (DPA)	−2684	−11238
diphenyl phthalate (DPP)	−2072	−8675
diphenylurea	−2227	−9324
diphenylurethane	−2739	−11468
ethyleneglycol dinitrate	+1757	+7357
ethylphenylurethane	−1639	−6862
glycol	−889	−3722
graphite	−3370	−14110
lead acetyl salicylate	− 857	−3588
lead ethylhexanoate	−1200	−5024
lead salicylate	−752	−3149
lead stearate	−2000	−8374
lead sulfate	+150	+ 628
methyl methacrylate (MMA)	−1671	−6996
Metriol trinitrate (MTN)	+1189	−4978
mineral jelly	−3302	−13825
nitrocellulose, 13.3 % N	+1053	+4409
nitrocellulose, 13.0 % N	+1022	+4279
nitrocellulose, 12.5 % N	+ 942	+3944
nitrocellulose, 12.0 % N	+ 871	+3647
nitrocellulose, 11.5 % N	+ 802	+3358
nitroglycerine (NG)	+1785	+7474
nitroguanidine (picrite)	+721	+3019
PETN	+1465	+6134
pentaerythrol trinitrate	+1233	+5163
polyethylene glycol (PEG)	−1593	−6670
poly methacrylate (PMA)	−1404	−5879
polyvinyl nitrate (PVN)	+ 910	+3810
potassium nitrate	+1434	+6004
potassium perchlorate	+1667	+6980
potassium sulfate	+300	+1256
TNT	+491	+2056
triacetin (TA)	−1284	−5376
triethyleneglycol dinitrate (TEGN)	+750	+3140

The values refer to water in the liquid state as a reaction product.

Heat Sensitivity

thermische Sensibilität; sensitiveness to Heat; sensibilité au chauffage

Heat sensitivity is determined by testing the flammability of explosives brought into contact with glowing objects, flame, sparks, the initiating flame of a black powder safety fuse, a red-hot iron rod, or a flame.

RID (*R*èglement *I*nternational concernant le Transport des Marchandises *D*angereuses) describes a method, in which a sample of about 500 g of the explosive, accommodated in a metal can of given dimensions, is exposed to a wood fire, and its behavior (combustion, intense decomposition or detonation) is observed.

In response to a suggestion made by *Koenen* (Bundesanstalt für Materialprüfung, Berlin, Germany), these tests, which are carried out with the purpose of evaluating the safety during transport, were improved as described below; the method is known as the "steel sleeve test" (Koenen test).

Koenen Test Procedure

The sample substance is introduced into a cylindrical steel sleeve (25 mm dia.×24 mm dia.×75 mm) up to a height of 60 mm, and the capsule is closed with a nozzle plate with a central hole of a given diameter. The diameter of the hole can vary between 1 and 20 mm; when the plate is not employed, the effect is equivalent to that of a 24-mm hole. The charged sleeve is placed inside a protective box and is simultaneously heated by four burners; the time elapsed up to incipient combustion and the duration of the combustion itself are measured with a stop watch. The plate perforation diameter is varied, and the limiting perforation diameter corresponding to an explosion caused by accumulation of pressure inside the steel sleeve is determined. Explosion is understood to mean fragmentation of the sleeve into three or more fragments or into a greater number of smaller fragments.

In this way, reproducible numerical data are obtained which allow classification of different explosives according to the explosion danger they represent.

The parameter which is reported is the largest diameter of the circular perforation in mm (limiting diameter) at which at least one explosion occurs in the course of three successive trials.

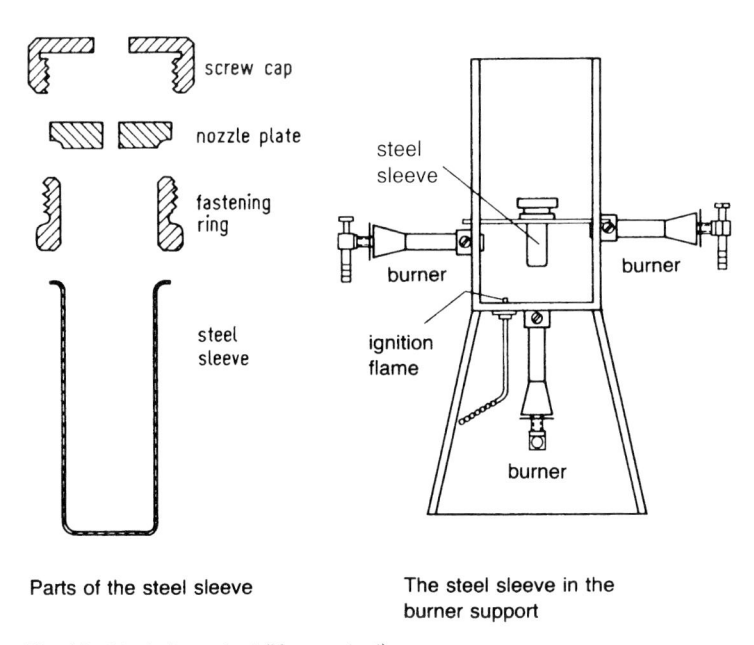

Fig. 15. Steel sleeve test (Koenen test)

Table 17. Results of steel sleeve test

Explosive Material	Limiting Diameter mm	Time until Ignition s	Time of Combustion s
A. *Homogeneous Explosives*			
nitroglycerine	24	13	0
nitroglycol	24	12	10
nitrocellulose, 13.4 % N	20	3	0
nitrocellulose, 12.0 % N	16	3	0
Hexogen	8	8	5
ammonium perchlorate	8	21	0
PETN	6	7	0
Tetryl	6	12	4
TNT	5	52	29
picric acid	4	37	16
dinitrotoluene	1	49	21
ammonium nitrate	1	43	29
B. *Industrial Explosives*blasting			
gelatin	24	8	0
guhr-dynamite	24	13	0
Gelignite	20	7	0
ammonium nitrate gelatin	14	10	0
ammonium-nitrate-based powder-form explosives	1.5–2.5	25	40
nitrocarbonitrates	2	25	4
ANFO blasting agents	1.5	33	5
gelatinous permitted explosives	14	12	0
ion-exchanged powder-form permitte explosives	1	35	5

Heptryl

N-(2,4,6 Trinitrophenyl-N-nitramino)-trimethylolmethane Trinitrate;
Trinitrate de trinitrophényl-nitramino-triméthylolméthane

$$O_2N-\underset{NO_2}{\overset{NO_2}{\bigcirc}}-N-\underset{NO_2}{C}\begin{matrix}CH_2-O-NO_2\\CH_2-O-NO_2\\CH_2-O-NO_2\end{matrix}$$

yellow crystals
empirical formula: $C_{10}H_8N_8O_{17}$
molecular weight: 512.24
energy of formation: -96.8 kcal/kg $= -405.0$ kJ/kg

oxygen balance: −21.9%
nitrogen content: 21.9%
volume of explosion gases: 787 l/kg
specific energy: 128.6 mt/kg = 1261 kJ/kg
melting point (decomp.): 154 °C = 309°F
deflagration point: 180 °C = 356°F
heat of combustion: 2265.9 kcal/kg

Heptryl is comparable in power and sensitivity to PETN. It can be prepared by nitrating 2,4-dinitroanilinotrimethylolmethane with mixed nitric-sulfuric acid and purified by reprecipitation from acetone.

HEX

Abbreviation for high energy explosive. The HEX series comprises modifications of → *Torpex*.

Hexal

Mixture of Hexogen, aluminum powder and added wax as phlegmatizer. It is used, press-molded, as a filling of anti-aircraft gunshells. Owing to the aluminum component, both an incendiary and an explosive effect are obtained.

Hexamethylene Diisocyanate

Hexamethylendiisocyanat; diisocyanate d'hexaméthylène

$O=C=N-(CH_2)_6-N=C=O$

colorless liquid
empirical formula: $C_8H_{12}N_2O_2$
molecular weight: 168.2
energy of formation: −468 kcal/kg = −1961 kJ/kg
enthalpy of formation: −496 kcal/kg = 2078 kJ/kg
oxygen balance: −205.4%
nitrogen content: 16.66%
density 20/4: 1.0528 g/cm^3
boiling point at 0.013 bar: 124 °C = 255°F

The compound acts as a hydroxy curing agent in the formation of polyurethane binders of → *Composite Propellants*; → also *Casting of Propellants*.

Hexamethylenetetramine Dinitrate

*Hexametylentetramindinitrat;
dinitrate d'hexaméthyléne tétramine*

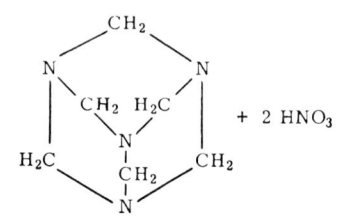

$$\begin{array}{c}\text{CH}_2\\ \text{N}\diagup\quad\diagdown\text{N}\\ |\quad\text{CH}_2\ \text{H}_2\text{C}\quad|\\ \quad\diagdown\text{N}\diagup\\ \text{H}_2\text{C}\quad|\quad\text{CH}_2\\ \quad\text{CH}_2\\ \quad|\\ \quad\text{N}\end{array} + 2\ HNO_3$$

colorless crystals
empirical formula: $C_6H_{14}N_6O_6$
molecular weight: 266.2
energy of formation: -309.9 kcal/kg = -1296.6 kJ/kg
enthalpy of formation:
    -338.8 kcal/kg = -1417.7 kJ/kg
oxygen balance: -78.3%
nitrogen content: 31.57%
volume of explosion gases: 1081 l/kg
heat of explosion
    (H_2O liq.): 631 kcal/kg = 2642 kJ/kg
    (H_2O gas): 582 kcal/kg = 2434 kJ/kg
specific energy: 76.4 mt/kg = 749 kJ/kg
melting point (decomposition): 158 °C = 316°F
lead block test: 220 cm^3/10 g
impact sensitivity: 1.5 kpm = 15 Nm
friction sensitivity: at 24 kp = 240 N pistil load reaction

This salt is soluble in water, but is insoluble in alcohol, ether, chloroform, and acetone.

Hexamethylenetetramine dinitrate can be prepared from hexamethylenetetramine and nitric acid of medium concentration; it is an important precursor of Hexogen manufactured by the Bachmann method.

Hexamethylenetriperoxide Diamine

Hexamethylentriperoxiddiamin; hexaméthylènetriperoxyde diamine; HMTD

$$\text{N}\begin{array}{c}\diagup\text{CH}_2\text{-O-O-CH}_2\diagdown\\ \text{-CH}_2\text{-O-O-CH}_2\text{-}\\ \diagdown\text{CH}_2\text{-O-O-CH}_2\diagup\end{array}\text{N}$$

colorless crystals
empirical formula: $C_6H_{12}N_2O_6$
molecular weight: 208.1

energy of formation: −384.3 kcal/kg = −1608 kJ/kg
enthalpy of formation:
  −413.7 kcal/kg = −1731 kJ/kg
oxygen balance: −92.2%
nitrogen content: 13.46%
volume of explosion gases: 1075 l/kg
heat of explosion
  (H_2O liq.): 825 kcal/kg = 3450 kJ/kg
  (H_2O gas): 762 kcal/kg = 3188 kJ/kg
specific energy: 87.3 mt/kg = 856 kJ/kg
density: 1.57 g/cm^3
lead block test: 330 cm^3/10 g
detonation velocity: 4500 m/s = 15000 ft/s
deflagration point: 200 °C = 390 °F
beginning of decomposition: 150 °C = 300 °F
impact sensitivity: 0.06 kp m = 0.6 N m
friction sensitivity: at 0.01 kp = 0.1 N pistil load reaction

This peroxide is practically insoluble in water and in common organic solvents. It is prepared from hexamethylenetetramine and hydrogen peroxide in the presence of citric acid, with efficient cooling.

It is an effective initiating explosive; nevertheless, it cannot be employed in practice owing to its poor storage properties. The thermal and mechanical stability is low.

Hexanitroazobenzene

Hexanitroazobenzol; hexanitroazobenzène

O_2N—⟨⟩—N=N—⟨⟩—NO_2 (with NO_2 groups ortho and para)

orange red crystals
empirical formula: $C_{12}H_4N_8O_{12}$
molecular weight: 452.2
oxygen balance: −49.7%
nitrogen content: 24.78%
melting point: 221 °C = 430 °F

This compound can be prepared from dinitrochlorobenzene and hydrazine. The tetranitrohydrazobenzene, which is obtained as an intermediate product, is treated with mixed acid, yielding hexanitroazobenzene by simultaneous oxidation and nitration. It is a more powerful explosive than hexanitrodiphenylamine.

2,4,6,2',4',6'-Hexanitrobiphenyl

Hexanitrobiphenyl; hexanitrobiphényle

$$O_2N-\underset{NO_2}{\overset{NO_2}{\bigcirc}}-\underset{NO_2}{\overset{NO_2}{\bigcirc}}-NO_2$$

pale yellow crystals
empirical formula: $C_{12}H_4N_6O_{12}$
molecular weight: 424.2
oxygen balance: −52.8 %
nitrogen content: 19.81 %
density: 1.6 g/cm^3
melting point: 263 °C = 505 °F
lead block test: 344 cm^3/10 g
deflagration point: 320 °C = 610 °F

Hexanitrobiphenyl is insoluble in water, but is soluble in alcohol, benzene, and toluene. It is a rather heat-intensive explosive.

2,4,6,2',4',6'-Hexanitrodiphenylamine

dipicrylamine; Hexanitrodiphenylamin; hexanitrodiphenylamine; Hexyl; hexite; HNDPhA: HNDP

$$O_2N-\underset{NO_2}{\overset{NO_2}{\bigcirc}}-NH-\underset{NO_2}{\overset{NO_2}{\bigcirc}}-NO_2$$

yellow crystals
empirical formula: $C_{12}H_5N_7O_{12}$
molecular weight: 439.2
energy of formation: +38.7 kcal/kg = +162 kJ/kg
enthalpy of formation: +22.5 kcal/kg = +94.3 kJ/kg
oxygen balance: −52.8 %
nitrogen content: 22.33 %
volume of explosion gases: 791 l/kg
heat of explosion
  (H_2O liq.): 974 kcal/kg = 4075 kJ/kg
  (H_2O gas): 957 kcal/kg = 4004 kJ/kg
specific energy: 112 mt/kg = 1098 kJ/kg
density: 1.64 g/cm^3
melting point: 240–241 °C = 464–466 °F
  (decomposition)
lead block test: 325 cm^3/10 g

detonation velocity, confined:
  7200 m/s = 23600 ft/s at ρ = 1.60 g/cm^3
deflagration point: 250 °C = 480 °F
impact sensitivity: 0.75 kp m = 7.5 N m
friction sensitivity: up to 353 N
  no reaction
critical diameter of steel sleeve test: 5 mm

This explosive is toxic (the dust attacks the skin and mucous membranes) and light-sensitive. It is insoluble in water and most organic solvents. It forms sensitive acid salts.

It is prepared by nitration of *asym*-dinitrodiphenylamine with concentrated nitric acid. *Asym*-Dinitrodiphenylamine is formed by condensation of dinitrochlorobenzene with aniline.

Its stability and brisance, as well as its sensitivity, are somewhat higher than those of picric acid.

Hexanitrodiphenylamine has been employed in underwater explosives in the form of pourable mixtures with TNT and aluminum powder. Since hexanitrodiphenylamine is toxic and is strongly colored, such mixtures are replaced by better ones (→ *Torpex;* → HBX).

By itself hexanitrodiphenylamine is an explosive with a relatively low sensitivity to heat.

The compound has been used as a precipitant for potassium.

Specifications

melting point: not less than	230 °C = 446 °F
insolubles in 1 : 3 pyridine acetone mixture: not more than	0.1 %

Hexanitrodiphenylaminoethyl Nitrate

Hexanitrodiphenylaminoethylnitrat; nitrato d'hexanitrodiphénylearninoéthyle

$$O_2N-C_6H_2(O_2N)(NO_2)-N(CH_2NO_2 \cdot CH_2-O-NO_2)-C_6H_2(NO_2)_2-NO_2$$

pale yellow platelets
empirical formula: $C_{14}H_8N_8O_{15}$
molecular weight: 528.3
oxygen balance: -51.5%
nitrogen content: 21.21 %
melting point: 184 °C = 363 °F
deflagration point: 390–400 °C = 735–750 °F

Hexanitrodiphenylglycerol Mononitrate

Heptanitrophenylglycerin; mononitrate d'hexanitrodiphényleglycérine

$$\begin{array}{l} CH_2\text{-}O\text{-}Ar(NO_2)_3 \\ CH\text{-}ONO_2 \\ CH_2\text{-}O\text{-}Ar(NO_2)_3 \end{array}$$

yellow crystals
empirical formula: $C_{15}H_9N_7O_{17}$
molecular weight: 559.3
oxygen balance: -50.1%
nitrogen content: 17.22%
melting point: 160–175 °C = 320–347 °F
lead block test: 355 cm³/10 g
impact sensitivity: 2.3 kp m = 23 N m

This compound is soluble in glacial acetic acid, sparingly soluble in alcohol, and insoluble in water.

It is prepared by dissolving glyceryl diphenyl ether in nitric acid and pouring the resulting solution into mixed acid.

2,4,6,2',4',6'-Hexanitrodiphenyl oxide

Hexanitrodiphenyloxid; hexanitrodiphényloxyde

$$O_2N\text{-}C_6H_2(NO_2)_2\text{-}O\text{-}C_6H_2(NO_2)_2\text{-}NO_2$$

yellow crystals
empirical formula: $C_{12}H_4N_6O_{13}$
molecular weight: 440.2
oxygen balance: -47.3%
nitrogen content: 19.09%
density: 1.70 g/cm³
melting point: 269 °C = 516 °F
lead block test: 373 cm³/10 g
detonation velocity, confined:
 7180 m/s = 23 600 ft/s at $\rho = 1.65$ g/cm³
impact sensitivity: 0.8 kp m = 8 N m

Hexanitrodiphenyl oxide is insoluble in water, but is sparingly soluble in alcohol and ether. It is a very stable compound, which is less sensitive to impact, but is a more powerful explosive than picric acid. It is prepared by nitrating dinitro-, trinitro-, tetranitro- and pentanitro-substituted diphenyl ether with mixed acid.

2,4,6,2',4',6'-Hexanitrodiphenylsulfide

Hexanitrodiphenylsulfid; Picrylsulfid; hexanitrodiphenylsulfide

$$O_2N-\underset{NO_2}{\overset{NO_2}{\bigcirc}}-S-\underset{NO_2}{\overset{NO_2}{\bigcirc}}-NO_2$$

reddish-yellow granular powder
empirical formula: $C_{12}H_4N_6O_{12}$
molecular weight: 456.2
oxygen balance: -56.1%
nitrogen content: 18.42%
density: 1.65 g/cm^3
melting point: $234\,°C = 453\,°F$
lead block test: 320 cm^3/10 g
detonation velocity, confined:
  7000 m/s = $23\,000$ ft/s at $\rho = 1.61$ g/cm^3
deflagration point: $305-320\,°C = 580-610\,°F$
impact sensitivity: 0.5 kp m = 6 N m

This explosive is not toxic, and its technological blasting performance resembles that of hexanitrodiphenylamine. It is sparingly soluble in alcohol and ether, but is readily soluble in glacial acetic acid and acetone.

It is prepared by reacting trinitrochlorobenzene with sodium thiosulfate in alkaline solution. It is relatively heat-insensitive.

2,4,6,2',4',6'-Hexanitrodiphenylsulfone

Hexanitrosulfobenzid; hexanitrodiphenylsulfone

$$O_2N-\underset{O_2N}{\overset{O_2N}{\bigcirc}}-SO_2-\underset{NO_2}{\overset{NO_2}{\bigcirc}}-NO_2$$

pale yellow crystals
empirical formula: $C_{12}H_4N_6O_{14}S$
molecular weight: 488.2
oxygen balance: -45.8%

nitrogen content: 17.22 %
melting point: 307 °C = 585 °F

Hexanitrodiphenylsulfone is soluble in acetone, but sparingly soluble in benzene and toluene. Its stability is satisfactory. It is prepared by oxidation of hexanitrodiphenylsulfide.

Hexanitroethane

Hexanitroethan; hexanitroéthane; HNE

$$\begin{array}{c} O_2N \\ O_2N-C-C-NO_2 \\ O_2N \end{array} \begin{array}{c} NO_2 \\ NO_2 \\ NO_2 \end{array}$$

colorless powder
empirical formula: $C_2N_6O_{12}$
molecular weight: 300.1
energy of formation: +101.8 kcal/kg = +425.9 kJ/kg
enthalpy of formation: +63.3 kcal/kg = +264.9 kJ/kg
oxygen balance: +42.7 %
nitrogen content: 28.01 %
volume of explosion gases: 734 l/kg
heat of explosion: 689 kcal/kg = 2884 kJ/kg
specific energy: 80.5 mt/kg = 789 kJ/kg
density: 1.85 g/cm^3
melting point: 147 °C = 297 °F
vapor pressure is relatively high.
transformation point: 17 °C = 63 °F
lead block test: 245 cm^3/10 g
detonation velocity, confined:
    4950 m/s = 16 240 ft/s at ρ = 0.91 g/cm^3
deflagration point: 175 °C = 347 °F
friction sensitivity: 240 N

Hexanitrohexaazaisowurtzitane

Hexanitrohexaazaisowurtzitan; HNIW; CL20
2,4,6,8,10,12-(hexanitro-hexaaza)-tetracyclododecane

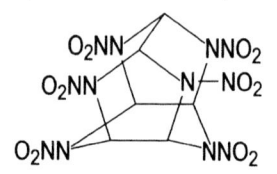

empirical formula: $C_6H_6N_{12}O_{12}$
molecular weight: 438.19

energy of formation: +240.3 kcal/kg = +1005.3 kJ/kg
enthalpy of formation: +220.0 kcal/kg = +920.5 kJ/kg
oxygen balance: −10.95%
nitrogen content: 38.3%
heat of explosion
  (H_2O liq.): 1509 kcal/kg = 6314 kJ/kg
  (H_2O gas): 1454 kcal/kg = 6084 kJ/kg
specific energy: 134.9 mt/kg = 1323 kJ/kg
density: 2.04 g/cm^3
melting point.: > 195 °C (decomposition)
impact sensitivity: 0.4 kp m = 4 Nm
friction sensibility: 4.9 kp = 48 N

Hexanitrohexaazaisowurtzitane is obtained by condensing glyoxal with benzylamine to yield hexabenzylhexaazaisowurtzitane. Next the benzyl groups are replaced under reducing conditions by easily removable substituents such as acetyl or silyl groups. Nitration to form hexanitrohexaazaisowurtzitane takes place in the final reaction step. Hexanitrohexaazaisowurtzitane exists in various crystal modifications, only the ε-modification is being of interest because of its high density and detonation velocity of more than 9000 m/s.

Being one of the most energy-rich organic explosives, CL 20 is attractive for many energetic systems.

Hexanitrooxanilide

Hexanitrodiphenyloxamid; HNO

O_2N-⟨⟩-NH-CO-CO-NH-⟨⟩-NO_2 (with O_2N substituents)

empirical formula: $C_{14}H_6N_8O_{14}$
molecular weight: 510.1
oxygen balance: −53.3%
nitrogen content: 21.97%
melting point: 295−300 °C = 565−570 °F
decomposition temperature: 304 °C = 579 °F

This compound is prepared by nitration of oxanilide. It is of interest as being relatively stable at high temperatures. The decomposition reaction above 304 °C is endothermic.

Hexanitrostilbene

Hexanitrostilben; hexanitrostilbène

$$O_2N-\underset{O_2N}{\overset{O_2N}{\bigcirc}}-\underset{H}{\overset{H}{C=C}}-\underset{NO_2}{\overset{NO_2}{\bigcirc}}-NO_2$$

yellow crystals
empirical formula: $C_{14}H_6N_6O_{12}$
molecular weight: 450.1
energy of formation: +57.3 kcal/kg = +239.8 kJ/kg
enthalpy of formation: +41.5 kcal/kg = +173.8 kJ/kg
oxygen balance: −67.6 %
nitrogen content: 18.67 %
volume of explosion gases: 766 l/kg
heat of explosion
   (H_2O liq.): 977 kcal/kg = 4088 kJ/kg
   (H_2O gas): 958 kcal/kg = 4008 kJ/kg
density: 1.74 g/cm^3
melting point: 318 °C = 604 °F (decomposition)
lead block test: 301 cm^3/10 g
impact sensitivity: 0.5 kp m = 5 N m
friction sensitivity:
   > 240 N

Hexanitrostilbene is manufactured as an additive to cast TNT, to improve the fine crystalline structure.

Hexogen

cyclo-1,3,5-trimethylene-2,4,6-trinitramine; Cyclonite; Trimethylentrinitramin; hexogène; RDX; T 4

$$O_2N-N\underset{H_2C\diagdown_N\diagup CH_2}{\overset{H_2}{\overset{C}{\diagup}\diagdown}N-NO_2}$$
$$\overset{|}{NO_2}$$

colorless crystals
empirical formula: $C_3H_6N_6O_6$
molecular weight: 222.1
energy of formation: +401.8 kJ/kg
enthalpy of formation: +301.4 kJ/kg
oxygen balance: −21.6 %
nitrogen content: 37.84 %
volume of explosion gases: 903 l/kg

heat of explosion
  calculated*)
  (H_2O liq.): 5647 kJ/kg
  (H_2O gas): 5297 kJ/kg
heat of detonation**)
  (H_2O liq.): 6322 kJ/kg
specific energy:
  1375 kJ/kg
density: 1.82 g/cm^3
melting point: 204 °C
heat of fusion: 161 kJ/kg

vapor pressure:

Pressure millibar	Temperature °C
0.00054	110
0.0014	121
0.0034	131
0.0053	138.5

lead block test: 480 cm^3/10 g
detonation velocity, confined:
  8750 m/s at ρ = 1.76 g/cm^3
impact sensitivity: 7.5 N m
friction sensitivity: 120 N pistil load
critical diameter of steel sleeve test: 8 mm

Hexogen is soluble in acetone, insoluble in water and sparingly soluble in ether and ethanol. Cyclohexanone, nitrobenzene and glycol are solvents at elevated temperatures.

Hexogen is currently probably the most important high-brisance explosive; its brisant power is high owing to its high density and high detonation velocity. It is relatively insensitive (as compared to, say → *PETN*, which is an explosive of a similar strength); it is very stable. Its performance properties are only slightly inferior to those of the homologous → *Octogen* (HMX).

The "classical" method of production (*Henning*, 1898) is the nitration of hexamethylene tetramine ($C_6H_{12}N_4$) to Hexogen ($C_3H_6O_6N_6$) using concentrated nitric acid; the concentrated reaction mixture is poured into iced water, and the product precipitates out. The structural formula shows that three methylene groups must be destroyed or split off by

---

  * computed by "ICT-Thermodynamic-Code".
  ** value quoted from *Brigitta M. Dobratz*, Properties of Chemical Explosives and Explosive Simulants, University of California, Livermore.

oxidation. As soon as this problem and the attendant dangers had been mastered, industrial-scale production became possible, and during the Second World War Hexogen was manufactured in large quantities on both sides, using several mutually independent chemical methods.

S-H process (inventor: *Schnurr*): continuous nitration of hexamethylenetetramine using highly concentrated nitric acid, accompanied by a decomposition reaction under liberation of nitrous gases, without destruction of the Hexogen formed. The reaction mixture is then filtered to separate the product from the waste acid, followed by stabilization of the product by boiling under pressure and, if required, recrystallization.

K process (inventor: *Knöffler*): an increased yield is obtained by the addition of ammonium nitrate to the nitration mixture of hexamethylene tetramine and nitric acid, followed by warming. The formaldehyde as a by-product forms more hexamethylenetetramine with the added ammonium nitrate and is converted by the nitric acid into Hexogen.

KA process (inventors: *Knöffler* and *Apel;* in USA: *Bachmann*): hexamethylenetetramine dinitrate is reacted with ammonium nitrate and a small amount of nitric acid in an acetic anhydride medium. Hexogen is formed in a similar manner as in the E process. The waste acetic acid thus formed is concentrated, subjected to the so-called ketene process, recycled, and the regenerated acetic anhydride is re-used.

E process (inventor: *Eble*): paraformaldehyde and ammonium nitrate are reacted in an acetic anhydride medium with formation of Hexogen (precursor of KA process).

W process (inventor: *Wolfram*): potassium amidosulfonate and formaldehyde are reacted to give potassium methyleneamidosulfonate ($CH_2 = N-SO_3K$), which is then nitrated to Cyclonite by a nitric acid-sulfuric acid mixture.

Phlegmatized and pressed Hexogen is used as a highly brisant material for the manufacture of → *Booster* and → *Hollow Charges.* Non-phlegmatized Hexogen in combination with TNT is also used as a pourable mixture for hollow charges and brisant explosive charges (→ *Compositions B*); mixtures of Cyclonite with aluminum powder are used as torpedo charges *(Hexotonal, Torpex, Trialen).* Hexogen may also be used as an additive in the manufacture of smokeless powders.

In manufacturing explosive charges which are required to have a certain mechanical strength or rubber-elastic toughness, Hexogen is incorporated into curable plastic materials such as polyurethanes, polybutadiene or polysulfide and is poured into molds (→ *Plastic Explosives*).

Specifications

melting point: at least	200 °C
for products prepared by the acetic anhydride method, at least	190 °C
acidity, as HNO_3: not more than	0.05 %
acetone-insolubles: not more than	0.025 %
ashes: not more than	0.03 %
sandy matter:	none

HMX

Homocyclonite, the U.S. name for → *Octogen*.

Hollow Charge

Hohlladung → *Shaped Charge*.

Hot Spots

This term denotes the increase of the detonation sensibility of explosives by finely dispersed air bubbles. The loss in sensitivity to detonation of gelatinous nitroglycerine explosives by long storage has been known since the time of Alfred Nobel; it is due to the loss or coagulation of the air bubbles that may have been left in the explosive by the manufacturing process. This effect can be explained by the adiabatic compression and heating of the air inclusions as the detonation wave is passing (→ *Detonation, wave theory*) and is termed "hot spots". This effect was used to make the recently developed cap sensitive → *Emulsion Slurries*. Conservation and independence from pressure of the air inclusions can be achieved by so-called → *Microballoons*.

Hot Storage Tests

Warmlagertest; épreuves de chaleur

These tests are applied to accelerate the decomposition of an explosive material, which is usually very slow at normal temperatures, able to evaluate the stability and the expected service life of the material from the identity and the amount of the decomposition products. Various procedures, applicable at different temperatures, may be employed for this purpose.

1. Methods in which the escaping nitrous gases can be recognized visually or by noting the color change of a strip of dyed filter paper. The former methods include the qualitative tests at 132, 100, 75, and 65.5 °C (270, 212, 167, and 150 °F). These tests include the U.S. supervision test, the methyl violet test, the *Abel* test, and the *Vieille* test.

2. Methods involving quantitative determination of the gases evolved. Here we distinguish between tests for the determination of acidic products (nitrous gases) only, such as the *Bergmann-Junk* test and methods which determine all the decomposition products, including manometric methods and weight loss methods.

3. Methods which give information on the extent of decomposition of the explosive material (and thus also on its stability), based on the identity and the amount of the decomposition products of the stabilizer formed during the storage. These include polarographic, thin-layer chromatographic and spectrophotometric methods.

4. Methods providing information on the stability of the explosive based on the heat of decomposition evolved during storage (silvered vessel test).

5. Methods in which stability can be estimated from the physical degradation of a nitrocellulose gel (viscometric measurements).

The tests actually employed vary with the kind of explosive tested (explosives, single-base, double-base or triple-base powders, or solid propellants) and the temporal and thermal exposure to be expected (railway transportation or many years' storage under varying climatic conditions). In the case of propellants about to be transported by train, only short-time testing is required. However, to obtain an estimate of the expected service life is required, the so-called long-time tests must be performed at 75 °C (167 °F) and below. The duration of such a storage is up to 24 months, depending on the propellant type. Short-time tests – the *Bergmann-Junk* test, the Dutch test, the methyl violet test, the *Vieille* test and, very rarely, the *Abel* test – are mostly employed in routine control of propellants of known composition, i.e., propellants whose expected service life may be assumed to be known. In selecting the test to be applied, the composition of the propellant and the kind and amounts of the resulting decomposition products must also be considered.

Contrary to the common propellants, which contain nitrates, the so-called composite propellants cannot be tested in the conventional manner owing to the relatively high chemical stability of the incorporated oxidants, e.g., ammonium perchlorate. In such cases the stability criterion of the propellants is the condition of the binder and its chemical and physical change.

HU-Zünder

HU-detonators have a high safety against static electricity, stray currents and energy from lightning discharge. They are safe against 4 A and 1100 mJ/ohm. All-fire current is 25 A, all-fire energy 2500 mJ/ohm. They are manufactured by ORICA Germany GmbH (formerly DYNAMIT NOBEL), as instantaneous detonators and with 20 ms and 30 ms short period delay, 18 delays each, and 24 delays of 250 ms long period delay.

HU-Zündmaschinen: the corresponding blasting machines are produced by WASAGCHEMIE Sythen, Haltern, Germany.

Hydan

Smokeless binary liquid explosive on hydrazine hydrate and → ammonium nitrate $(NH_2)_2 \cdot H_2O/NH_4NO_3$ -basis developed at Dynamit Nobel Wien in 1994. Some characteristic values for a mixture 50/50.

Hydan II:			
	oxygen balance	[%]	−4,0
	heat of explosion	[kJ/kg]	3879
	fume volume	[l/kg]	1112
	explosion temperature	[°C]	2400
	specific energy	[mt/kg]	112,3
	brisance (Kast)	$[10^6]$	105,3
	specific weight ρ	[g/cm³]	1,36
	detonation velocity	[m/s]	7150*

Despite a favourable price, high security and stability when stored separately these explosive mixtures got no civilian market due to the handling problems of hydrazine hydrate (corrosive, toxic). It can be used as a liquid, cold burning monergole propellant (→ Monergol) with low smoke signature. Japanese and US- institutions have worked on similar mixtures for the use in ship artillery (→ Liquid Propellants).

Hybrids

lithergoles

Hybrids is the name given in rocket technology to systems in which a solid fuel in the form of a case-bonded charge with a central perfora-

---

* Ignition with 2 g → Pentrit -Booster.

tion is reacted with a liquid oxidant. Hybrids with solid oxidant and liquid fuel also exist. Hybrids can be thrust-controlled during combustion and can even be re-ignited if hypergolic components are incorporated in the formulation of the fuel charge.

Hydrazine

Hydrazin; hydrazine

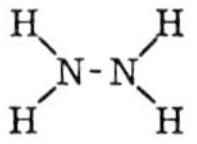

$$\begin{array}{c} H \\ \diagdown \\ N-N \\ \diagup \\ H \end{array} \begin{array}{c} H \\ \diagup \\ \\ \diagdown \\ H \end{array}$$

colorless liquid
empirical formula: H_4N_2
molecular weight: 32.05
energy of formation: +433.1 kcal/kg = +1812 kJ/kg
enthalpy of formation: +377.5 kcal/kg = +1580 kJ/kg
oxygen balance: −99.9 %
nitrogen content: 87.41 %
density: 1.004 g/cm^3

Hydrazine and alkylhydrazines are important propellants in rocket engines, especially for flight control rockets which are actuated only for short periods of time during space travel. In the presence of special catalysts, hydrazine can be made to decompose within milliseconds; → also *Dimethylhydrazine.* Hydrazine and its derivates are toxic.

Hydrazine Nitrate

Hydrazinnitrat; nitrate d'hydrazine

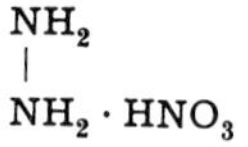

$$\begin{array}{c} NH_2 \\ | \\ NH_2 \cdot HNO_3 \end{array}$$

colorless crystals
empirical formula: $H_5N_3O_3$
molecular weight: 95.1
energy of formation: −586.4 kcal/kg = −2453 kJ/kg
enthalpy of formation: −620.7 kcal/kg = −2597 kJ/kg
oxygen balance: −8.6 %
nitrogen content: 44.20 %
volume of explosion gases: 1001 l/kg
heat of explosion
   (H_2O liq.): 1154 kcal/kg = 4827 kJ/kg
   (H_2O gas): 893 kcal/kg = 3735 kJ/kg
specific energy: 108 mt/kg = 1059 kJ/kg
density: 1.64 g/cm^3

melting point:
　　stable modification: 70.7 °C = 159.3 °F
　　unstable modification: 62.1 °C = 143.8 °F
lead block test: 408 cm^3/10 g
detonation velocity, confined:
　　8690 m/s 28500 ft/s at ρ = 1.60 g/cm^3
decomposition temperature: 229 °C = 444 °F
impact sensitivity: 0.75 kp m = 7.4 N m
critical diameter of steel sleeve test: 6 mm

Hydrazine nitrate is readily soluble in water.

The high detonation velocity of the salt is interesting. Mixtures with → Octogen, pressed to high density, reach more than 9000 m/s.

Hydrazine Perchlorate

perchlorate d'hydrazine

$$\begin{array}{c} NH_2 \\ | \\ NH_2 \cdot HClO_4 \end{array}$$

colorless crystals
empirical formula: $H_5N_2O_4Cl$
molecular weight: 132.5
energy of formation: −291 kcal/kg = −1216 kJ/kg
enthalpy of formation: −318 kcal/kg = −1331 kJ/kg
oxygen balance: +24.1 %
nitrogen content: 21.14 %
volume of explosion gases: 838 l/kg
heat of explosion
　　(H$_2$O liq.): 882 kcal/kg = 3690 kJ/kg
　　(H$_2$O gas): 725 kcal/kg = 3033 kJ/kg
density: 1.83 g/cm^3
melting point: 144 °C = 291 °F
lead block test: 362 cm^3/10 g
deflagration point: 272 °C = 522 °F
impact sensitivity: 0.2 kp m = 2 N m
friction sensitivity: at 1 kp = 10 N pistil load no reaction
critical diameter of steel sleeve test: 20 mm

The product is thus very sensitive.

Hygroscopicity

Hygroskopizität; hygroscopicité

Tendency of a substance to absorb moisture from its surroundings; specifically, absorption of water vapor from atmosphere.

Hypergolic

Liquid Propellant system based on two or more substances capable of spontaneous ignition on contact.

IATA

Means "International Air Transport association Dangerous Goods Regulations" and contains regulations for the transport of dangerous goods by air.

ICAO TI

Means "International Civil Aviation Organisation Technical Instructions for the Safe Transport of Dangerous Goods by Air" and contains the conditions under which it is permissible to transport dangerous goods by commercial aircraft.

ICT

Fraunhofer-Institut für Chemische Technologie
D-76327 Pfinztal-Berghausen
(www.ict.fraunhofer.de)

German research institute for propellants and explosives and organizer of international meetings at Karlsruhe, Germany.

Igniter

Anzünder, allumeur

A pyrotechnic and/or propellant device used to initiate burning of propellant.

Igniter Cord

Anzündlitze; corde d'allumage

An igniter cord is a safety fuse which burns at a fast rate (6–30 s/m) and with an open flame. The cord can be ignited by an open flame or with by a conventional safety fuse (guide fuse) a connector. Its function is to ignite the cords in the desired sequence.

Igniter Cord Connector

Anzündlitzenverbinder

Igniter cord connectors ensure a safe transmission of the sparking combustion of the igniter cord into the gunpowder core of a connected safety fuse.

Igniter Safety Mechanism

Zündsicherung; dispositif de securite d'allumage

Device for interrupting (safing) or aligning (arming) an initiation train of an explosive device, i.e., a rocket motor or gas generator.

Igniter Train

Anzünd-Kette; chaine d'allinmage (d'amorcage)

Step-by-step arrangement of charges in pyrotechnic or propellant by which the initial fire from the primer is transmitted and intensified until it reaches and sets off the main charge. Also called *burning train* or *explosive train*.

Ignitibility

Zündwilligkeit; inflammabilité

Statement of ease with which burning of a substance may be initiated.

Ignition System

Zündanlage; système d'allumage

Arrangement of components used to initiate combustion of propellant charge of gas generator (→ *Ignitor Train*).

Illuminant Composition

Leuchtsatz; composition lumineuse

A mixture of materials used in the candle of a pyrotechnic device to produce a high intensity light as its principal function. Materials used include a fuel (-reducing agent), an oxidizing agent, a binder plus a color intensifier and waterproofing agent. The mixture is loaded under pressure in an container to form the illuminant charge. Basic formulations contain sodium nitrate, magnesium and binder.

IMDG Code

Abbreviation for "International Maritime Dangerous Goods Code". It contains the regulations for the transport of dangerous goods by ocean-going ships, inter alia about their classification, packaging and stowing.

IMO

Abbreviation for International Maritime Organization, London, with the International Maritime Dangerous Goods (IMDG Code) contains texts of international conventions on classification, compatibility, packing, storage, etc. during transportation by sea; explosives and primers belong to class 1 of the code.

IME: Institute of Makers of Explosives

A non-profit trade association representing leading U.S. producers of commercial explosive materials and dedicated to safety in the manufacture, transportation, storage, and use of explosive materials.

Immobilization

Festlegung; immobilisation

Method of fixing propellant grain in definite position relative to generator case.

Impact Sensitivity

Schlagempfindlichkeit; sensitiveness to impact, sensibilité à l'impact

The sensitiveness to impact of solid, liquid, or gelatinous explosives is tested by the fallhammer method. Samples of the explosives are subjected to the action of falling weights of different sizes. The parameter to be determined is the height of fall at which a sufficient amount of impact energy is transmitted to the sample for it to decompose or to explode.

The US standard procedures are:

(a) Impact sensitivity test for solids: a sample (approximately 0.02 g) of explosive is subjected to the action of a falling weight, usually 2 kg. A 20-milligram sample of explosive is always used in the Bureau of Mines (BM) apparatus when testing solid explosives. The weight of the sample used in the Picatinny Arsenal (PA) apparatus is indicated in each case. The impact test value is the minimum height at which at least one of 10 trials results in explosion. In the BM apparatus, the explosive is held between two flat, parallel hardened steel surfaces; in the PA apparatus it is placed in the depression of a small steel die-cup, capped by a thin brass cover, in the center of which a slotted-vented-cylindrical steel plug is placed, with the slotted side downwards. In the BM apparatus, the impact impulse is transmitted to the sample by the upper flat surface; in the PA, by the vented plug. The main differences between the two tests are that the PA test involves greater confinement, distributes the translational impulse over a smaller area (due to the inclined sides of the diecup cavity), and involves a frictional component (against the inclined sides).

The test value obtained with the PA apparatus depends greatly on the sample density. This value indicates the hazard to be expected on subjecting the particular sample to an impact blow, but is of value in assessing a material's inherent sensitivity only if the apparent density (charge weight) is recorded along with the impact test value. The samples are screened between 50 and 100 mesh, U.S. where single-component explosives are involved, and through 50 mesh for mixtures.

(b) Impact sensitivity test for liquids: the PA Impact Test for liquids is run in the same way as for solids. The die-cup is filled, and the top of the liquid meniscus is adjusted to coincide with the plane of the top rim of the die-cup. To date, this visual observation has been found adequate to assure that the liquid does not wet the die-cup rim after the brass cup has been set in place. Thus far, the reproducibility of data obtained in this way indicates that variations in sample size obtained are not significant.

In the case of the BM apparatus, the procedure that was described for solids is used with the following variations:

1. The weight of explosives tested is 0.007 g.

2. A disc of desiccated filter paper (Whatman No. 1) 9.5 mm ∅ is laid on each drop, on the anvil, and then the plunger is lowered onto the sample absorbed in the filter paper.

The fallhammer method was modified by the German Bundesanstalt für Materialprüfung (BAM), so as to obtain better reproducible data*). The sample is placed in a confinement device, which consists of two coaxial cylinders placed one on top of the other and guided by a ring. The cylinders have a diameter of $10_{-0.005}^{-0.003}$ mm and a height of 10 mm, while the ring has an external diameter of 16 mm, a heigh of 13 mm and a bore of $10_{+0.01}^{+0.005}$ mm; all parts, cylinders and rings, must have the

---

* *Koenen* and *Ide*, Explosivstoffe, Vol. 9, pp. 4 and 30 (1961).

Impact Sensitivity

same hardness*). Cylinders and rings are renewed for each falling test procedure. If the sample is a powder or a paste, the upper cylinder is slightly pressed into the charged confinement device as far as it will go without flattening the sample. If liquids are tested, the distance between the cylinders is 2 mm. The charged device is put on the anvil of the fallhammer apparatus, and the falling weight, guided by two steel rods, is unlocked. For sensitive explosives such as primary explosives, a small fallhammer is used for insensitive explosives a large hammer. The small hammer involves the use of fall weight of up to 1000 g, while the fall weights utilized with the large hammer are 1, 5 and 10 kg. The fall heights are 10–50 cm for the 1-kg weight, 15–50 cm for the 5-kg weight and 35–50 cm for the 10-kg weight. This method is the recommended test method in the UN-recommendations for the transport of dangerous goods and it is standardized as EN 13631-4 as a so-called Harmonized European Standard.

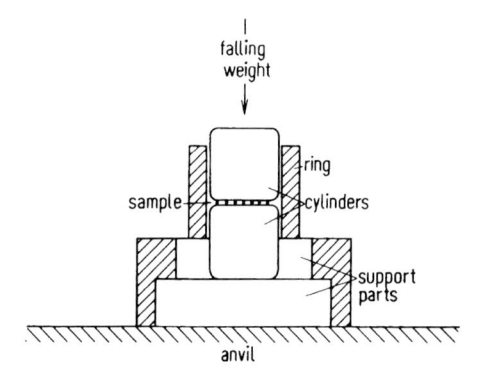

Fig. 16. Fallhammer confinement device

The influence of friction test results is thus elliminated.

---

* The ground and hardened cylinders and rings are standard parts for ball bearings; they are available on the market.

Table 18. Impact sensitivities given as the product of fall weight and fall height (kp m). In the following Table the kp m values are listed at which at least one of six tested samples explodes.

Explosive	Fall Weight kp	N	Fall Height m	Fall Energy kp m	N m
A. Homogeneous explosives					
nitroglycol	0.1	1	0.2	0.02	0.2
nitroglycerine	0.1	1	0.2	002	0.2
Tetrazene	1	10	0.2	0.2	2
mercury fulminate	1	10	0.2	0.2	2
PETN	1	10	0.3	0.3	3
Tetryl	1	10	0.3	0.3	3
nitrocellulose 13.4 % N	1	10	0.3	0.33	
nitrocellulose 12.2 % N	1	10	0.4	0.44	
lead azide	5	50	0.15	0.75	7.5
Hexogen	5	50	0.15	0.75	7.5
picric acid	5	50	0.15	0.75	7.5
TNT	5	50	0.30	1.5	15
lead styphnate	5	50	0.30	1.5	15
ammonium perchlorate	5	50	0.50	2.525	
dinitrobenzene	10	100	0.50	5	50
B. Industrial explosives					
Guhr dynamite	1	10	0.10	0.1	1
Gelignite	1	10	0.10	0.1	1
seismic gelatins	1	10	0.10	0.1	1
blasting gelatin	1	10	0.20	0.2	2
ammonium nitrate nitroglycol gelatin	1	10	0.20	0.2	2
gelatinous permitted explosives	1	10	0.30	0.3	3
nitroglycerine sensitized powders and permitted explosives	5	50	0.20	1.0	10
powder-form explosives without nitroglycerine	5	50	0.40	2.0	20

For explosives of high critical diameter, *Eld, D.* and *Johansson, C. H.* described an impact testing method*) by shooting the explosive sample (unconfined; filled in bakelite tubes 30 mm ⌀; 30 mm length and covered with paper or plastic foil) with brass cylinders (15 mm ⌀;

---
* *Eld, D.* and *Johansson, C. H.*: Shooting test with plane surface for determining the sensitivity of explosives EXPLOSIVSTOFFE 11, 97 (1963) and Johansson-Persson, Detonics of High Explosives, Academic Press, London and New York (1970), p. 108 ff.

15 mm length; 19 g) and varying their velocity. They are accelerated in a gun by means of compressed air or gunpowders. The front of them is plane, the back concave for better flight stability. A pendulum bearing a shock acceptance plate, hanging at about one yard from the shooting line, and the observation of the appearance of light and of smell are used to determine reaction of the explosive sample.

Results:

Table 19. Low → *Critical Diameter* (<40 mm)

Explosive	Projectile Velocity m/s	Pendulum °	Smell	Light
DYNAMITE 1.34 g/cm^3	186	15	+	+
	143	16	+	+
	116	15	+	+
	88	19	+	+
	73	19	+	+
	62	15	+	+
	55	12	+	+
	53	17	+	+
limit:				
	49	0	−	−
	49	0	−	−
	47	0	−	−
	45	0	−	−
	43	0	−	−
Test Explosive 0.84 g/cm^3	377	17	+	+
60% PETN/40% NaCl	316	19	+	+
	238	15	+	+
	222	12	+	+
	195	0	−	−
	189	14	+	+
limit:				
	174	0	−	−
	174	0	−	−
	170	0	−	−
	164	0	−	−

Table 20. High critical diameter (<132>40 mm)

Explosive	Projectile Velocity m/s	Pendulum °	Smell	Light
ANFO 0.91 g/cm³	1110	3	+	+
(94% ammonium nitrate;	765	2	+	+
6% liquid hydrocarbon)	675	3	+	+
	520	2	+	+
	460	1	+	+
	440	1	+	+
	435	0	+	–
	415	0	+	+
limit:				
	390	0	–	–
	380	0	–	–
	350	0	–	–
	320	0	–	–

For impact sensitivity of confined explosive charges thrown against a steel target → *Susan Test*.

Impulse

Product of thrust in pounds by time in seconds (also → *Specific Impulse*).

Incendiary

Designates a highly exothermic composition or material that is primarily used to start fires.

Inert

Descriptive of condition of device that contains no explosive, pyrotechnic, or chemical agent.

To Inflame

Ignite; anzünden; inflammer; allumer

The mode of ignition affects the manner in which an explosive reacts, detonating (→ *Detonation*) or deflagrating (→ *Deflagration*). The effect of flame ignition differs from that of a brisant initiation produced by a blasting cap or by a booster. The non-brisant ignition is termed inflammation.

The sensitivity of explosives to inflammation varies widely. Black powder can be exploded by a spark from a spark-producing tool; smokeless powders are ignited by the brief flame jet produced by striking a percussion cap. On the other hand, the combustion of an ion-exchanged → *Permitted Explosive*, ignited by a gas flame, is extinguished as soon as the flame source is removed.

→ *Initiating Explosives* always detonate when inflamed.

Inhibited Propellant

Oberflächenbehandelter Treibsatz; propellant traité de surface

A propellant grain in which a portion of the surface area has been treated to control the burning.

Initiating Explosives

primary explosives; Initialsprengstoffe;
explosifs d'amorçage; explosifs primaires

Primary explosives can detonate by the action of a relatively weak mechanical shock or by a spark; if used in the form of blasting caps, they initiate the main explosive. They are also filled in percussion caps mixed with friction agents and other components.

An initiating explosive must be highly brisant and must have a high triggering velocity. The most important primary explosives are mercury fulminate, lead azide, lead trinitroresorcinate, silver azide, diazodinitrophenol, tetrazene, and the heavy metal salts of 5-nitrotetrazole, which is used as an additive in primers. Initiating charges must be transported only if they are already pressed into capsules; the latter are usually made of aluminum, and sometimes of copper, while plastic capsules are used for special purposes (→ *Blasting Caps;* → Bridgewire Detonators; → Bullet Hit Squib).

Initiation

Zündung; mise à feu

Initiation means to set off explosive charges. The decomposition of an explosive can undergo → *Deflagration* (subsonic propagation rate) or → *Detonation* (supersonic propagation rate), depending on the manner and intensity of the ignition and on amount of → *Confinement*.

Non-brisant, i.e. flame ignition, is known as "inflammation". Brisant initiators include blasting caps, electric detonators, → *Primers* and → *Detonating Cords*. The initiating shock can be intensified by interposition of → *Boosters*, when the charge is insensitive.

Im rocketry, initiation means the functioning of the first element in an → *Igniter Train*.

Initiator

A device used as a primary stimulus component in all explosive or pyrotechnic devices, such as detonator primer or squib, which, on receipt of proper mechanical or electrical impulse, produces burning or detonating action. Generally contains a small quantity of sensitive explosive (→ *Squib;* → Detonator; → Primer; → Bullet Hit Squib).

Instantaneous, Detonator

Momentzünder; détonateur instantané

A detonator that has a firing time essentially of zero seconds as compared to delay detonators with firing times of from several milliseconds to several seconds.

Insulation

Isolierung; isolement

Thermal barrier designed to prevent excessive heat transfer from hot combustion products to case of rocket.

Ion Propellants

Ionentreibstoffe

In vacuo, i.e., under space travel conditions, ions, which are atomic carriers of electric charges, can be accelerated by electric fields and bunched to give a single beam. The discharge velocity thus attained is of a higher order than that of gaseous products from chemical reactions. For this reason, very high values of specific impulse can be produced.

The preferable ion propellant is cesium owing to its high molecular weight and to the fact that it is easily ionized.

Iron Acetylacetonate

Eisenacetylacetonat; acétylacetonate de fer

$$\left[H_3C - \underset{\underset{O}{\|}}{C} - CH = \underset{\underset{CH_3}{|}}{C} - O - \right]_3 Fe$$

empirical formula: $C_{15}H_{21}O_6Fe$
molecular weight: 353.2
energy of formation: -836 kcal/kg $= -3498$ kJ/kg
enthalpy of formation: -859 kcal/kg $= -3593$ kJ/kg
oxygen balance: $+163.1\%$
density: 1.34 g/cm^3

Iron(III) acetylacetonate is a curing catalyst for polymethane binders with combustion-modifying abilities for → *AP containing* → *Composite Propellants*.

ISL

Institut Franco-Allemand de Recherches de St. Louis,
Deutsch-Französisches Forschungsinstitut St. Louis, France

German-French research institute especially for ballistics and detonation physics.

Isosorbitol Dinitrate

Isosorbitdinitrat; dinitrate d'isosorbitol;

white microcrystals
empirical formula: $C_6H_8N_2O_8$
molecular weight: 236.1
oxygen balance: -54.2%
nitrogen content: 11.87%
melting point: 70 °C = 158 °F (decomposition)
lead block test: 311 cm^3/10 g
detonation velocity, confined:
    5300 m/s at $\rho = 1.08$ g/cm^3
deflagration point: 173 °C = 343 °F

impact sensitivity: 1.5 kp m = 15 N m
friction sensitivity: over 16 kp = 160 N
　pistil load crackling

Isosorbitol dinitrate serves as an effective cardial medicine (in low percentage mixture with milk sugar). The pure substance is a strong explosive; it is more effective than → *PETN*.

Jet Tappers

Abstichladungen; ouvreuses explosives de percée

Jet tappers are used in tapping Siemens-Martin (open hearth) furnaces. They are hollow charges, which are insulated from heat by earthenware jackets; when detonated, the tapping channel is produced. Other shaped charges are used to break up blast furnace hangups.

Kelly

A hollow bar attached to the top of the drill column in rotary drilling; also called grief joint, kelly joint, kelly stem, or kelly bar.

Lambrex

Trade name of a cartridged slurry blasting agent distributed in Austria by Austin Power GmbH (formerly Dynamit Nobel Wien):

Cartridged:	Density: g/cm^3	Oxygen balance: %	Gas Volume: l/kg	Specific energy: kJ/kg	Velocity of Detonation: m/s
LAMBREX 1	1,2	+2,3	910	765	5500 (unconfined)
LAMBREX 2	1,2	+0,3	871	853	5400 (unconfined)
LAMBREX 2 CONTOUR	1,05	0,0	856	804	4200 (unconfined)

Lambrit

Trade name of an ANFO blasting agent distributed in Austria by Austin Powder GmbH (formerly Dynamit Nobel Wien).

density: 0,8 g/cm^3
oxygen balance: −1,6%
gas volume: 981 l/kg
specific energy: 1001 kJ/kg
velocity of detonation: 3700 m/s
(steel tube confinement 52/60/1500 mm)

Large Hole Blasting

Großbohrloch-Sprengverfahren; sautage à grand trou

In large-scale blasting processes in open pit mining and quarrying, rows of nearly vertical boreholes are drilled parallel to the quarry face; the diameter of each borehole is 3–8 in. (in Germany more often 3–4 in.), while the borehole length is over 12 m. The holes are filled with explosive and stemmed. → *Free-flowing Explosives* or pumped → *Slurries* can be applied.

Lead Acetylsalicylate

Bleiacetylsalicylat; acétylsalicylate de plomb

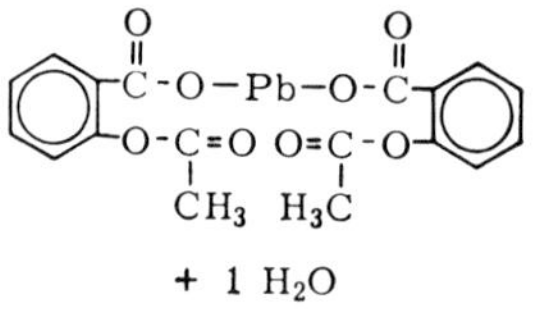

$$\text{C}_6\text{H}_4(\text{O-C(=O)CH}_3)\text{-C(=O)-O-Pb-O-C(=O)-}\text{C}_6\text{H}_4(\text{O-C(=O)CH}_3)$$

+ 1 H$_2$O

colorless, fine crystals
empirical formula: $C_{18}H_{14}O_8Pb \cdot H_2O$
molecular weight: 583.5
energy of formation: −810 kcal/kg = −3391 kJ/kg
enthalpy of formation: −823 kcal/kg = −3444 kJ/kg
oxygen balance: −98.7%

Lead acetylsalicylate is a combustion-modifying additive, especially so in → *Double Base* rocket propellants.

Lead Azide

Bleiazid; azoture de plomb

$Pb(N_3)_2$

colorless crystals; microcrystalline
granules, if dextrinated
molecular weight: 291.3

energy of formation: +397.5 cal/kg = +1663.3 kJ/kg
enthalpy of formation: +391.4 cal/kg = +1637.7 kJ/kg
oxygen balance: −5.5 %
nitrogen content: 28.85 %
volume of explosion gases: 231 l/kg
explosion heat: 391 kcal/kg = 1638 kJ/kg
density: 4.8 g/cm^3
lead block test: 110 cm^3/10 g
detonation velocity, confined:
   4500 m/s = 14 800 ft/s at ρ = 3.8 g/cm^3
   5300 m/s = 17 400 ft/s at ρ = 4.6 g/cm^3
deflagration point: 320–360 °C = 600–680 °F
impact sensitivity:
   pure product: 0.25–0.4 kp m = 2.5–4 N m
   dextrinated: 0.3–0.65 kp m = 3–6.5 N m
friction sensitivity:
   at 0.01–1 kp = 0.1–1 N pistil load explosion

Lead azide is insoluble in water, is resistant to heat and moisture, and is not too hygroscopic. It is prepared by reacting aqueous solutions of sodium azide and lead nitrate with each other. During the preparation, the formation of large crystals must be avoided, since the breakup of the crystalline needles may produce an explosion. Accordingly, technical grade product is mostly manufactured which contains 92–96 % Pb(N$_3$)$_2$, and is precipitated in the presence of dextrin, polyvinyl alcohol, or other substances which interfere with crystal growth. Lead azide is employed as an initiating explosive in blasting caps. When used as a primary charge, it is effective in smaller quantities than mercury fulminate, has a higher triggering rate, and, unlike mercury fulminate, cannot be dead-pressed by even relatively low pressures. In order to improve its flammability, an easily flammable additive, such as lead trinitroresorcinate, is added. Lead azide is decomposed by atmospheric CO_2, with evolution of hydrazoic acid.

Lead azide detonators for use in coal mining have copper capsules; for all other blastings, aluminum caps are used.

Specifications

net content (by determination as PbCrO$_4$): not less than	91.5 %
moisture: not more than	0.3 %
mechanical impurities:	none
water solubles: not more than	1 %
lead content: at least	68 %
copper:	none
reaction:	neutral, no acid
bulk density: at least	1.1 g/cm^3
deflagration point: not below	300 °C = 572 °F

Lead Block Test

Bleiblockausbauchung; essai au bloc de plomb, coefficient d'utilisation pratique, c. u. p.

The *Trauzl* lead block test is a comparative method for the determination of the → *Strength* of an explosive. Ten grams of the test sample, wrapped in tinfoil, are introduced into the central borehole (125 mm deep, 25 mm in diameter) of a massive soft lead cylinder, 200 mm in diameter and 200 mm long. A copper blasting cap No. 8 with an electric primer is introduced into the center of the explosive charge, and the remaining free space is filled with quartz sand of standard grain size. After the explosion, the volume of the resulting bulge is determined by filling it with water. A volume of 61 cm^3, which is the original volume of the cavity, is deducted from the result thus obtained.

In France the lead block performance value is given by the coefficient d'utilisation pratique (c. u. p.): if m_x is the mass of the tested explosive, which gives exactly the same excavation as 15 g of picric acid, the ratio

$$\frac{15}{m_x} \cdot 100 = \% \text{ c. u. p.}$$

is the coefficient d'utilisation pratique. Also, 10 g of picric acid can be applied as a standard comparison explosive. For the relationship with other testing procedures → *Strength*.

Another modification of the lead block test is recommended by BAM (Bundesanstalt für Materialprüfung, Germany). The test sample is prepared as follows: a special instrument wraps the sample in tinfoil and molds it into a cylinder of 11 ml capacity (24.5 mm in diameter,

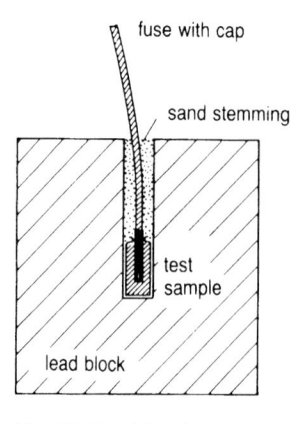

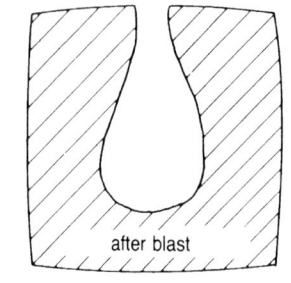

Fig. 17. Lead block test

25 mm in height, with a coaxial cavity 7 mm in diameter and 20 mm long for the blasting cap), whereby the resulting density should be only slightly higher than the pour (bulk) density. Liquids are filled into thin-walled cylindrical glass ampoules or, in special cases, directly into the cavity of the lead block.

The initiation is effected with an electric copper blasting cap No. 8 containing 0.4 g of high pressed (380 kp/cm^2) and 0.2 g of low pressed → *PETN* as the secondary charge and 0.3 g of lead azide as the initiating charge.

Table 21. Lead block excavation values.

A. Homogeneous Explosives

Explosive	Test Value cm^3/10 g	Explosive	Test Value cm^3/10 g
nitroglycol	610	picric acid	315
methylnitrate	600	trinitroaniline	311
nitroglycerine	530	TNT	300
PETN	520	urea nitrate	272
RDX	483	dinitrophenol	243
nitromethane	458	dinitrobenzene	242
ethylnitrate	422	DNT	240
Tetryl	410	guanidine nitrate	240
nitrocellulose 13.4 % N	373	ammonium perchlorate	194
ethylenediamine dinitrate	350	ammonium nitrate	178

B. Industrial Explosives

Explosive	Density g/cm^3	Test Value cm^3/10 g
blasting gelatin	1.55	600
gulii dynamite	1.35	412
Gelignite 65 % nitroglycerine	1.53	430
ammonium-nitrate-based gelatins, 40 % nitroglycerine	1.47	430
powder-form ammonium-nitrate-based explosives	1.0	370
ANFO	0.9	316
gelatinous permitted explosive	1.69	130
ion-exchanged permitted explosive	1.25	85

The empty space above the test sample is filled with dried, screened quartz sand (grain size 0.5 mm), as in the original method.

The volume of the excavation is determined by filling it with water; after 61 ml have been deducted from the result, the net bulge corresponding to the weight of the compressed sample is obtained. In accordance with the international convention, this magnitude is recalculated to a 10-g sample.

The European Commission for the Standardization of Testing of Explosive Materials*) recalculated the results for a 10-ml test sample, using a calibration curve established by *Kurbalinga* and *Kondrikov*, as modified by *Ahrens;* the reported value refers to the mixture of PETN with potassium chloride which gives the same result as the test sample under identical experimental conditions.

Since this regulation is still recent, the values given in the following table, as well as the values given under the appropriate headings of the individual explosive materials, are still based on the older method, in which a 10-g sample is employed. Other conventional methods for the determination of the explosive strength are the ballistic mortar test and the sand test.

For further details, including descriptions of other tests, → *Strength*.

Lead Dioxide

Bleidioxid; dioxide de plomb

$$O = Pb = O$$

dark brown powder
empirical formula: PbO_2
molecular weight: 239.2
energy of formation: -274.7 kcal/kg = -1149.4 kJ/kg
enthalpy of formation: -277.2 kcal/kg = -1159.8 kJ/kg
oxygen balance: +6.7%
density: 9.38 g/cm^3

Lead dioxide serves as an oxidizer in primer and pyrotechnic compositions, and in crackling stars.

---

* Now: International Study Group for the Standardization of the Methods of Testing Explosives. Secretary: Dr. Per-Anders Persson, Swedish Detonic Research Foundation, Box 32058, S 12611 Stockholm, Sweden.

Lead Ethylhexanoate

Bleiethylhexanoat; éthylhexanoate de plomb

$$\left[CH_3-CH_2-CH_2-CH_2-\underset{C_2H_5}{CH}-C\underset{O^-}{\overset{O}{\diagup\!\!\!\diagdown}} \right]_2 Pb$$

technical product: brownish, nearly amorphous
empirical formula: $C_{16}H_{30}O_4Pb$
molecular weight: 493.6
energy of formation: −703 kcal/kg = −2940 kJ/kg
enthalpy of formation: −724 kcal/kg = −3027 kJ/kg
oxygen balance: −142.6 %

Lead ethylhexanxoate is a combustion-modifying additive, especially in → *Double Base Propellants* for rockets.

Lead-free Priming Compositions

Air contamination with health-impairing pollutants gave rise to the demand for sport ammunition free from lead, barium and mercury.

→ *SINTOX Primer Composition.*

Lead Nitrate

Bleinitrat; nitrate de plomb

$Pb(NO_3)_2$

colorless crystals
molecular weight: 331.2
energy of formation: −318.9 kcal/kg = −1334.4 kJ/kg
enthalpy of formation: −326.1 kcal/kg = −1364.3 kJ/kg
oxygen balance: +24.2 %
nitrogen content: 8.46 %
density: 4.53 g/cm^3
beginning of decomposition: 200 °C = 390 °F

Lead nitrate is employed as an oxidizer in initiating mixtures in which a particularly high density is required.

Lead Picrate

Bleipikrat; picrate de plomb

$$O_2N-\underset{NO_2}{\overset{NO_2}{\bigcirc}}-O-Pb-O-\underset{O_2N}{\overset{O_2N}{\bigcirc}}-NO_2$$

yellow crystals
empirical formula: $C_{12}H_4N_6O_{14}Pb$
molecular weight: 663.3
oxygen balance: −31.4 %
nitrogen content: 12.7 %

Lead picrate is insoluble in water, ether, chloroform, benzene and toluene, and sparingly soluble in acetone and alcohol. It is prepared by precipitation with a solution of lead nitrate in a solution of sodium picrate and picric acid.

It can be used as an active component in initiating mixtures, e.g. for electrical squibs in bridgewire detonators. It is more powerful and more sensitive than → Lead Styphnate.

The unintentional formation of picrates by reaction of picric acid with the surrounding metals must be strictly avoided.

Lead Styphnate

lead trinitroresorcinate; Bleitrizinat; trinifroresorcinate de plomb

$$O_2N-\underset{NO_2}{\overset{}{\bigcirc}}\overset{NO_2}{\underset{O}{\overset{O}{<}}}Pb \quad + 1\ H_2O$$

orange-yellow to dark brown crystals
empirical formula: $C_6H_3N_3O_9Pb$
molecular weight: 468.3
energy of formation: −417.6 kcal/kg = −1747.2 kJ/kg
enthalpy of formation:
−427.1 kcal/kg = −1786.9 kJ/kg
oxygen balance: −18.8 %
nitrogen content: 8.97 %
density: 3.0 g/cm^3
lead block test: 130 cm^3/10 g
detonation velocity, confined:
  5200 m/s = 17 000 ft/s at ρ = 2.9 g/cm^3

heat of explosion: 347 kcal/kg = 1453 kJ/kg
deflagration point: 275–280 °C = 527–535 °F
impact sensitivity: 0.25–0.5 kp m = 2.5–5 N m

Lead trinitroresorcinate is practically insoluble in water (0.04 %), and is sparingly soluble in acetone and ethanol; it is insoluble in ether, chloroform, benzene and toluene. It is prepared by precipitation with a solution of lead nitrate from a solution of magnesium trinitroresorcinate, while maintaining certain concentration relationships and working in a given temperature and pH range, with stirring, in a reaction vessel which can be heated or cooled. The magnesium trinitroresorcinate solution required for the precipitation of lead trinitroresorcinate is obtained as a brown-to-black solution in a dissolving vessel by reacting an aqueous suspension of trinitroresorcinol with magnesium oxide powder while stirring.

Lead trinitroresorcinate is mostly employed as an initiating explosive in the form of a mixture with lead azide forming the detonator charge; it is particularly suited for this purpose, since it has a high ignition sensitivity, and its hygroscopicity is low. It is also employed as the main component of "sinoxide" charges in non-eroding percussion caps; these charges also contain the usual additives and a low percentage of tetrazene.

In the absence of any admixtures, lead trinitroresorcinate readily acquires an electrostatic charge, easily causing explosion.

Specifications

net content: not less than	98 %
moisture: not more than	0.15 %
lead content (determination as $PbCrO_4$):	43.2–44.3 %
heavy metals other than lead: not more than	0.05 %
Ca + Mg: not more than	0.5 %
Na: not more than	0.07 %
pH:	5–7
nitrogen content: at least	8.8 %
bulk density:	1.3–1.5 g/cm^3
deflagration point: not below	270 °C = 518 °F

Leading Lines

Leading Wires; Zündkabel; ligne de fir

The wire(s) connecting the electrical power source with electric blasting cap circuit.

Leg Wires

Zünderdrähte; fils du défonaleur

The two single wires or one duplex wire extending out from an electric blasting cap.

Linear Burning Rate

Linear Regression Rate; lineare Brenngeschwindigkeit; velocité de combustion linéaire

Distance normal to any burning surface of pyrotechnic or propellant burned through in unit time; → *Burning Rate.*

Liquid Explosives

flüssige Sprengstoffe; explosifs liquides

Numerous explosive materials are liquid. This applies primarily to several nitric acid esters such as → *Nitroglycerin,* → *Nitroglycol,* → *Diethyleneglycol Dinitrate,* → *Triethyleneglycol Dinitrate,* → *Butanetriol Trinitrate* and many more. Most of them are so highly sensitive to impact that they are converted to the less sensitive solid state, e.g., by gelatinization with nitrocellulose; as is well known, such processes formed the subject of the pioneering patents of *Alfred Nobel.* It was shown by *Roth* that the impact sensitivity of explosive liquids is considerably enhanced if they contain air bubbles. Nitrocellulose gelatinization increases the minimum explosion-producing impact energy in fallhammer tests performed on nitroglycerine from 0.02 to 0.2 kpm.

→ *Nitromethane* is considerably less sensitive. The volatility of the compound is high, and the handling of the constituent explosion-producing liquids is complicated. Nevertheless, nitromethane was used in the USA for preliminary studies to the big nuclear explosions ("pregondola" etc.). It has also been used in stimulation explosions carried out in gas wells and oil wells. PLX ("Picatinny Liquid Explosive") consists of 95% nitromethane and 5% ethylenediamine.

It has been proposed that liquid oxidizers (highly concentrated nitric acid, nitrogen tetroxide, tetranitromethane) be incorporated into the explosive mixture only on the actuation site or in the weapon itself so as to produce an approximately equalized oxygen balance and thus attain a higher degree of transport safety. Well known liquid explosives include "Panklastites" (nitrogen tetroxide with nitrobenzene, benzene, toluene, or gasoline) and "Hellhoffites" (concentrated nitric acid with dinitrobenzene or dinitrochlorobenzene). The mixture, under the name "Boloron", was still a recommended procedure in Austria after the Second World War. Similar explosives are known as → *Dithekite.* The explosive strength of these mixtures is very high. However, since the components are corrosive, their handling is very unpleasant; when mixed together, the product becomes highly sensitive. For all these reasons they are no longer employed in practice.

For mud-like ammonium nitrate explosives → *Slurries* and → *Emulsion Slurries*.

Liquid Oxygen Explosives

Flüssigluft-Sprengstoffe; Oxyliquit;
explosifs à l'oxygène liquide

Liquid oxygen explosives are made by impregnating cartridges made of combustible absorbent materials such as wood dust, cork meal, peat dust, → *Carbene*, etc., with liquid oxygen. They must be exploded immediately after the impregnation and loading, which are carried out in situ. They are energy-rich and cheap, but their manner of utilization does not permit rational working, such as detonating a large number of charges in one priming circuit. They are, accordingly, hardly ever employed in practice.

Liquid Propellants

Flüssigtreibstoffe; propergols liquides;
(Monergol; Hypergolic)

Combinations of pairs of liquids which react with each other (fuels and oxidizers in the most general sense of the word) which release energy in the form of hot gaseous reaction products; the → *Gas Jet Velocity* builds up the → *Thrust*. The caloric yield and the possible magnitude of the specific impulse may be higher than in one-component systems – i. e., higher than those of → *Monergols*, homogeneous propellants, and composite propellants.

Examples of fuels are alcohol, hydrocarbons, aniline, hydrazine, dimethylhydrazine, liquid hydrogen, liquid ammonia.

Examples of oxidizers are liquid oxygen, nitric acid, concentrated H_2O_2, N_2O_4, liquid fluorine, nitrogen trifluoride, chlorine trifluoride.

Certain pairs of the reacting liquids are → *Hypergolic*.

The liquid-propellant technique was developed for rocketry, but it is also considered today for small caliber cannons (e.g. 30 mm⌀).

Lithium Nitrate

Lithiumnitrat; nitrate de lithium

$LiNO_3$

molecular weight: 68.95
oxygen balance: +58.1
nitrogen content: 20.32 %
density: 2.38 g/cm^3
melting point: 256 °C = 493 °F

Lithium nitrate is soluble in water and is highly hygroscopic. Its only use is as a flame-coloring oxidizer in pyrotechnical formulations.

Lithium Perchlorate

Lithiumperchlorat; perchlorate de lithium

$LiClO_4$

colorless crystals
molecular weight: 106.4
oxygen balance: +60.2 %
density: 2.43 g/cm^3
melting point: 239.0 °C = 462 °F
decomposition point: 380 °C = 716 °F

Lithium perchlorate is soluble in water and alcohol and is very hygroscopic. The hydrated salt melts at 95 °C (203 °F).

Lithium perchlorate is prepared by saturating perchloric acid with lithium hydroxide or lithium carbonate. It is also used in batteries.

Loading Density

Ladedichte; densité de chargement

The ratio between the weight of the explosive and the explosion volume, i. e., the space, in which the explosive is detonated. In a similar manner, the loading density of a powder is the ratio between the maximum weight of the powder and the space into which it is loaded.

Loading density is a very important parameter, both in propellant powders (owing to the necessity of ensuring the strongest possible propellant effect in the loaded chamber, whose shape and size are mostly limited by the design of the weapon) and in brisant explosives (→ *Brisance*).

It is often essential to attain the maximum possible loading density (especially in shaped charges). This is done by casting and pressing methods, such as Vacuum, sedimentation, and mold-casting processes.

LOVA

An abbreviation for low-vulnerability ammunition. This term is descriptive of the trend towards choosing substances for both explosive and propellant charges which are as intensitive as possible even if losses in effectivity have to be accepted. The development of → *Shaped Charges* has made it possible to hit stored ammunition with simple tactical weapons even behind armoured walls.

The sensitivity of high-brisance explosives, e.g. → *Hexogen*, can be reduced by embedding them in rubberlike plastics (→ *Plastic Explosives*).

Low-sensitivity propellants, too, are based on plastic compounded nitramines. Another example of an insensitive explosive and propellant is → *Nitroguanidine.*

LOVA Gun-Propellant

LOVA-Treibladungspulver; LOVA-poudre

Since 1970, in addition to the various well-known → *Gun Powders*, LOVA gun propellants have been developed and used in the production of propellants. The acronym LOVA stands for (LOw Vulnerability Ammunition) which has come to represent a type of gun propellant.

This name expresses the unique characteristics of this type of munitions and those of gun propellants. That is, under external influences from the bullet casing, a hollow charge or fire, or a possible reaction, can at most, lead to combustion, but not to → *Deflagration* or → *Detonation*. Nevertheless, the ballistic capability of traditional gun powders must be equaled and surpassed.

To meet this challenge, one can used as an energy carrier standard → *Explosives* imbedded in a matrix of → *Energetic* or inert *Binders* so that the energy carrier loses its explosive properties while allowing for a regulated combustion.

The most widely-used energy carries are → *Hexogen* and → *Octogen* and to an extent → *Triaminoguanidine Nitrate*. Depending on the desired purpose, → *Nitroguanidine, Guanidinnitrate* and → *Ammonium Perchlorate* can also be used.

As a binder system polymers are utilized. If the binders contain energy or gas-producing molecular groups ($-NO_2$, $-N_3$), one classifies the binders as → *Energetic Binders* (e.g. polynitropolyphenylene, glycidyl azide polymer, polyvinyl nitrate and nitrocellulose). If these substances are not present, then the binders are classified as inert binders.

Depending on available processing methods, binder types such as thermoset material, thermoplast or gelatinizers can be used. They can then be formed and cured by chemical or physical means.

For thermoset materials, reactive polymers such as polyesters or polybutadiene derivatives combined with curing agents (e.g. isocyanates) are utilized. For thermoplasts one uses long-chained, partially branched polyether (Movital) or polymeric flouridated hydrocarbons (Fluorel). An example of a gelatin binder type is celluloseacetobutyrat (CAB), which is normally used in combination with nitrocellulose.

The production of LOVA powders is dependant on the chosen binder type. When thermoset materials are used, the system of energy carrier/binders/curing agents is kneaded together. The same is true when gelatines are used, however in this case, gelatinizing solvents (usually alcohol and ether) are added.

Thermoplasts, after being combined with energy carriers, are processed on hot rollers into a plastic material. The subsequent shaping is achieved by means of hydraulic mold presses and cutting machines. Depending on the binder type, the resulting powder kernels are cured (thermoset material), cooled (thermoplast), or dried by the removal of solvents (gelatin).

The possible forms of LOVA powders correspond to those of traditional → *Gunpowder* and are adapted according to the desired ballistic characteristics.

LX

Code of Lawrence Livermore National Laboratory for designated formulations in production. Examples*) are:

LX	Synonym	HMX %	Additive	confined %	Detonation Velocity, m/s	ft/s	at $\rho =$ g/cm^3
−04−1	PBHV−85/15	85	Viton A	15	8460	27 740	1.86
−07−2	RX−04−BA	90	Viton A	10	8640	28 330	1.87
−09−0	RX−09−CB	93	"DNPA"	7	8810	28 890	1.84
−10−0	RX−05−DE	95	Viton A	5	8820	28 920	1.86
−11−0	RX−04−P1	80	Viton A	20	8320	27 280	1.87
−14−0		95	Estane	5	8837	28 970	1.83

* Data quoted from the publication UCRL-51319 of the U.S. Department of Commerce: Properties of Chemical Explosives and Explosive Stimulants, edited and compiled by *Brigitta M. Dobratz*, University of California (1974).

Magazine

Sprengstofflager; dépôt

Any building or structure approved for the storage of explosive materials.

Magazine Keeper; Lagerverwalter; agent du dépôt

A person responsible for the safe storage of explosive materials, including the proper maintenance of explosive materials, storage magazines and areas.

Magazine, Surface; Übertage-Lager; dépôt superficiel

A specially designed structure for the storage of explosive materials aboveground.

Magazine, Underground; Untertage-Lager; dépôt souterrain

A specially designed structure for the storage of explosive materials underground.

Mannitol Hexanitrate

nitromannitol; Nitromannit; hexanitrate de mannitol; Hexanitromannit; MHN

$$\begin{array}{l}CH_2\text{-}O\text{-}NO_2\\CH\text{-}O\text{-}NO_2\\CH\text{-}O\text{-}NO_2\\CH\text{-}O\text{-}NO_2\\CH\text{-}O\text{-}NO_2\\CH_2\text{-}O\text{-}NO_2\end{array}$$

colorless needles
empirical formula: $C_6H_8N_6O_{18}$
molecular weight: 452.2
energy of formation: -336.2 kcal/kg = -1406.8 kJ/kg
enthalpy of formation:
   -357.2 kcal/kg = -1494.4 kJ/kg
oxygen balance: +7.1 %
nitrogen content: 18.59 %
volume of explosion gases: 694 l/kg
heat of explosion
   (H_2O gas): 1399 kcal/kg = 5855 kJ/kg
specific energy: 110 mt/kg = 1078 kJ/kg
density: 1.604 g/cm^3
melting point: 112 °C = 234 °F
lead block test: 510 cm^3/10 g

detonation velocity, confined:
  8260 m/s = 27 100 ft/s at ρ = 1.73 g/cm^3
deflagration point: 185 °C = 365 °F
impact sensitivity: 0.08 kp m = 0.8 N m

Nitromannitol is insoluble in water, but is soluble in acetone, ether and hot alcohol; it is difficult to stabilize.

It is prepared by dissolving mannitol in cold concentrated nitric acid and precipitating it with cold concentrated sulfuric acid. The crude product is washed with a dilute bicarbonate solution and with water, and is then recrystallized from hot alcohol.

In the USA nitromannitol was used as a filling for → *Blasting Caps*.

MAPO

Abbreviation for methylaziridine phosphine oxide, a bonding agent for AP containing → *Composite Propellants*.

$$\left[\begin{array}{c} H_3C-\overset{H}{C} \\ \diagdown N \\ H_2C\diagup \end{array} \right]_3 \equiv P=O$$

empirical formula: $C_9H_{18}N_3OP$
molecular weight: 215.14
density: 1.08 g/cm^3
boiling point at 0.004 bar: 120 °C = 248 °F

Mass Explosion Risk

Massen-Explosionsfähigkeit, Massen-Explosionsgefährlichkeit;
danger d'explosion en masse

A term describing the behavior of explosive materials and items (chiefly ammunition) in bulk. The question to be answered is whether a local explosion or fire will or will not detonate the entire bulk of the explosive (a truckload, or even a shipload of explosives). A number of tests have been laid down, in which first a parcel, then a case, and finally a caseload – in the form in which it is to be dispatched – are primed or inflamed in the manner in which this is to be done in actual service. When testing a caseload, the cases are arranged in predetermined locations, covered by inert cases of identical construction, and ignited.

Mass explosion risk does not depend solely on the properties of the explosive material, but also on the stacking height (in extreme cases, an entire shipload), on the nature of the confinement (e.g., buildings,

Mass Ratio

Massenverhältnis; relation des masses

In rocket technology, the ratio between the initial mass of the rocket and its final mass, after the propellant has burnt out. The relation between the end-velocity of a rocket projectile (theoretical value, without considering friction by the atmosphere) and the mass ratio is described by the equation

$$v_b = I_s \cdot g \cdot \ln \frac{1}{1 - \frac{M_e}{M_i}}$$

v_b: velocity of projectile at the end of burning,
I_s: specific impulse,
g: gravitation constant,
M_e: mass of the rocket projectile after propellant has burnt out,
M_i: mass of charged missile at beginning of burning.

Other keywords in this connection: → *Rocket Motor*; → *Solid Propellant Rocket*; → *Specific Impulse*.

Mercury Fulminate

Knallquecksilber; fulminate de mercure

$Hg=(ONC)_2$

colorless crystals
empirical formula: $C_2N_2O_2Hg$
molecular weight: 284.6
energy of formation: +229 kcal/kg = +958 kJ/kg
enthalpy of formation: +225 kcal/kg = +941 kJ/kg
oxygen balance: −11.2 %
nitrogen content: 9.84 %
heat of explosion: 415 kcal/kg = 1735 kJ/kg
density: 4.42 g/cm^3
deflagration point: 165 °C = 330 °F
impact sensitivity: 0.1−0.2 kp m = 1−2 N m

Mercury fulminate is toxic and is practically insoluble in water. When dry, it is highly sensitive to shock, impact, and friction, and is easily detonated by sparks and flames. It can be phlegmatized by the addition of oils, fats, or paraffin, and also by press-molding under very high pressure.

Mercury fulminate is prepared by dissolving mercury in nitric acid, after which the solution is poured into 95 % ethanol. After a short time, vigorous gas evolution takes place and crystals are formed. When the reaction is complete, the crystals are filtered by suction and washed until neutral. The mercury fulminate product is obtained as small, brown to grey pyramid-shaped crystals; the color is caused by the presence of colloidal mercury.

If small amounts of copper and hydrochloric acid are added to the reaction mixture, a white product is obtained. Mercury fulminate is stored under water. It is dried at 40 °C shortly before use. Owing to its excellent priming power, its high brisance, and to the fact that it can easily be detonated, mercury fulminate was the initial explosive most frequently used prior to the appearance of lead azide. It was used in compressed form in the manufacture of blasting caps and percussion caps. The material, the shells, and the caps are made of copper.

Mesa Burning

Mesa-Abbrand

→ Burning Rate.

Metadinitrobenzene

m-Dinitrobenzol; métadinitrobenzéne

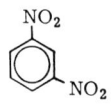

$$\text{NO}_2\text{-C}_6\text{H}_4\text{-NO}_2$$

pale yellow needles
empirical formula: $C_6H_4N_2O_4$
molecular weight: 168.1
energy of formation: -21.1 kcal/kg = -88.1 kJ/kg
enthalpy of formation: -38.7 kcal/kg = -161.8 kJ/kg
oxygen balance: -95.2 %
nitrogen content: 16.67 %
volume of explosion gases: 907 l/kg
heat of explosion
  (H_2O liq.): 637 kcal/kg = 2666 kJ/kg
  (H_2O gas): 633 kcal/kg = 2646 kJ/kg
specific energy: 79.7 mt/kg = 781 kJ/kg
density: 1.5 g/cm^3
solidification point: 89.6 °C = 193.3 °F
lead block test: 242 cm^3/10 g

detonation velocity, confined:
  6100 m/s = 20000 ft/s at ρ = 1.47 g/cm^3
deflagration point:
  evaporation at 291 °C = 556 °F;
  no deflagration
impact sensitivity: 4 kp m = 39 N m
friction sensitivity:
  up to 36 kp = 353 N no reaction
  critical diameter of steel sleeve test: 1 mm

Dinitrobenzene is sparingly soluble in water. It is prepared by direct nitration of benzene or nitrobenzene. It is an insensitive explosive. For purposes of official transport regulations, the sensitivity and the reactivity of dinitrobenzene are just on the limit between high-explosive and the non-dangerous zone.

Dinitrobenzene is toxic and produces cyanosis.

The maximum permissible concentration of this compound in the air at the workplace is 1 mg/m^3. The compound is of no interest as an explosive, since toluene – from which → *TNT* is produced – is available in any desired amount.

Methylamine Nitrate

Methylaminnitrat; nitrate de méthylamine; MAN

$CH_3-NH_2 \cdot HNO_3$

colorless crystals
empirical formula: $CH_6N_2O_3$
molecular weight: 94.1
energy of formation: −862 kcal/kg = −3604 kJ/kg
enthalpy of formation: −896 kcal/kg = −3748 kJ/kg
oxygen balance: −34 %
nitrogen content: 20.77 %
volume of explosion gases: 1191 l/kg
heat of explosion
  (H$_2$O liq.): 821 kcal/kg = 3437 kJ/kg
  (H$_2$O gas): 645 kcal/kg = 2698 kJ/kg
specific energy: 95.3 mt/kg = 934 kJ/kg
density: 1.422 g/cm^3
melting point: 111 °C = 232 °F
lead block test: 325 cm^3/10 g

Methylamine nitrate is considerably more hygroscopic than ammonium nitrate. Its sensitivity to impact is very low. It can be employed as a flux component in castable ammonium nitrate mixtures, but requires incorporation of brisant components. Methylamine nitrate is also employed as a component in → *Slurries*.

Methyl Nitrate

Methylnitrat; nitrate de méthyle

CH_3-O-NO_2

colorless volatile liquid
empirical formula: CH_3NO_3
molecular weight: 77.0
energy of formation: −456.7 kcal/kg = −1911 kJ/kg
enthalpy of formation:
  −483.6 kcal/kg = −2023.6 kJ/kg
oxygen balance: −10.4%
nitrogen content: 18.19%
volume of explosion gases: 873 l/kg
heat of explosion
  (H_2O liq.): 1613 kcal/kg = 6748 kJ/kg
  (H_2O gas): 1446 kcal/kg = 6051 kJ/kg
specific energy: 123 mt/kg = 1210 kJ/kg
density: 1.217 g/cm^3
boiling point: 65 °C = 149 °F
lead block test: 610 cm^3
detonation velocity, confined:
  6300 m/s = 20 700 ft/s at ρ = 1.217 g/cm^3
deflagration: evaporation, no deflagration
impact sensitivity: 0.02 kp m = 0.2 N m
friction sensitivity:
up to 36 kp = 353 pistil load no reaction
critical diameter of steel sleeve test: 18 mm

Methyl nitrate is a highly volatile liquid, and its brisance is about equal to that of nitroglycerine. Its vapors are both flammable and explosive and produce headaches. Methyl nitrate dissolves nitrocellulose, yielding a gel, from which it rapidly evaporates.

It can be prepared by introducing methyl alcohol into a nitrating mixture at a low temperature or by distilling methanol with medium-concentrated nitric acid.

Methylphenylurethane

Methylphenylurethan; Ethyl-N,N-Phenylmethylcarbamat; méthylphénylurethane

$$O=C \diagup_{OC_2H_5}^{N<_{CH_3}^{C_6H_5}}$$

colorless liquid
empirical formula: $C_{10}H_{13}O_2N$
molecular weight: 179.2
boiling point: 250 °C = 482 °F
refractive index n_D^{20}: 1.51558
energy of formation: −538.5 kcal/kg = 2253 kJ/kg
enthalpy of formation:
 −564.7 kcal/kg = −2363 kJ/kg
oxygen balance: −218.7 %
nitrogen content: 7.82 %

Methylphenylurethane is a gelatinizing → *Stabilizer* especially for → *Double Base Propellants*.

Specifications

density (20/4):	1.071–1.090 g/cm³
boiling analysis:	248–255 °C
	= 478–491 °F
reaction:	neutral

Methyl Violet Test

In this test, which was developed in the USA about 50 years ago from another test known as the German test (testing for visible nitrous gases at 135 °C), visual inspection of the nitrous gases is replaced by testing with a strip of paper, impregnated with methyl violet. This test is performed at 134.5 °C for nitrocellulose and single-base powders and at 120 °C for multi-base propellants. At the end of the test the violet dye changes color to blue-green and then to salmon-pink. For single-base powder, this color change should not take place after less than 30 minutes and for a multibase powder after less than 60 minutes. Only highly unstable powders can be detected by this test, therefore the latter is now rarely used.

Metriol Trinitrate

*methyltrimethylolmethane trinitrate; Metrioltrinitrat;
Nitropentaglycerin; trinitrate de triméthylolméthylméthane; TMETN*

$$CH_3-\underset{\underset{CH_2-O-NO_2}{|}}{\overset{\overset{CH_2-O-NO_2}{|}}{C}}-CH_2-O-NO_2$$

pale-colored, oily substance
empirical formula: $C_5H_9N_3O_9$
molecular weight: 255.1
energy of formation: -373.8 kcal/kg = -1564.1 kJ/kg
enthalpy of formation:
  -398.2 kcal/kg = -1666.1 kJ/kg
oxygen balance: -34.5%
nitrogen content: 16.47%
volume of explosion gases: 966 l/kg
heat of explosion
  (H_2O liq.): 1182 kcal/kg = 4945 kJ/kg
  (H_2O gas): 1087 kcal/kg = 4549 kJ/kg
specific energy: 124 mt/kg = 1215 kJ/kg
density: 1.460 g/cm^3
solidification point: -15 °C = +5 °F
lead block test: 400 cm^3/10 g
deflagration point: 182 °C = 360 °F
impact sensitivity: 0.02 kp m = 0.2 N m

The oil is practically insoluble in water and is chemically very stable. When mixed with nitrocellulose, it can be gelatinized on rollers only to a moderate extent and only at an elevated temperature. The volatility of the trinitrate is low.

Metriol trinitrate can be prepared by nitration of methyltrimethylolmethane (Metriol) with mixed acid. Metriol is prepared by condensation of propanal with formaldehyde in a manner similar to that employed in the synthesis of pentaerythritol.

During the Second World War, mixtures of metriol trinitrate with triglycol dinitrate (a good gelatinizer of nitrocellulose) were processed together with nitrocellulose to produce tropic-proof propellants. They were also distinguished by good mechanical strenght for employment in rocket motors. TMETN is an excellent plasticizer for GAP or other energetic polymer binders.

Microballoons

Microspheres

Microballoons are bubbles with an average diameter of 40 µm (range 10–100 µm). They were originally used as a filler to control the density of plastic products. They are available as glass or plastic material.

Microballoons have proved to produce a fine gas bubble distribution in low-sensitivity explosives, particularly in emulsion slurries. Finely distributed gas bubbles considerably increase the sensitivity to detonation ("hot spots"). In the form of microballoons, gas distribution stabilises; gas distributions without enclosure may experience a loss in effectiveness as a result of coagulation into coarse bubbles, or by escape.

Millisecond Delay Blasting

Millisekunden-Sprengen; tir à microretard

The explosive charges are successively initiated at time intervals as short as 20–100 milliseconds with the aid of millisecond detonators (→ *Bridgewire Detonators*, → *Dynatronic*).

Experience shows that rock fragmentation is better with this technique, and a smaller amount of explosive is required to produce the same blasting effect since there is better mutual support of the charges.

Minex

A cast explosive charge used in the USA consisting of RDX, TNT, ammonium nitrate, and aluminum powder.

Miniaturized Detonating Cord

Mild detonating fuse; nicht-sprengkräftige detonierende Zündschnur; cordeau détonant miniaturé

Detonating cord with a core load of 5 or less grains of explosive per foot (≤ 0.1 g/m).

Minol

A pourable mixture of TNT, ammonium nitrate, and aluminum powder (40:40:20).

    casting density: 1.70 g/cm^3
    detonation velocity at casting density,
    confined: 6000 m/s = 19 700 ft/s

Misfire

Versager; raté

An explosive material charge that fails to detonate after an attempt of initiation.

Missile

Raketenflugkörper; roquette

The integral functional unit consisting of initiator and igniter devices, rocket engine, guiding equipment, and useful payload. → *Rocket Motor.* Missiles are, in principle, guided rocket projectiles.

Mock Explosives*)

Sprengstoff-Attrappen; factices

Mocks are nonexplosive simulants for high explosives. They duplicate the properties for test purposes without hazard. The required properties to copy need different mocks, e.g., for physical properties, for density, or for thermal behavior.

Monergol

In rocket technology the name for liquid or gelled homogeneous propellants, which require no other reaction partner for the formation of gaseous reaction products. Gas formation can be due to catalytic decomposition (on concentrated H_2O_2 or anhydrous hydrazine) or to an intramolecular reaction, e.g., by decomposition of propylnitrate generating N_2, CO, CO_2, NO, etc., → *Liquid Propellants.*

---

* For more details see *Dobratz, B. M.*, Properties of Chemical Explosives and Explosive Simulants, UCRL-51319, Rev. 1, University of California.

Motor

Motor; moteur

Generic term for solid propellant gas generator or rocket.

MOX

Abbreviation for metal oxidizer explosives (USA). Compositions:

Table 22.

MOX	1 %	2 B %	3 B %	4 B %	6 B %
ammonium perchlorate	35	35	–	–	–
aluminum (fine grain)	26.2	52.4	47	47	49.2
magnesium (fine grain)	26.2	–	–	–	–
Tetryl	9.7	–	–	–	–
RDX	–	5.8	29.1	29.1	28.7
TNT	–	3.9	2.0	2.0	–
potassium nitrate	–	–	18	–	–
barium nitrate	–	–	–	18	–
copper oxide	–	–	–	–	19.7
wax	–	–	0.9	0.9	0.9
calcium stearate	1.9	1.9	2.0	2.0	–
graphite	1.0	1.0	1.0	1.0	1.5

Muckpile

Haufwerk; déblai

The pile of broken burden resulting from a blast.

Mud Cap

Auflegerladung; pétardage

Mud caps are explosive charges which have a strong destructive effect even when not placed in a confining borehole. They are used for the demolition of boulders and concrete structures. (Synonymous with *Adobe Charge* and *Bulldoze*).

A highly brisant explosive is required for this purpose. Mud caps are usually covered with mud in order to enhance their brisance. It is often desirable to use charges of definite shape (→ *Shaped Charges*; → *Cutting Charges*).

Munroe Effect

The effect of shaped charges is known in the USA as the Munroe effect after *Munroe*, who described it in 1888. The terms "cavity effect" and "lined cavity effect" are also employed (→ *Shaped Charges*).

Muzzle Flash

Mündungsfeuer; lueur à 1 a bouche

Muzzle flash is the appearance of a flame at the muzzle of a barrel or a tube during the shot. The flash is a secondary effect which takes place when the still flammable explosion gases (CO, H_2) become mixed with air on emerging from the barrel.

The reasons for the appearance of the flash are not yet fully clear; it is also unclear why the flash can be supressed by introducing certain additives to the powder (probably catalytic termination of chain reactions). It is certain that muzzle flash is promoted by the high temperature of the combustion gases, the high gas pressure and the high velocity of the gas emerging from the muzzle.

In a given weapon, fast-burning powders have a lower tendency to flash than slow-burning powders. Weapons with a high ballistic performance (high projectile velocity and high gas pressure) give a larger flash, which is more difficult to suppress than that given by firearms with a lower performance.

In general, alkali metal salts damp muzzle flash better than alkaline earth salts. It is also fairly certain that the flash-damping effect in the alkali metal group increases from lithium to cesium. In the First World War, bags filled with sodium chloride placed in front of the propellant charge, was used as a muzzle flash damper.

Subsequently, potassium salts, in particular the sulfate, nitrate and bitartrate, proved to be more effective. Other muzzle flash dampers, used with varying degrees of success, are oxalates, phosphates, and bicarbonates.

Nano Energetic Materials

Nano sized energetic materials possesses desirable combustion characteristics such as high heats of combustion and fast energy release rates. Because of their capability to enhance performance, various metals have been introduced in solid propellants formulations, gel propellants, and solid fuels. Besides, shortened ignition delay and burn times, enhanced heat transfer rates, greater flexibility in design and the use of nano materials as gelling agent stimulate actual research activities.

Neutral Burning

Burning of a propellant grain in which a reacting surface area remains approximately constant during combustion (→ *Burning Rate, Progressive Burning* and *Regressive Burning*).

Nitroaminoguanidine

*Nitraminoguanidin; N'-Nitro-N-aminoguandine,
1-Amino-3-nitroguanidine, NaGu*

$$HN=C\underset{NH-NO_2}{\overset{NH-NH_2}{\diagup}}$$

empirical formula: $CH_5N_5O_2$
molecular weight: 119,09
energy of formation: +74.2 kcal/kg = +310.2 kJ/kg
enthalpy of formation: +44.3 kcal/kg = +185.5 kJ/kg
oxygen balance: −33,6 %
nitrogen content: 58,2 %
heat of explosion
  (H_2O liq.): 895.2 kcal/kg = 3746 kJ/kg
  (H_2O gas): 816.9 kacl/kg = 3418 kJ/kg
specific energy: 114.5 mt/kg = 1124 kJ/kg
density: 1,71 g/cm^3
deflagration point: 188 °C
impact sensitivity: 0,3 kpm − 3 Nm
friction sensibillty: 25 kp = 240 N
critical diameter of steel sleeve test: 12 mm

Nitroaminoguanidine is obtained by reacting → *nitroguanidine* with hydrazine in aqueous solution. Nitroaminoguanidine has gained a certain attractiveness as a reduced carbon monoxide propellant because of its ready ignitability and its burn-up properties.

Nitrocarbonitrate

NCN; N.C.N.

Nitrocarbonitrates are relatively low-sensitive explosives, usually based on ammonium nitrate, which do not contain any high explosives such as nitroglycerine or TNT. The components are named by *nitro*: dinitrotoluene; by *carbo*: solid carbon carriers as fuel: by *nitrate*: ammonium nitrate.

NCN as a classification for dangerous goods has been removed by the US Department of Transportation and replaced by "Blasting Agent" as a shipping name (→ Blasting Agents).

Nitrocellulose

Nitrocellulose; nitrocellulose; NC

white fibers
empirical formula of the structural unit: $C_{12}H_{14}N_6O_{22}$
nitrogen content referring to the unattainable*) nitration
 grade = 14.14 %
nitrogen content, practical maximum value: ca. 13.4 %
molecular weight of the structure unit:
 324.2 + % N/14.14 270

energy of formation and enthalpy of formation in relation to the nitrogen content:

% N	Energy of Formation		Enthalpy of Formation	
	kcal/kg	kJ/kg	kcal/kg	kJ/kg
13.3	−556.1	−2327	−577.4	−2416
13.0	−574.6	−2404	−596.1	−2494
12.5	−605.6	−2534	−627.2	−2624
12.0	−636.5	−2663	−658.4	−2755
11.5	−667.4	−2793	−689.6	−2885
11.0	−698.4	−2922	−720.7	−3015

* Nitrogen content > 13.5 % may be reached in the laboratory by use of acid mixtures with anhydrous phosphoric acid as a component.

the following data refer to 13.3% N:
oxygen balance: −28.7%
volume of explosion gases: 871 l/kg
heat of explosion
   (H_2O liq.): 1031 kcal/kg = 4312 kJ/kg
   (H_2O gas): 954 kcal/kg = 3991 kJ/kg
density: 1.67 g/cm^3 (maximum value attainable by pressing: 1.3 g/cm^3)
lead block test: 370 cm^3/10 g
impact sensitivity: 0.3 kp m = 3 N m
friction sensitivity:
   up to 36 kp = 353 N pistil load no reaction
critical diameter of steel sleeve test: 20 mm

Nitrocellulose is the commonly employed designation for nitrate esters of cellulose (cellulose nitrates). Nitrocellulose is prepared by the action of a nitrating mixture (a mixture of nitric and sulfuric acids) on well-cleaned cotton linters or on high-quality cellulose prepared from wood pulp. The concentration and composition of the nitrating mixture determine the resulting degree of esterification, which is measured by determining the nitrogen content of the product.

The crude nitration product is first centrifuged to remove the bulk of the acid, after which it is stabilized by preliminary and final boiling operations. The spent acid is adjusted by addition of concentrated nitric acid and anhydrous sulfuric acid and recycled for further nitration operations. The original form and appearance of the cellulose remains unchanged during the nitration. Subsequent boiling of the nitrocellulose under pressure finally yields a product with the desired viscosity level. The nitrated fibers are cut to a definite fiber length in hollanders or refiners. Apart from the numerous types of lacquer nitrocelluloses, which include ester- and alcohol-soluble products with a nitrogen content of 10.3–12.3% at all viscosity levels used in technology, standard nitrocellulose types are manufactured and blended to the desired nitrogen content. Blasting soluble nitrocotton (dynamite nitrocotton; 12.2–12.3% N) is held at a high viscosity to maintain good gelatinizing properties.

All nitrocelluloses are soluble in acetone. The viscosity of the solutions is very variable. (For its adjustment by pressure boiling see above.)

Nitrocellulose is transported in tightly closed drums or in pasteboard drums lined with plastic bags inside, which contain at least 25% of a moisturizing agent (water, alcohol, isopropanol, butanol, etc.). Spherical NC particles are precipitated from solution under vigorous stirring, and preferably used for manufacturing of cast or composite double base propellants.

Specifications

The specified nitrogen content, solubility in alcohol, ether-alcohol mixtures and esters, as well as viscosity, etc., all vary in various types of nitrocellulose. The nitrogen content should not vary more than +0.2 % from the specified value. The following specifications are valid for all types:
Bergmann-Junk-Test at 132 °C (270 °F):

not more than	2.5 cm^3/g NO
ashes: not more than	0.4 %
insolubles in acetone:	
not more than	0.4 %
alkali, as $CaCO_3$: not more than	0.05 %
sulfate, as H_2SO_4: not more than	0.05 %
$HgCl_2$:	none

Nitrocellulose for gelatinous explosives must gelatinize nitroglycerine completely within 5 minutes at 60 °C (140 °F).
Linters (cotton fibers) as raw material
Properties
$(C_6H_{10}O_5)_n$
white fibers
molecular weight of structure unity: 162.14

Specifications

α-cellulose content (insoluble in 17.5 % NaOH):	
at least	96 %
fat; resin (solubles in CH_2Cl_2):	
not more than	0.2 %
moisture: not more than	7.0 %
ash content: not more than	0.4 %

appearance:
homogenous, white or pale yellow,
free of impurities (knots; rests of capsules)

Nitroethane

Nitroethan; nitroéthane

$$CH_3\text{-}CH_2\text{-}NO_2$$

colorless liquid
empirical formula: $C_2H_5NO_2$
molecular weight: 75.07
energy of formation: −426.7 kcal/kg = −1785.3 kJ/kg
enthalpy of formation:
    −458.3 kcal/kg = −1917.4 kJ/kg
oxygen balance: −95.9 %
nitrogen content: 18.66 %

heat of explosion
  (H_2O liq.): 403 kcal/kg = 1686 kJ/kg
  (H_2O gas): 384 kcal/kg = 1608 kJ/kg
specific energy: 75.3 mt/kg = 738 kJ/kg
density: 1.053 g/cm^3
boiling point: 114 °C = 237 °F

Nitroparaffins are produced by vapor phase nitration with nitric acid vapor. Nitroethane is also prepared in this way. The individual reaction products (nitromethane, nitroethane, nitropropane) must then be separated by distillation.

All these products can be reacted with formaldehyde; polyhydric nitroalcohols are obtained, which can be esterified with nitric acid.

Nitroethylpropanediol Dinitrate

Nitroethylpropandioldinitrat;
dinitrate d'éthyl-nitropropandiol

$$H_5C_2-\underset{\underset{CH_2-O-NO_2}{|}}{\overset{\overset{CH_2-O-NO_2}{|}}{C}}-NO_2$$

empirical formula: $C_5H_9N_3O_8$
molecular weight: 239.2
oxygen balance: −43.5 %
nitrogen content: 17.57 %
volume of explosion gases: 1032 l/kg
heat of explosion (H_2O liq.): 1037 kcal/kg = 4340 kJ/kg
specific energy: 122 mt/kg = 1193 kJ/kg

The product is prepared by condensing → *Nitropropane* with formaldehyde and by nitration of the resulting nitroethylpropanediol.

Nitroglycerine

glycerol trinitrate; Nitroglycerin; nitroglycérine;
trinitrate de glycérine; NG; Ngl.

$$\begin{array}{l} CH_2-O-NO_2 \\ CH-O-NO_2 \\ CH_2-O-NO_2 \end{array}$$

yellow oil
empirical formula: $C_3H_5N_3O_9$
molecular weight: 227.1
energy of formation: −368.0 kcal/kg = −1539.8 kJ/kg
enthalpy of formation:
  −390.2 kcal/kg = −1632.4 kJ/kg

oxygen balance: +3.5%
nitrogen content: 18.50%
volume of explosion gases: 716 l/kg
heat of explosion
  (H_2O liq.): 1594 kcal/kg = 6671 kJ/kg
  (H_2O gas): 1485 kcal/kg = 6214 kJ/kg
specific energy: 106.6 mt/kg = 1045 kJ/kg
density: 1.591 g/cm^3
solidification point:
  +13.2 °C = 55.8 °F (stable modification)
  +2.2 °C = 35.6 °F (unstable modification)
specific heat: 0.32 kcal/kg = 1.3 kJ/kg
vapor pressure:

Pressure millibar	Temperature °C	°F
0.00033	20	68
0.0097	50	122
0.13	80	176
0.31	90	194

lead block test: 520 cm^3/10 g
detonation velocity, confined:
  7600 m/s = 25000 ft/s at ρ = 1.59 g/cm^3
impact sensitivity: 0.02 kp m = 0.2 N m
friction sensitivity:
  up to 36 kp = 353 N pistil load no reaction
critical diameter of steel sleeve test: 24 mm

Nitroglycerine is almost insoluble in water, but is soluble in most organic solvents; it is sparingly soluble in carbon disulfide. It readily dissolves a large number of aromatic nitro compounds and forms gels with soluble guncotton. Its volatility is negligible, but is still sufficient to cause headaches.

The acid-free product is very stable, but exceedingly sensitive to impact. The transportation of nitroglycerine and similar nitrate esters is permitted only in the form of solutions in non-explosive solvents or as mixtures with fine-powdered inert materials containing not more than 5% nitroglycerine. To avoid dangers, internal transport within the factories is made by water injection (→ *Water-driven Injector Transport*). Transport of pure nitroglycerine and similar products outside factory premises is difficult; in the U.S., special vessels have been developed in which the oil is bubble-free covered with water without air bubbles which raise the impact sensitivity considerably. The nitroglycerine produced is ideally processed immediately to the products (e.g., explosives; double base powders).

Nitroglycerine is prepared by running highly concentrated, almost anhydrous, and nearly chemically pure glycerin (dynamite glycerin) into a highly concentrated mixture of nitric and sulfuric acids, with constantly efficient cooling and stirring. At the end of the reaction the nitroglycerine acid mixture is given to a separator, where the nitroglycerine separates by gravity. Following washing processes with water and an alkaline soda solution remove the diluted residual acid.

Since nitroglycerine is dangerous to handle, its industrial production by continuous method has always been of the highest interest, since it is always desirable to have the smallest possible quantity of the product in any particular manufacturing stage. Accordingly, several competing methods (*Schmidt, Meissner, Biazzi*, KONTINITRO), have been developed, each method being characterized by a different approach to the problem of safety. The most recent procedures involve the reaction of glycerin and acid in injectors (Nitroglycerine AB).

Nitroglycerine is one of the most important and most frequently used components of explosive materials; together with nitroglycol, it is the major component of gelatinous industrial explosives. In combination with nitrocellulose and stabilizers, it is the principal component of powders, gun propellants and smokeless solid rocket propellants (→ *double base propellants*).

Care has to be taken for complete removal of acid residues from nitroglycerine, because they may enhance the exothermic autocatalytic decomposition process, from which severe explosions have occurred in the path.

Specifications

1. **Nitroglycerine** as a component of explosives	
Nitrogen content: not less than	18.35 %
Abel test at (82.2 °C) 180 °F:	
not less than	10 min
2. **Nitroglycerine** as a component of propellants	
nitrogen content: not less than	18.40 %
moisture: not more than	0.5 %
alkalinity, as Na_2CO_3: not more than	0.002 %
acidity, as HNO_3: not more than	0.002 %
3. **Glycerol** as a raw material	
smell:	not offensively pungent
color:	clear, as pale as possible
reaction to litmus:	neutral
inorganic impurities:	none

reducing matter (ammoniacal AgNO$_3$ test):	traces only
fatty acids:	traces only
ash content:	max. 0.03 %
water content:	max. 0.50 %
net content (oxidation value):	min. 98.8 %
density, d_4^{20}:	1.259– 1.261 g/cm^3
refractive index n_D^{20}:	1.4707–1.4735
acidity: not more than	0.3 ml n/10 NaOH/100 ml
alkalinity: not more than	0.3 ml n/10 HCl/100 ml

Nitroglycerine Propellants

→ *Double Base Propellants* and → *Gunpowder.*

Nitroglycide

glycide nitrate; Nitroglycid; nitrate de glycide

```
  CH₂-O-NO₂
  |
  CH
O⟨  |
  CH₂
```

water-white liquid
empirical formula: $C_3H_5NO_4$
molecular weight: 119.1
oxygen balance: –60.5 %
nitrogen content: 11.76 %
density: 1.332 g/cm^3
lead block test: 310 cm^3/10 g
deflagration point: 195–200 °C = 383–390 °F
impact sensitivity: 0.2 kp m = 2 N m

Nitroglycide is soluble in alcohol, ether, acetone, and water; it is highly volatile.

This nitrate ester of glycide is prepared from dinitroglycerine by splitting off one HNO$_3$ molecule with concentrated alkali. It is the anhydride of glycerin mononitrate.

Nitroglyceride is precursor for → *Polyglyn.*

Nitroglycol

ethyleneglycol dinitrate; Nitroglykol;
dinitrate de glycol; EGDN

$$CH_2\text{-}O\text{-}NO_2$$
$$|$$
$$CH_2\text{-}O\text{-}NO_2$$

colorless, oily liquid
empirical formula: $C_2H_4N_2O_6$
molecular weight: 152.1
energy of formation: -358.2 kcal/kg = -1498.7 kJ/kg
enthalpy of formation: -381.6 kcal/kg = -1596.4 kJ/kg
oxygen balance: $\pm 0\%$
nitrogen content: 18.42%
volume of explosion gases: 737 l/kg
heat of explosion
 (H_2O liq.): 1742 kcal/kg = 7289 kJ/kg
 (H_2O gas): 1612 kcal/kg = 6743 kJ/kg
specific energy: 121 mt/kg = 1190 kJ/kg
density: 1.48 g/cm^3
solidification point: $-20\,°C = -4\,°F$
vapor pressure:

Pressure millibar	Temperature °C	°F
0.006	0	32
0.05	20	68
0.35	40	104
1.7	60	140
7.8	80	176
29	100	212

lead block test: 620 cm^3/10 g
detonation velocity, confined:
 7300 m/s = 24 000 ft/s at $\rho = 1.48$ g/cm^3
deflagration point: 217 °C = 423 °F
impact sensitivity: 0.02 kp m = 0.2 N m
friction sensitivity:
 at 36 kp = 353 N pistil load no reaction
critical diameter of steel sleeve test: 24 mm

Nitroglycol is not hygroscopic, is sparingly soluble in water and readily soluble in common organic solvents; its properties and performance characteristics are practically the same as those of nitroglycerine; it is 150 times more volatile and about four times more soluble in water; it is less viscous and gelatinizes nitrocellulose more rapidly than nitroglycerine.

Glycol can be nitrated in the same vessels as glycerin – in batches or continuously. The same applies to its separation and washing, which are in fact easier since nitroglycol is less viscous.

Nitroglycol is utilized in mixtures with nitroglycerine, since it markedly depresses the freezing temperature of the latter compound. Ammongelits contain only nitroglycol alone as the main explosive component, and therefore only freeze in winter below −20 °C.

The vapor pressure of nitroglycol is markedly higher than that of nitroglycerine; for this reason nitroglycol cannot be used in propellant formulations.

Like all nitrate esters, nitroglycol strongly affects blood circulation; its maximum permitted concentration at the workplace is 1.5 mg/m^3.

Specifications

nitrogen content: not below	18.30%
Abel test: not under	15 min
alkali (Na$_2$CO$_3$): for use in industrial explosives	no limit

Specifications for glycol (raw material)

net content (determination by oxidation with dichromate): at least	98%
density (20/4):	1.1130– 1.1134 g/cm^3
content of diglycol and triglycol (residue of vacuum destillation): not more than	2.5%
moisture: not more than	0.5%
glow residue: not more than	0.02%
chlorides:	none
reaction:	neutral
reducing components (test with NH$_3$–AgNO$_3$):	none
test nitration: no red fumes, yield: at least	230%

Nitroguanidine; Picrite

Nitroguanidin; nitroguanidine; Nigu; NQ

$$NH=C\begin{matrix}NH_2 \\ NH-NO_2\end{matrix}$$

white, fiber-like crystals
empirical formula: CH$_4$N$_4$O$_2$

molecular weight: 104.1
energy of formation: −184.9 kcal/kg = −773.4 kJ/kg
enthalpy of formation: −213.3 kcal/kg = −893.0 kJ/kg
oxygen balance: −30.7 %
nitrogen content: 53.83 %
volume of explosion gases: 1042 l/kg
heat of explosion
   (H_2O liq.): 734 kcal/kg = 3071 kJ/kg
   (H_2O gas): 653 kcal/kg = 2730 kJ/kg
specific energy: 95 mt/kg = 932 kJ/kg
density: 1.71 g/cm^3
melting point: 232 °C = 450 °F (decomposition)
specific heat: 0.078 kcal/kg = 0.33 kJ/kg
lead block test: 305 cm^3/10 g
detonation velocity, confined at max. density:
   8200 m/s = 26 900 ft/s
deflagration point:
   at melting point decomposition;
   no deflagration
impact sensitivity:
   up to 5 kp m = 49 N m no reaction
friction sensitivity:
   up to 36 kp = 353 N pistil load no reaction
critical diameter of steel sleeve test:
   at 1 mm ⌀ no reaction

Nitroguanidine is soluble in hot water, practically insoluble in cold water, very sparingly soluble in alcohol, insoluble in ether, and readily soluble in alkali. It is not very sensitive to shock or impact. It has excellent chemical stability.

Guanidine nitrate, which has been prepared from dicyanodiamide and ammonium nitrate, is dehydrated under formation of nitroguanidine, when treated with concentrated sulfuric acid. Nitroguanidine can be incorporated into nitrocellulose powder, nitroglycerine powder, or diglycol dinitrate powder; it is not dissolved in the powder gel, but is embedded in it as a fine dispersion. These "cold" (calorie-poor) powders erode gunbarrels to a much lesser extent than do the conventional "hot" powders.

Nitroguanidine has the advantage of quenching muzzle flash, but smoke formation is somewhat more intensive.

Nitroguanidine is also of interest as an insensitive high explosive (→ *LOVA*); its energy is low, but density and detonation velocity are high.

Specifications

    white, free flowing, crystalline powder

Type 1
grain size:	4.3–6.0 H
net content: not less than	98%
acidity as H_2SO_4: not more than	0.60%

Type 2
grain size: not more than	3.3 H
net content: not less than	99%
acidity as H_2SO_4: not more than	0.06%

Both types
ash content: not more than	0.30%
acid content, as H_2SO_4: not more than	0.06%
volatile matter: not more than	0.25%
sulfates: not more than	0.20%
water insolubles: not more than	0.20%
pH:	4.5–7.0

Nitroisobutylglycerol Trinitrate

nib-glycerin trinitrate; Nitroisobutylglycerintrinitrat trimethylolnitromethane trinitrate; trinitrate de nitroisobintylglycérine; NIBTN

$$NO_2-C\begin{matrix}CH_2-O-NO_2\\-CH_2-O-NO_2\\CH_2-O-NO_2\end{matrix}$$

yellow viscous oil
empirical formula: $C_4N_6N_4O_{11}$
molecular weight: 286.1
energy of formation: -169.1 kcal/kg = -707.5 kJ/kg
enthalpy of formation:
   -190.8 kcal/kg = -797.5 kJ/kg
oxygen balance: $\pm 0\%$
nitrogen content: 19.58%
volume of explosion gases: 705 l/kg
heat of explosion
   (H_2O liq.): 1831 kcal/kg = 7661 kJ/kg
   (H_2O gas): 1727 kcal/kg = 7226 kJ/kg
specific energy: 125 mt/kg = 1225 kJ/kg
density: 1.68 g/cm^3
solidification point: -35 °C = -31 °F
lead block test: 540 cm^3/10 g

detonation velocity, confined:
  7600 m/s = 24 900 ft/s at ρ = 1.68 g/cm^3
deflagration point: 185 °C = 365 °F
impact sensitivity: 0.2 kp m = 2 N m

The compound is less volatile than nitroglycerine, practically insoluble in water and petroleum ether, soluble in alcohol, acetone, ether, benzene, and chloroform, and is a good gelatinizer of guncotton. Its explosive strength is close to that of nitroglycerine.

It is prepared by condensation of formaldehyde with nitromethane and by nitration of the nitroisobutylglycerine product under the same conditions as nitroglycerine. The nitration and stabilization procedures are very difficult because of decomposition reactions.

While being of interest to the explosives industry, since it has an ideal oxygen balance, its stabilization in practice has proven to be impossible.

Nitromethane

Nitromethan; nitrométhane; NM

CH_3NO_2

colorless liquid
molecular weight: 61.0
energy of formation: −413.7 kcal/kg = −1731 kJ/kg
enthalpy of formation: −442.8 kcal/kg = −1852.8 kJ/kg
oxygen balance: −39.3 %
nitrogen content: 22.96 %
volume of explosion gases: 1059 l/kg
heat of explosion
  (H_2O liq.): 1152 kcal/kg = 4821 kJ/kg
  (H_2O gas): 1028 kcal/kg = 4299 kJ/kg
specific energy: 127 mt/kg = 1245 kJ/kg
density: 1.1385 g/cm^3
solidification point: −29 °C = −20 °F
boiling point: 101.2 °C = 214.2 °F
heat of vaporization: 151 kcal/kg = 631 kJ/kg

vapor pressure:

Pressure millibar	Temperature °C	°F
1.3	−29	−20 (solidification point)
10	0	32
32	20	68
140	50	122
283	80	176
1010	101.2	214.2 (boiling point)

lead block test: 400 cm^3/10 g
detonation velocity, confined:
 6290 m/s = 20 700 ft/s at ρ = 1.138 g/cm^3

Nitromethane is sparingly soluble in water. The compound is of industrial interest as a solvent rather than as an explosive. Its technical synthesis involves nitration of methane with nitric acid above 400 °C in the vapor phase.

It was used in the USA for underground model explosions ("Pre-Gondola"), in preparation for the → *Nuclear Charge* technique. Nitromethane was also employed in stimulation blasting in oil and gas wells. PLX (*P*icatinny *L*iquid *E*xplosive) is a mixture of nitromethane with 5 % ethylenediamine and is used to clean up mine fields.

Nitromethane is of interest both as a monergolic and as a liquid fuel for rockets.

Nitromethylpropanediol Dinitrate

methylnitropropanediol dinitrate; Nitromethylpropandioldinitrat; dinitrate de nitromethylpropanediol

$$\begin{array}{l} CH_2\text{-}O\text{-}NO_2 \\ |NO_2 \\ C \\ |CH_3 \\ CH_2\text{-}O\text{-}NO_2 \end{array}$$

empirical formula: $C_4H_7N_3O_8$
molecular weight: 225.1
oxygen balance: −24.9 %
nitrogen content: 18.67 %
volume of explosion gases: 890 l/kg
heat of explosion

(H_2O liq.): 1266 kcal/kg = 5295 kJ/kg
(H_2O gas): 1163 kcal/kg = 4866 kJ/kg
specific energy: 126.5 mt/kg = 1240 kJ/kg

The product is prepared by condensation of → *Nitroethane* with formaldehyde and subsequent nitration of nitromethylpropanediol.

Nitroparaffins

are aliphatic hydrocarbons with NO_2-groups attached directly to carbon atoms. They are mainly obtained by nitration in a gaseous state; → *Nitromethane;* → *Nitroethane;* → *Trinitromethane;* → *Tetranitromethane.*

Nitroparaffins can be reacted with formaldehyde to obtain nitroalcohols, which can be further esterified with nitric acid (→ e.g. *Nitroisobutylglycerol Trinitrate*).

Nitrostarch

Nitrostärke; nitrate d'amidon

$$[C_6H_7O_2(ONO_2)_3]_n$$

pale yellow powder
empirical formula of the structural unit: $C_6H_7N_3O_9$
oxygen balance at 12.2% N: −35%
density: 1.6 g/cm^3
maximum value attainable by pressing: 1.1 g/cm^3
lead block test: 356 cm^3/10 g
deflagration point: 183 °C = 361 °F
impact sensitivity: 1.1 kp m = 11 Nm

Nitrostarch is insoluble in water and ether, but is soluble in ether alcohol mixtures and in acetone.

Nitrostarch, with various nitrogen contents (12−13.3%), is prepared by nitration of starch with nitric acid or nitrating mixtures. The resulting crude product is washed in cold water and is then dried at 35−40 °C.

Nitrostarch resembles nitrocellulose in several respects, but, owing to its poor stability, difficulty in preparation and hygroscopicity, it is not used anywhere outside the USA. "Headache-free" industrial explosives are based on nitrostarch.

Nitro-Sugar

Nitrozucker, Zuckernitrat; nitrate de sucre

Nitro-sugar in its pure form is too unstable to be utilized in practice; however, during the First World War, a liquid explosive named "Nitrohydren" was prepared by nitration of solutions of cane sugar in glycerol and was then further processed into explosives and gunpowders. However, these mixtures are much more difficult to stabilize than nitroglycerine alone and are no longer of interest, since glycerin is freely available.

Nitrotoluene

Nitrotoluol; nitrotoluène

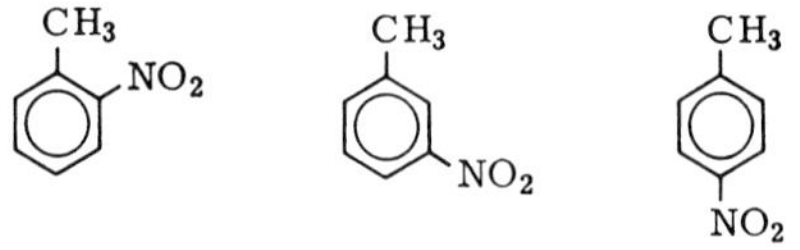

pale yellow liquid
empirical formula: $C_7H_7O_2N$
molecular weight: 137.1
oxygen balance: -180.9%
nitrogen content: 10.22%

Nitrotoluene is of importance as an intermediate or precursor for in the manufacture of TNT. There are three isomers, of which only the ortho- and para-isomers can yield pure 2,4,6-trinitrotoluene. "Mononitration" of toluene yields mostly the orthocompound, as well as 4% of the meta-, and about 33% of the para-compound.

It is often advantageous to separate the isomers from each other (by distillation or by freezing out) in the mononitro stage.

3-Nitro-1,2,4-triazole-5-one

Oxynitrotriazole, NTO, ONTA

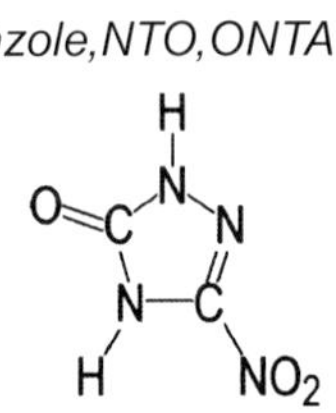

colorless crystals
empirical formula: $C_2H_2N_4O_3$
molecular weight: 130,1

energy of formation: − 164,69 kcal/kg = − 689,10 kJ/kg
enthalpie of formation: − 185,14 kcal/kg = − 774,60 kJ/kg
oxygen balance: − 24,6 %
nitrogen content: 43,07 %
volume of detonation gases: 855 l/kg
heat of explosion
   (H_2O liq.): 752,4 kcal/kg = 3148 kJ/kg
   (H_2O gas): 715,4 kcal/kg = 2993 kJ/kg
specific energy: 96,4 mt/kg = 945,4 kJ/kg
density: 1,91 g/cm^3
detonation velocity: unconfined 7860 m/s at ρ =
   1,80 g/cm^3
confined 7940 m/s at ρ = 1,77 g/cm^3
deflagration point: > 270 °C = > 540 °F
impact sensitivity: ≥ 120 Nm
friction sensitivity: at 36 kp = 353 N pistil load no reaction

NTO is synthesised in a two step process by reacting semicarbazide HCl with formic acid to obtain 1,2,4 triazole-5-one and followed by nitration to NTO.

NTO is used as an component in IHE (insensitive high explosives)

NTO is like EDNA an acetic nitramine. It forms easily salts with organic bases like melamine guanidine and ethylene diamine.

Nitrourea

Nitroharnstoff; nitro-urée

$$O=C\begin{matrix}NH-NO_2\\NH_2\end{matrix}$$

colorless crystals
empirical formula: $CH_3N_3O_3$
molecular weight: 105.1
energy of formation: −617.2 kcal/kg = −2582.4 kJ/kg
enthalpy of formation:
   −642.5 kcal/kg = −2688.4 kJ/kg
oxygen balance: −7.6 %
nitrogen content: 39.98 %
volume of explosion gases: 853 l/kg
heat of explosion (H_2O liq.): 895 kcal/kg = 3745 kJ/kg
specific energy: 93.0 mt/kg = 912 kJ/kg
melting point: 159 °C = 318 °F (decomposition)
beginning of decomposition: 80 °C = 176 °F

Nitrourea is soluble in benzene, ether, and chloroform; it is decomposed by water.

It is synthesized by dehydration of urea nitrate with sulfuric acid.

Nobelit®

Registered trademark for water-in-oil emulsion explosives manufactured and distributed by Orica Germany GmbH. Outstanding features of this non-Ngl explosives generation are complete water-resistance, safe handling, long shelf life, drastic reduction of toxic fumes and excellent performance marked by high detonating velocities of up to 6000 m/sec. (19000 ft/sec.). Nobelit® series 100, 200 and 300 are available as pumpable and packaged blasting agents and in cap-sensitive cartridges from 25 mm diameter and upward.

No-Fire Current

Grenz-Stromstärke; intensite de courant de non-allumage

Maximum current that can be continuously applied to bridgewire circuit without igniting prime material (Note: Continued applications of this current may degrade prime and "dud" the unit).

Non-electric Delay Device

Detonationsverzögerer; detonateur avec retard

A detonator with an integral delay element used in conjunction with, and capable of being initiated by, a detonating impulse.

Nonel

Trade name of a new "non electric device" for the firing of explosive charges. The basic unit consists, of detonating cords of a plastic hose (3 mm ⌀), the inner wall of which is coated with a thin layer of explosive instead of electrical wires. A shock wave initiated by a special initiator passes through the tube with a speed of approx. 2000 m/s. The spectator observes this shock wave process as a flash in the hose. The plastic tube is not destroyed by the shock.

In order to initiative a charge, the Nonel line must be combined with a conventional detonator. Branching is possible.

The device is distributed by NITRO NOBEL, Sweden; and DYNO, Norway. Its applications are of interest in electrically endangered areas (e.g., by thunderstorms and stray currents).

Nozzle

Düse; tuyère

Mechanical device designed to control the characteristics of a HYPERLINK "http://encyclopedia.thefreedictionary.com/fluid" fluid flow from a chamber into an outer medium.

In rocket technology it is intended to increase the HYPERLINK "http://encyclopedia.thefreedictionary.com/kinetic+energy" kinetic energy of exhaust fumes by reduction of pessure energy or HYPERLINK "http://encyclopedia.thefreedictionary.com/internal+energy" internal energy. A HYPERLINK "http://encyclopedia.thefreedictionary.com/de+Laval+nozzle" Laval nozzle, as often used, has a convergent section followed by a divergent section. Convergent nozzles accelerate fluids even up to the sonic velocity at the smallest cross section if the nozzle pressure ratio enough. Divergent nozzles decelerate subsonic flowing fluids and accelerate sonic or supersonic fluids.

Obturate

Verschluss; dispositif de clôture

To stop or close an opening so as to prevent escape of gas or vapor, to seal as in delay elements.

Octogen

cyclotetramethylene tetranitramine; Homocyclonit; cyclotétraméthyléne tétranitramine; Octogéne; HMX

$$\begin{array}{c} NO_2 \\ | \\ H_2C-N-CH_2 \\ | \qquad \quad | \\ O_2N-N \qquad N-NO_2 \\ | \qquad \quad | \\ H_2C-N-CH_2 \\ | \\ NO_2 \end{array}$$

colorless crystals
empirical formula: $C_4H_8N_8O_8$
molecular weight: 296.2
energy of formation: +84.5 kcal/kg = +353.6 kJ/kg
enthalpy of formation: +60.5 kcal/kg = +253.3 kJ/kg
oxygen balance: −21.6 %
nitrogen content: 37.83 %
volume of explosion gases: 902 l/kg

heat of explosion
   (H_2O gas): 1255 kcal/kg = 5249 kJ/kg } calculated*)
   (H_2O liq.): 1338 kcal/kg = 5599 kJ/kg
heat of detonation
   (H_2O liq.): 1480 kcal/kg = 6197 kJ/kg experimental**)
specific energy: 139 mt/kg = 1367 kJ/kg
density:
   α-modification: 1.87 g/cm^3
   β-modification: 1.96 g/cm^3
   γ-modification: 1.82 g/cm^3
   δ-modification: 1.78 g/cm^3
melting point: 275 °C = 527 °F
modification transition temperatures:
   $\alpha \rightarrow \delta$: 193–201 °C = 379–394 °F
   $\beta \rightarrow \delta$: 167–183 °C = 333–361 °F
   $\gamma \rightarrow \delta$: 167–182 °C = 333–359 °F
   $\alpha \rightarrow \beta$: 116 °C = 241 °F
   $\beta \rightarrow \gamma$: 154 °C = 309 °F
transition enthalpies:
   $\alpha \rightarrow \delta$: 5.98 kcal/kg = 25.0 kJ/kg
   $\beta \rightarrow \delta$: 7.90 kcal/kg = 33.1 kJ/kg
   $\gamma \rightarrow \delta$: 2.26 kcal/kg = 9.46 kJ/kg
   $\beta \rightarrow \gamma$: 5.64 kcal/kg = 23.6 kJ/kg
   $\alpha \rightarrow \gamma$: 3.71 kcal/kg = 15.5 kJ/kg
   $\alpha \rightarrow \beta$: 1.92 kcal/kg = 8.04 kJ/kg
specific heat, β-modification:
   0.3 kcal/kg at 80 °C = 176 °F
lead block test: 480 cm^3/10 g
detonation velocity, confined, β-mod.:
   9100 m/s = 29 800 ft/s at ρ = 1.9 g/cm^3
deflagration point: 287 °C = 549 °F
impact sensitivity: 0.75 kp m = 7.4 N m
friction sensitivity: at 12 kp = 120 N
pistil load: reaction
critical diameter of steel sleeve test: 8 mm

Octogen appears in four modifications, of which only the β-modification displays a particularly high density and hence also a particularly fast detonation rate.

It is practically insoluble in water. Its solubilities in other solvents resemble those of → Hexogen.

---

* computed by the "ICT-Thermodynamic-Code".
** value quoted from *Brigitta M. Debratz*, Properties of Chemical Explosives and Explosive Simulants, University of California, Livermore.

The compound is formed as a by-product from the manufacture of Hexogen by the Bachmann process (from hexamethylenetetramine, ammonium nitrate, nitric acid, and acetic anhydride). It is obtained as the sole product, when 1,5-methylene-3,7-dinitro-1,3,5,7-tetrazacyclooctane is treated with acetic anhydride, ammonium nitrate, and nitric acid.

The above starting material is formed when acetic anhydride is made to react on hexamethylenetetramine dinitrate.

In high-power charges, especially in shaped charges, Octogen performs better than Hexogen.

The δ-phase, which may occur after heating is much more sensitive against impact. friction and dectrostatic impulse.

Specifications

net content of β-modification:	
grade A, not less than	93 %
grade B, not less than	98 %
melting point: not less than	270 °C = 518 °F
acetone-insolubles:	
not more than	0.05 %
ashes: not more than	0.03 %
acidity, as CH_3COOH:	
not more than	0.02 %

Octol

A mixture of Octogen (HMX) and TNT 70/30 and 75/25. Performance values:

	70/30	75/25	
detonation velocity, confined:	8377	8643	m/s
at $\rho =$	1.80	1,81	g/cm^3
volume of explosion gases:	827	825	l/kg
heat of explosion (H_2O liq.):	1112	1147	kcal/kg
	4651	4789	kJ/kg

Oxidizer

Sauerstoffträger; comburant

All explosive materials contain oxygen, which is needed for the explosive reaction to take place. The oxygen can be introduced by chemical reactions (nitration) or by mechanical incorporation of materials containing bound oxygen. The most important solid-state oxidizers are nitrates, especially → *Ammonium Nitrate* and → *Sodium Nitrate* for explosives; → *Potassium Nitrate* for → *Black Powder* and ion exchanged → *Permitted Explosives*; potassium chlorate for → *Chlorate Explosives* and for pyrotechnical compositions; → *Ammonium Perchlorate* (APC) for → *Composite Propellants*.

Important liquid oxidizers for liquid fuel rocket motors include liquid oxygen (LOX), highly concentrated nitric acid, liquid N_2O_4, liquid fluorine, and halogen fluorides. See also → *Oxygen Balance*.

Oxygen Balance

Sauerstoffwert; bilan d'oxygène

The amount of oxygen, expressed in weight percent, liberated as a result of complete conversion of the explosive material to CO_2, H_2O, SO_2, Al_2O_3, etc. ("positive" oxygen balance). If the amount of oxygen bound in the explosive is insufficient for the complete oxidation reaction ("negative" oxygen balance), the deficient amount of the oxygen needed to complete the reaction is reported with a negative sign. This negative oxygen balance can be calculated in exactly the same manner for non-explosive fuels.

Examples:
TNT ($C_7H_5N_3O_6$)   −74 %
nitroglycerine ($C_3H_5N_3O_9$)   + 3.5 %
ammonium nitrate (NH_4NO_3) +20 %

Table 23. Oxygen balance of explosives and explosive components.

Material	Available O_2, %	Material	Available O_2, %
aluminum	− 89.0	ammonium chloride	− 44.9
ammonium nitrate	+ 20.0	ammonium perchlorate	+ 34.0
ammonium picrate	− 52.0	barium nitrate	+ 30.6
dinitrobenzene	− 95.3	dinitrotoluene	−114.4
wood meal, purified	−137.0	potassium chlorate	+ 39.2
potassium nitrate	+ 39.6	carbon	−266.7
sodium chlorate	+ 45.0	sodium nitrate	+ 47.0
nitroglycerine	+ 3.5	nitroguanidine	− 30.8
nitrocellulose (guncotton)	− 28.6	nitrocellulose (soluble guncotton)	− 38.7
picric acid	− 45.4	sulfur	−100.0
Tetryl	− 47.4	trinitroresorcinol	− 35.9
TNT	− 74.0		

The most favorable composition for an explosive can be easily calculated from the oxygen values of its components. Commercial explosives must have an oxygen balance close to zero in order to minimize the amount of toxic gases, particularly carbon monoxide, and nitrous gases, which are evolved in the fumes.

Further data are found under the respective compounds described in this book; also, → *Thermodynamic Calculation of Decomposition Reactions.*

Paraffin

$$CH_3\text{-}(CH_2)_x\text{-}CH_3$$

Paraffin serves to impregnate explosive cartridges against moisture. The technical product may contain ceresin, wax, or fat.

Specifications

solidification point: not below	50–55 °C (122–131 °F)
inflammation point: not below	200 °C (392 °F)
volatile matter: not more than	1%
glow residue	none
insolubles in toluene: not more than	0.03%
solution in ether, CS_2, ligroin	clear, without residue
acidity, as CH_3COOH: not more than	0.005%
alkalinity; test with concentrated sulfuric acid:	none no alteration, no darkoning of the acid
saponification index:	zero
iodine index:	low to zero
adhesion test:	negative

Parallel Connection

Parallelschaltung; branchement en parallèle

In multiple blastings with electric priming, → *Bridgewire Detonators* are usually connected in series to the priming line. If the boreholes are very wet, and there is a real danger of voltage loss, the charges are connected in parallel. Since only a very small fraction of the electric energy employed is then actuated in the primer bridges (the bulk of the energy is dissipated in the lead wires), parallel connections require special high-energy-supplying blasting machines.

Paste

Pulverrohmasse; galette

A nitrocellulose-nitroglycerine mixture for the solvent-free manufacture of → *Double Base Propellants*. It is obtained by introducing nitroglycerine (or diglycol dinitrate or similar nitrate esters) into a stirred nitrocellulose suspension in water. The mixture is then centrifuged or filtered off; it contains about 35 % water; its appearance resembles that of moist nitrocellulose. The paste, containing materials such as stabilizers and gelatinizers, is manufactured to the double base powder by hot rolling and pressing without application of solvents.

PBX

Abbreviation for plastic-bonded explosives: also → *LX*.
Pressed explosives:

PBX-9010:	90 % RDX, 10 % Kel F*)
PBX-9011:	90 % HMX, 10 % Estane
PBX-9404-03:	94 % HMX, 3 % NC, 3 % chloroethylphosphate
PBX-9205:	92 % RDX, 6 % polystyrene, 2 % ethylhexylphthalate
PBX-9501	95 % HMX, 2.5 % dinitropropyl acrylate-furmarate, 2.5 % estane
PBXN-1:	68 % RDX, 20 % Al, 12 % nylon
PBXN-2:	95 % HMX, 5 % nylon
PBXN-3:	86 % HMX, 14 % nylon
PBXN-4	94 % DATNB, 6 % nylon
PBXN-5:	95 % HMX, 5 % Viton A
PBXN-6:	95 % RDX, 5 % Viton A

Extruded explosive:

PBXN-201:	83 % RDX, 12 % Viton A, 5 % Teflon

Cast explosives:

PBXN-101:	82 % HMX, 18 % Laminac
PBXN-102:	59 % HMX, 23 % Al, 18 % Laminac

---

* Kel F: chlorotrifluoropolyethylene; Sylgard: silicone resin.

Injection molded explosive:
    PBXC-303   80% PETN, 20% Sylgard 183*)

PE

Abbreviation for "plastic explosives". They consist of high brisance explosives such as RDX or PETN, plasticized with vaseline or other plasticizers. Depending on the additives they contain, the plastic explosives are denoted as PE-1, PE-2 or PE-3 (→ also *Plastic Explosives* and *PBX*).

Pellet Powder

Black powder pressed into cylindrical pellets 2 inches in length and 1 1/4 to 2 inches in diameter.

In the United Kingdom pellet powder is the term used for rounded black powder for hunting ammunition.

Pellets

Explosives in the form of round-shaped granules, e.g., of TNT, used for filling the residual vacant spaces in boreholes.

Pentaerythritol Trinitrate

Pentaerylhrittrinitrat; trinitrate de pentaérythrite; PETRIN

$$HOH_2C-C\begin{smallmatrix}CH_2-O-NO_2\\-CH_2-O-NO_2\\CH_2\;O-NO_2\end{smallmatrix}$$

empirical formula: $C_5H_9N_3O_{10}$
molecular weight: 271.1
energy of formation: -470.2 kcal/kg = -1967 kJ/kg
enthalpy of formation: -494.2 kcal/kg = -2069 kJ/kg
oxygen balance: -26.5%
nitrogen content: 15.5%
density: 1.54 g/cm^3
volume of explosion gases: 902 l/kg
heat of explosion
   (H_2O liq.): 1250 kcal/kg = 5230 kJ/kg
   (H_2O gas): 1142 kcal/kg = 4777 kJ/kg
specific energy: 125 mt/kg = 1227 kJ/kg

The compound is prepared by cautious partial nitration of pentaerythritol.

The free hydroxyl group can react with an acid, e.g., acrylic acid; the polymer PETRIN acrylate serves as a binder.

Pentastit

Name for → *PETN* phlegmatized with 7 % wax.

> detonation velocity, confined:
>   7720 m/s = 23 700 ft/s at $\rho = 1.59$ g/cm^3
> deflagration point: 192–194 °C = 378 – 390 °F
> impact sensitivity: 3 kp m = 29 N m
> friction sensitivity: crackling at 24 kp = 240 N pistil load
> critical diameter of steel sleeve test:
> begins to explode at 4 mm ⌀

Pentolite

Pourable mixtures of → *TNT* and *PETN*, used for shaped charges and for cast boosters (for initiation of insensitive explosives, such as ANFO). A 50:50 mixture has a density of 1.65 g/cm^3; the detonation velocity is 7400 m/s.

Perchlorate Explosives

Perchlorat-Sprengstoffe; explosifs perchlorates

In these explosives, the main oxidizer is sodium, potassium, or ammonium perchlorate; the combustible components consist of organic nitro compounds, hydrocarbons, waxes, and other carbon carriers. Nowadays, these explosives are uneconomical and are no longer industrially produced.

A mixture of 75 % KClO$_4$ and 25 % asphalt pitch, melted together under the name of Galcit, was used as a rocket propellant and was thus a precursor of the modern → *Composite Propellants*.

Percussion Cap

Anzündhütchen; amorce

Percussion caps serve as primers for propellant charges. In mechanical percussion caps, a friction-sensitive or impact-sensitive priming charge (containing, e.g., mercury fulminate with chlorates or lead trinitroresorcinate with Tetrazene) is ignited by the mechanical action of a firing pin.

Percussion Primer = Percussion-actuated initiator.

Perforation of Oil Wells

Perforation von Erdölbohrlöchern; perforation des trous de sondage

In petroleum technology, shaped charges fired from special firing mechanisms (jet perforators) are lowered into the borehole down to the level of the oil horizon. Their purpose is to perforate the pipework and the cement work at the bottom of the borehole, so as to enable the oil to enter it.

Permissibles; Permitted Explosives

Wettersprengstoffe; explosifs antigrisouteux

1. Definition

Shotfiring in coal mines constitutes a risk in the presence of firedamp and coal dust. Permitted explosives are special compositions which produce short-lived detonation flames and do not ignite methane-air or coal-dust-air mixtures.

The methane oxidation

$$CH_4 + 2O_2 = CO_2 + 2 H_2O$$

needs an "induction period"*) before the reaction proceeds. If the time required for ignition by the detonation flames is shorter than the induction period, then ignition of firedamp will not occur. Thus, the composition of permitted explosives must ensure that any secondary reactions with a rather long duration, which follow the primary reaction in the detonation front, are suppressed and that slow → *Deflagration* reactions are avoided (→ *Audibert Tube*).

Such explosives are known as "permissibles" in the USA, as "permitted explosives" in the United Kingdom, as "Wettersprengstoffe" in Germany, as "explosifs antigrisouteux" in France, and as "explosifs S.G.P." (sécurité grisou poussières) in Belgium.

Safety measures to avoid ignition of firedamp uses salt (NaCl) which is included in the usual compositions of commercial explosives. It lowers the → *Explosion Temperature* and shortens the detonation flame.

---

* Contrary to the delayed ignition, the oxidation of hydrogen with the salt-pair aid of an ignition source, $2 H_2 + O_2 = 2 H_2O$, is instantaneous.

Higher safety grades are achieved in ionexchange explosives in which the ammonium and sodium (or potassium) ions are exchanged; instead of

NH_4NO_3 + (inert) NaCl = N_2 + 2 H_2O + 1/2O_2 + (inert) NaCl

the reaction is:

NH_4Cl + $NaNO_3$ (or KNO_3) = N_2 + 2 H_2O + 1/2O_2 + NaCl (or kcl).

Thus, a flame-extinguishing smoke of very fine salt particles is produced by the decomposition reaction itself. Combinations of salt-pair reactions and "classic" detonation reactions quenched by adding salt are possible.

Permitted explosives with a higher grade of safety are powder explosives. They contain a minimum percentage of nitroglycerine-nitroglycol to ensure reliable initiation and transmission of detonation and to exclude slow deflagration reactions. The mechanism of salt-pair detonation in confined and unconfined conditions is explained in → *Detonation, Selective Detonation*.

2. Testing galleries

Versuchsstrecken, Sprengstoffprüfstrecken; galeries d'essai.

All coal-mining countries have issued detailed regulations for the testing, approval, and use of explosives which are safe in firedamp. The main instrument for these tests is the testing gallery.

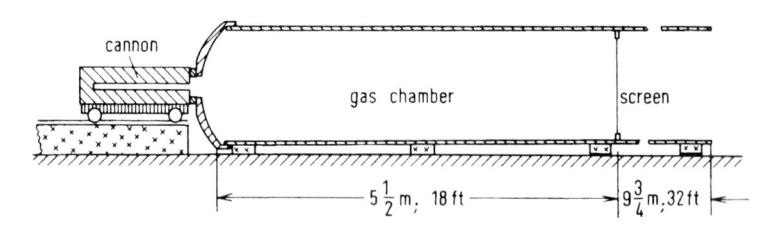

Fig. 18. Testing gallery with borehole cannon.

A test gallery consists of a steel cylinder which initates an underground roadway; the cross sectional area is about 2 m² (5 ft ∅; one end is closed by a shield of about 30 cm (1 ft) ∅ against which the cannon is placed. The other end of the chamber which has a volume of ca. 10 m³ (18 ft length) is closed by means of a paper screen. The remaining part of the tube length (10 m; 32 ft) behind the paper screen is left open to the atmosphere. (The gallery tube can be constructed in closed form if the noise of the test shots can be diminished.) After charging and positioning the cannon, the closed chamber is filled with a methane-air mixture (containing, e.g., 9.5% CH_4 to give the most dangerous composition), and the charge is fired. Whether or not ignition of the gas occurs is observed from a safe position.

Amongst the known types of mortars is the borehole cannon, as shown in Fig. 18. A steel cylinder about 1.5 m (5 ft) long and about 35 cm (1–1/8 ft) in diameter has in it a borehole of 55 mm (2–11/64 in.) diameter and 1.20 m (47 in.) length. The explosive to be tested is placed in the borehole, unstemmed or stemmed by a clay plug, and the detonator is introduced last in the hole (direct initiation). If the detonator is inserted first, followed by the train of cartridges, initiation is "inverse". The required test conditions can be severe; ignition of the gas mixture is more probable to occur using unstemmed charges and inverse initiation than with stemmed charges and direct initiation. The different mortars are designed to simulate different underground conditions. The borehole cannon in the testing gallery illustrates the action of a single shot in the roadway of gassy mines. The British break test and the slotted mortar in Poland imitate the exposure of a charge and, consequently, the more extended contact between the firing charge and the firedamp atmosphere where breaks in the strata intersect a shothole:

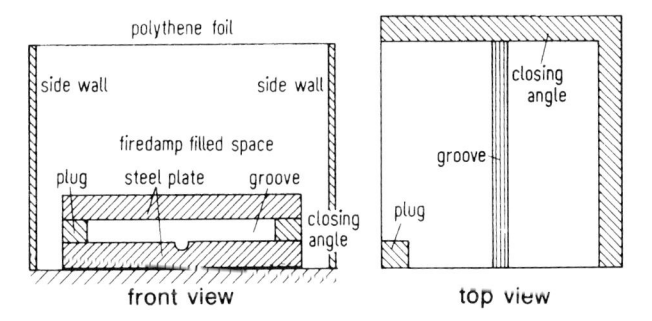

Fig. 19. Break test.

Two steel plates are held at a given distance by means of a closing angle and a plug. The lower plate has a groove for the cartridge train. The plate arrangement is covered with a polythene sheet laid upon two steel side walls; the gas-tight room is filled with the methane-air mixture after charging. The break test conditions are varied; permitted explosives which meet the most stringent test conditions belong to the British safety class P4.

The slotted mortar allows similar test procedures.

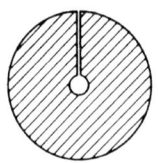

Fig. 20. Slotted mortar.

The slot does not extend over the whole length of the borehole and does not begin at the mouth of the hole.

A specially dangerous condition can arise when several shots are fired in one round by means of electric delay detonators. A preceding shot may then break the coal of another hole or even cut off the whole burden of the charge in question so that it is partly or completely exposed. This condition is simulated in the angle-mortar test.

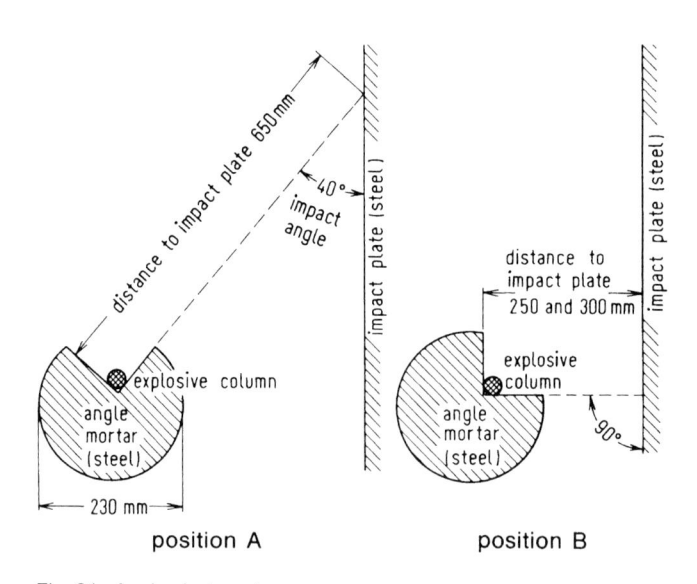

Fig. 21. Angle shot mortar.

A steel cylinder of 230 mm (9 in.) diameter and 2 m (~1/2 ft) in length with a right-angled groove is positioned in the gas chamber of a testing gallery against an impact steel plate at given distances and different impact angles, as shown in Fig. 21. Trains of several cartridges or of the full length of 2 m are placed in the groove of the angle and fired into the methane-air mixture.

Table 24. Areas of use and associated authorized application of German permitted explosives

Working Areas	CH$_4$ in the Mine Air	Type of Explosive	Safety Class
Working in stone without coal	0–0.5	non permitted explosives for rock blasting	–
(except upcasts)	0.5–1.0	permitted explosives	Class I
Working in stone with coal seams up to 0.2 m thickness (except upcasts)	0–0.5 0.5–1.0	permitted explosives permitted explosives	Class I Class II
Working in stone with coal bands of more than 0.2 m thickness gate-end roads (except upcasts)	0–0.3 0.3–0.5 0.5–1.0	permitted explosives permitted explosives permitted explosives	Class I Class II Class III
Rises and dips, gate roads, coal faces and adjacent rock in areas near coal faces, upcasts	less than 1.0	permitted explosives	Class III

3. Safety classes

The different mortar set-ups and other test arrangements can be varied to give a higher or lower probability of ignition; consequently, different safety grades for the explosives have been defined.

In France, there are three categories: explosifs roche, couche, and couche améliorés. They satisfy different requirements according to the borehole cannon test: short or long cannon, direct or inverse initiation, different thicknesses of stemming by means of steel plates.

In the United Kingdom, 5 groups are listed: group P1, the "classic" permitted explosives diluted with rock salt which must pass the least severe cannon test, direct initiation and stemmed; group P2, the now abandoned → *Sheathed Explosives*; group P3, the successor of Eq. S. (equivalent to sheathed) explosives; group P4, the class of highest safety, which meets the most severe break test conditions; and group P5, safe in cut-off conditions.

In Germany there are three classes: class I, the classic permissibles; class II, which are safe in the anglemortar test in position A, with charges of 40 cm in length in the groove; class III, the class of highest safety, which must give no ignition in the angle-mortar test in position B and with the groove filled over its full length with the explosive charge (2 m, 6–1/2 ft).

As an example for possible authorised applications a diagram of the use of the German permitted explosives is given in Table 24.

Peroxides

Organic peroxides may act as explosive. They are usually not manufactured for blasting purposes, but rather as catalysts for polymerization reactions. They are utilized in a safely phlegmatized condition. → *Tricycloacetone Peroxide* and → *Hexamethylenetriperoxide Diamine* display properties of primary explosives.

PETN

pentaerythritol tetranitrate; Nitropenta; tétranitrate de pentaerythrite; Pertitrit; corpent

$$O_2N\text{-}O\text{-}H_2C\diagdown C\diagup CH_2\text{-}O\text{-}NO_2$$
$$O_2N\text{-}O\text{-}H_2C\diagup \diagdown CH_2\text{-}O\text{-}NO_2$$

colorless crystals
empirical formula: $C_5H_8N_4O_{12}$
molecular weight: 316.1
energy of formation: -385.0 kcal/kg = -1610.7 kJ/kg
enthalpy of formation: -407.4 kcal/kg = -1704.7 kJ/kg
oxygen balance: -10.1%
nitrogen content: 17.72 %
volume of explosion gases: 780 l/kg
heat of explosion
  (H_2O gas): 1398 kcal/kg = 5850 kJ/kg ⎫ calculated*)
  (H_2O liq.): 1507 kcal/kg = 6306 kJ/kg ⎭
heat of detonation
  (H_2O liq.): 1510 kcal/kg = 6322 kJ/kg experimental**)
specific energy: 123 mt/kg = 1205 kJ/kg
density: 1.76 g/cm^3
melting point: 141.3 °C = 286.3 °F
heat of fusion: 36.4 kcal/kg = 152 kJ/kg
specific heat: 0.26 kcal/kg = 1.09 kJ/kg

---

* computed by the "ICT-Thermodynamic-Code".
** value quoted from *Brigitta M. Dobratz*, Properties of Chemical Explosives and Explosive Simulants, University of California, Livermore.

vapor pressure:

Pressure millibar	Temperature °C	°F	
0.0011	97.0	207	
0.0042	110.6	231	
0.015	121.0	250	
0.050	131.6	267	
0.094	138.8	282	(near melting point)

lead block test: 523 cm^3/10 g
detonation velocity, confined:
    8400 m/s = 27 600 ft/s at ρ = 1.7 g/cm^3
deflagration point: 202 °C = 396 F
impact sensitivity: 0.3 kp m = 3 N m
friction sensitivity: 6 kp = 60 N pistil load
critical diameter of steel sleeve test: 6 mm

PETN is very stable, insoluble in water, sparingly soluble in alcohol, ether, and benzene, and soluble in acetone and methyl acetate.

It is prepared by introducing pentaerythrol into concentrated nitric acid with efficient stirring and cooling.

The bulk of the tetranitrate thus formed crystallizes out of the acid. The solution is diluted to about 70% HNO$_3$ in order to precipitate the remainder of the product. The washed crude product is purified by reprecipitation from acetone.

PETN is one of the most powerful and most brisant explosives, its → *Stability* is satisfactory, and its → *Sensitivity* is moderate. It is used in high-efficiency blasting-cap fillings and detonation cords. If phlegmatized with a small amount of wax and pressed, it may be used to produce boosters and fillings for smaller caliber projectiles. PETN can also be incorporated into gelatinous, industrial explosives (e.g., for seismic prospecting).

Specifications

melting point: not below	140 °C (284 °F)
nitrogen content: not below	17.5%
Bergmann-Junk test at 132 °C (267 °F): not above	2 ml NO/g
deflagration point: not below	190 °C (374 °F)
acetone-insoluble matter: not more than	0.1%
acidity, as HNO$_3$: not more than	0.003%
alkalinity, as Na$_2$CO$_3$: not more than	0.003%

→ *Vacuum Test* at 120 °C (248 °F):
not more than   5 cm³

Pentaerythrol (raw material):
$C(CH_2OH)_4$
molecular weight: 136.15
melting point: 260.5 °C (501 °F)

Specifications

melting point:
beginning not below   230 °C (446 °F)
moisture: not more than   0.5 %
chlorides: none
not more than   0.5 %
reaction: neutral
reducing matter ($AgNO_3$-NH_3-test):
not more than   traces

Petroleum Jelly

Vaseline

This substance is used as a gunpowder stabilizer. It is believed that the stabilizing effect is due to the presence of unsaturated hydrocarbons, which are capable of binding any decomposition products formed.

Phlegmatization

The impact sensitivity and friction sensitivity of highly sensitive crystalline explosives (e.g., → *Hexogen* and → *PETN*) can be reduced to a considerable extent by the addition of small amounts of a phlegmatizer. This can be an organic polymer or plasticizer, which may act as a lubricant or elastifying agent. For pressed charges it was wax that serves as a desirable lubricant and as a binder. RDX, PETN, and → *Octogen* cannot be compacted by pressing, unless they contain phlegmatization additives. Wax can also be added to pourable mixtures if they contain aluminum powder (→ *Torpex*).

Picramic Acid

Dinitroaminophenol; acide picramique

$$\underset{NO_2}{\underset{|}{O_2N}}\text{—}\underset{}{\overset{OH}{\underset{}{C_6H_3}}}\text{—}NH_2$$

empirical formula: $C_6H_5N_3O_5$
molecular weight: 199.1
energy of formation: −279 kcal/kg = −1167 kJ/kg
enthalpy of formation: −298 kcal/kg = −1248 kJ/kg
oxygen balance: −76.3 %
nitrogen content: 21.11 %
volume of explosion gases: 847 l/kg
heat of explosion
  (H_2O liq.): 639 kcal/kg = 2674 kJ/kg
specific energy: 68.2 mt/kg = 669 kJ/kg
melting point: 169.9 °C = 337.8 °F
lead block test: 166 cm^3/10 g
deflagration point: 240 °C = 464 °F
impact sensitivity: 3.5 pm m = 34 N m
friction sensitivity: up to 36 kp = 353 N
  pistil load no reaction
critical diameter of steel sleeve test: 2.5 mm

Diazotization of picramic acid yields → *Diazodinitrophenol* (DDNP). Lead picramate and DDNP are → *Initiating Explosives*.

Picratol

A 52:48 mixture of ammonium picrate and TNT was used as a bomb filling in the Second World War.

casting density:	1.62 g/cm^3
detonation velocity,	
at casting density, confined:	22 600 ft/s

Picric Acid

2,4,6-trinitrophenol; Pikrinsäure; acide picrique

$$\underset{NO_2}{\underset{|}{O_2N\diagdown\overset{OH}{\overset{|}{\bigcirc}}\diagup NO_2}}$$

yellow crystals; (colorant)
empirical formula: $C_6H_3N_3O_7$
molecular weight:: 229.1
energy of formation: −242.5 kcal/kg = −1014.5 kJ/kg
enthalpy of formation: −259.3 kcal/kg = −1084.8 kJ/kg
oxygen balance: −45.4 %
nitrogen content: 18.34 %
volume of explosion gases: 826 l/kg
heat of explosion
  (H_2O liq.): 822 kcal/kg = 3437 kJ/kg
  (H_2O gas): 801 kcal/kg = 3350 kJ/kg
specific energy: 101 mt/kg = 995 kJ/kg
density: 1.767 g/cm^3
solidification point: 122.5 °C = 252.5 °F
heat of fusion: 18.2 kcal/kg = 76.2 kJ/kg
specific heat: 0.254 kcal/kg = 1.065 kJ/kg
vapor pressure:

Pressure millibar	Temperature °C	°F	
0.01	122	252	(melting point)
2.7	195	383	
67	255	491	

lead block test: 315 cm^3/10 g
detonation velocity, confined:
  7350 m/s at ρ = 1.7 g/cm^3
deflagration point: 300 °C = 570 °F
impact sensitivity: 0.75 kp m = 7.4 N m
friction sensitivity: up to 353 N
  pistil load no reaction
critical diameter of steel sleeve test: 4 mm

Picric acid is toxic, soluble in hot water, and readily soluble in alcohol, ether, benzene, and acetone.

The explosive power of picric acid is somewhat superior to that of → TNT, both as regards the strength and the detonation velocity. Picric acid is prepared by dissolving phenol in sulfuric acid and subsequent nitration of the resulting phenoldisulfonic acid with nitric acid or by further nitration of dinitrophenol (prepared from dinitrochlorobenzene). The crude product is purified by washing in water.

Picric acid was used as a grenade and mine filling. It needs a high pouring temperature, which is undesirable. However, the solidification point can be depressed by the addition of nitronaphthalene, dinitrobenzene or trinitrocresol.

A drawback of picric acid is its tendency to form impact-sensitive metal salts (picrates) when in direct contact with shell walls, etc.; → *TNT*.

Plastic Explosives

kunststoffgebundene Sprengstoff-Mischungen; explosif-liant plastique

High-brisance crystalline explosives, such as RDX or octogen, can be embedded in curable or polyadditive plastics such as polysulfides, polybutadiene, acrylic acid, polyurethane, etc. The mixture is then cured into the desired shape. Other components such as aluminum powder can also be incorporated. The products obtained can be of any desired size, and specified mechanical properties can be imparted to them, including rubber-like elasticity (→ *LX* and → *PBX*). They can also be shaped into foils.

"Plastic" also means mixtures of → *RDX* with vaseline or gelatinized liquid nitro compounds of plastiline-like consistency.

Also propellant charges for rockets and guns have also been developed by compounding solid explosives such as nitramines (e.g. → *Hexogen*) with plastics. Plastic explosives and plastic propellants are of interest, if low thermal and impact sensitivity is needed (→ *LOVA;* → *IHE*).

Plate Dent Test

is a brisance comparison test used in the USA for military explosives. There are two methods:

Method A – The charge is contained in a copper tube, having an internal diameter of 3/4-inch and a 1/16-inch wall. This loaded tube is placed vertically on a square piece of cold-rolled steel plate, 5/8-inch thick; 4-inch and 3-1/4-inch square plate gave the same results. The steel plate is in a horizontal position and rests in turn on a short length of heavy steel tubing 1-1/2 inches ID and 3 inches OD. The charge rests on the center of the plate, and the centers of the charge, plate, and supporting tube are in the same line. A 20 g charge of the explosive under test is boostered by a 5g pellet of tetryl, in turn initiated by a No. 8 detonator.

Method 13 – A 1-5/8-inch diameter, 5-inch long uncased charge is fired on a 1-3/4-inch thick, 5-square inch cold-rolled steel plate, with one or more similar plates as backing. The charge is initiated with a No. 8 detonator and two 1-5/8-inch diameter, 30-g Tetryl boosters.

Plate dent test value, or relative brisance =

$$\frac{\text{Sample Dent Depth}}{\text{Dent Depth for TNT at } 1.61 \text{ g/cm}^3} \times 100.$$

Plateau Combustion

Plateau-Abbrand

→ Burning Rate.

Pneumatic Placing

Druckluft-Ladeverfahren; chargement pneumatique

The loading of explosives or blasting agents into a borehole using compressed air as the loading force.

Poly-3-azidomethyl-3-methyl-oxetane

Poly-AMMO

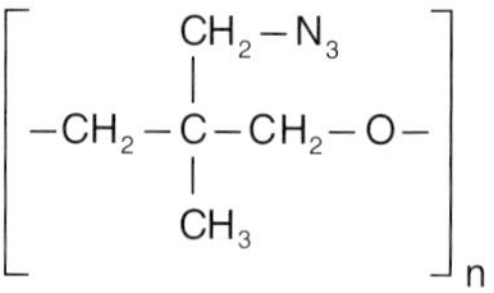

$$\left[-CH_2 - \underset{\underset{CH_3}{|}}{\overset{\overset{CH_2-N_3}{|}}{C}} - CH_2 - O - \right]_n$$

colorless oil to wax	
empirical formula of the structural unit:	$C_5H_9N_3O$
molecular weight of the structural unit:	127.15
mean molecular weight:	1000–3000
energy of formation:	471.88 kJ/kg
enthalpy of formation:	345.19 kJ/kg
oxygen balance:	−169.88 %
nitrogen content:	33.05 %
density:	1.17 g/cm^3
specific energy:	568.3 kJ/kg

Poly-AMMO is synthesized via cationic polymerisation from the monomer 3-azidomethyl-methyl-oxetane (AMMO). The polymerisation reaction is quenched with water to get polymer chains with hydroxyl endgroups which enable to react these pre-polymers later with isocyanate for curing reaction. Poly-AMMO is suggested as → energetic binder component in → composite propellants and is in the scope of actual research.

Poly-3,3-bis-(azidomethyl)-oxetane

Poly-BAMO

$$\left[-CH_2 - \underset{\underset{CH_2-N_3}{|}}{\overset{\overset{CH_2-N_3}{|}}{C}} - CH_2 - O - \right]_n$$

colorless solid
empirical formula of the structural
    unit: $C_5H_8N_6O$
molecular weight of the structural
    unit: 168.16
mean molecular weight: 1000–10000
energy of formation: 2517.7 kJ/kg
enthalpy of formation: 2460.8 kJ/kg
oxygen balance: −123.69 %
nitrogen content: 49.98 %
density: 1.25 g/cm^3
mech. sensitivity: 5.0 Nm (impact);
    288 N (friction)

Poly-BAMO is synthesized via cationic polymerisation from the monomer 3,3-bis(azidomethyl)l-oxctane (BAMO) The polymerisation reaction is quenched with water to get polymer chains with hydroxyl endgroups which enable to react these pre-polymers later with isocyanate for curing reaction. Poly-BAMO has one of the highest nitrogen content of the → energetic binder components and is suggested for → composite propellants. It is in the scope of actual research.

Polynitropolyphenylene

Polynitropolyphenylen; Polynitropolyphénylène; PNP

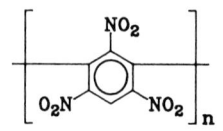

$$\left[\begin{array}{c} NO_2 \\ O_2N \underset{}{\underset{}{\bigcirc}} NO_2 \end{array} \right]_n$$

>greenish, yellow-brown amorphous powder
>empirical formula of structural unit: $C_6HN_3O_6$
>molecular weight of structural unit: 211.1
>mean molecular weight: 2350
>oxygen value: −49.3 %
>nitrogen content: 19.91 %
>explosion heat (H_2O liq.): 3200 kJ/kg = 764 kcal/kg
>density: 1.8−2.2 g/cm^3
>bulk density: 520 g/l
>deflagration temperature: 280°−304 °C
>impact sensitivity: 3−5 Nm = 0.2−0,5 kpm
>sensitivity to friction: at 360 N = 37 kp pin load, reaction
>marginal diameter, stell case test: 6 mm

Polynitropolyphenylene is obtained from the reaction of a solution of 1,3-dichloro-2,4,6-trinitrobenzene in nitrobenzene at 150°−180 °C with copper powder (Ullmann reaction).

The raw product obtained in this manner is first separated from copper chloride and then cleaned in several stages from solvent residues and low molecular weight elements. The resulting compound is a non-crystalline explosive of extremely high thermal resistance. In the field of → *LOVA* technology, it is used as an → *Energetic Binder* in high ignition temperature propellants.

Polypropylene Glycol

Polypropylenglykol; polypropylene glycol; PPG

$$HO-[(CH_2)_3-O-]_{34}H$$

>viscous liquid
>empirical formula: $C_{10}H_{20.2}O_{3.4}$
>molecular weight: 1992
>energy of formation: −853 kcal/kg = −3571 kJ/kg
>enthalpy of formation: −888 kcal/kg = −3718 kJ/kg
>oxygen balance: −218.4 %
>density (20/4): 1.003 g/cm^3

PPG serves as a prepolymer, which reacts with diisocyanates as curing agents to form polyurethanes used as a binder in → Composite Propellants

Polyvinyl Nitrate

Polyvinylnitrat; nitrate de polyvinyle; PVN

$$\left[\begin{array}{c} -CH_2-CH \\ | \\ O-NO_2 \end{array} \right]_n$$

yellowish-white powder
empirical formula of the structure unit: $C_2H_3NO_3$
molecular weight of the structure unit: 89.05
average molecular weight: 200000
energy of formation: -252.1 kcal/kg = -1054.8 kJ/kg
enthalpy of formation: -275.4 kcal/kg = -1152.1 kJ/kg
oxygen balance: -44.9%
nitrogen content: depends on nitration grade
volume of explosion gases: 958 l/kg
heat of explosion
  (H_2O liq.): 1143 kcal/kg = 4781 kJ/kg
  (H_2O gas): 1073 kcal/kg = 4490 kJ/kg
specific energy: 129 mt/kg = 1269 kJ/kg
density: 1.6 g/cm^3
softening point: 30–40 °C = 86–104 °F
detonation velocity:
7000 m/s = 23000 ft/s at ρ = 1.5 g/cm^3
deflagration point: 175 °C = 350 °F
impact sensitivity: 1.0 kpm = 10 Nm
friction sensitivity: at 20 kp = 196 N
  pistil load reaction
critical diameter of steel sleeve test: 8 mm

Polyvinyl nitrate is prepared by esterification of polyvinyl alcohol (PVA) using nitric acid or a nitrating mixture. Depending on the degree of saponification of polyvinyl alcohol, which is prepared from polyvinyl acetate, the products have varying nitrogen contents and rheological properties, depending on the manufacturing conditions manufacture and the degree of polymerization. PVN is a thermoplastic, macromolecular substance, with a softening zone which varies between 30 and 45 °C, depending on the molecular weight of the starting polyvinyl alcohol. Polyvinyl Nitrate is used in a plasticizer for TNT charges called X28M, which is a solution of 22% PVN in 78% mononitrotoluene MNT.

Porous Powder

Poröses Pulver; poudre poreux

Special powders for exercise ammunition with a large internal surface area and thus, a fast burning rate. The porosity is produced by adding a soluble salt to the powder being manufactured; the salt is then leached out again at a later stage.

Post Combustion

Nachflammen

Combustion of flammable fumes of a deflagrated or detonated explosive with a negative oxygen balance (→ also *Muzzle Flash*).

Potassium Chlorate

Kaliumchlorat; chlorate de potassium

$KClO_3$

colorless crystals
molecular weight: 122.6
oxygen balance: +39.2 %
density: 2.34 g/cm^3
melting point: 370 °C = 700 °F

Potassium chlorate is sparingly soluble in cold water, readily soluble in hot water, and insoluble in alcohol. It is the principal component of → *Chlorate Explosives* and is an important component of primer formulations and pyrotechnical compositions, in particular matchheads.

Potassium Nitrate

saltpetre; Kaliumnitrat; nitrate de potasse

KNO_3

colorless crystals
molecular weight: 101.1
energy of formation: −1157 kcal/kg = −4841 kJ/kg
enthalpy of formation: −1169 kcal/kg = −4891 kJ/kg
oxygen balance: +39.6 %
nitrogen content: 13.86 %
density: 2.10 g/cm^3
melting point: 314 °C = 597 °F

Potassium nitrate is readily soluble in water, sparingly soluble in alcohol, and insoluble in ether.

It is used as a component in pyrotechnical compositions, in inustrial explosives, and in black power.

Table 25. Specifications

	Class 1	Class 2	Class 3
net content (e.g., by N-determination) at least	99.5%	99.5%	99.5%
moisture: not more than	0.2%	0.2%	0.2%
water-insoluble: not more than	0.1%	0.1%	0.1%
grit:	none	none	none
acidity:	0	0	0
alkalinity:	0	0	0
chlorides as KCl: not more than	0.07%	0.07%	0.07%
chlorates and perchlorates, as K-salt: not more than	0.5%	0.5%	0.5%
$Al_2O_3 + Fe_2O_3$: not more than	0.5%	0.5%	–
CaO + MgO: not more than	0.5%	0.5%	0.5%
Na as Na_2O: not more than	0.25%	0.25%	–
nitrogen content: at least	13.77%	13.77%	13.77%

Potassium Perchlorate

Kaliumperchlorat; perchlorate de potassium

$KClO_4$

colorless crystals
molecular weight: 138.6
oxygen balance: +46.2%
density: 2.52 g/cm^3
melting point: 610 °C = 1130 °F
(decomposition begins at 400 °C = 750 °F)

Potassium perchlorate is insoluble in alcohol but soluble in water. It is prepared by reacting a soluble potassium salt with sodium perchlorate or perchloric acid. It is employed in pyrotechnics.

Specifications

	colorless odorless crystals	
	net content (KCl determination after reduction): not below	99%
	moisture: not more than	0.5%
	insolubles in water: not more than	0.1%
	solution in hot water:	clear
	chlorides as KCl: not more than	0.1%
	bromate as $KBrO_3$: not more than	0.1%
	chlorate as $KClO_3$: not more than	0.1%
	NH_4-, Na-, Mg- and Ca-salts:	none
	heavy metals	none
	pH	6.5 ± 0.5

Poudre B

French gunpowder. A single base nitrocellulose propellant stabilized by 1.5–2% diphenylamine. The sufix (e.g., Poudre B Ba) denotes:

Suffix	Shape of Powder Grain
Ba	short rods (bâtonnet)
Bd	bands
Cd	long rods (corde)
Di	disks
FP	flakes (paillette; obsolete denomination)
Pa	flakes (paillette)
Se	flattened balls (sphère écrasée)
SP	balls (sphère)
7 T	tubes with 7 holes
19 T	tubes with 19 holes
Tf	slotted tubes (tube fendu)
Tu	tubes (tubulaire)

Powder Form Explosives

Pulverförmige Sprengstoffe; explosifs pulverulents

Industrial explosives must be easy to shape, i.e., must have a gelatinous or powdery consistency in order to introduce the detonator or electric cap. Powder-form explosives are mostly based on ammonium nitrate and fuel components (e.g., aluminum).

The powders can be sensitized by the addition of nitroglycerine in small percentages. Non-cap sensitive powders (→ *Blasting Agents*) need a booster charge for safe initiation.

Certain types of powder-form explosives contain moisture repelling additives such as stearates; in paraffinated cartridges they can be applied even under wet conditions. Non-cartridged powder form explosives must be free-flowing (→ *ANFO*).

Ion exchanged → *Permitted Explosives* are based on so-called salt pairs (sodium nitrate ammonium chloride or potassium nitrate – ammonium chloride) and are thus also in powder form.

Pre-ignition

Vorzeitige Selbstentzündung; allumage spontané

Spontaneous and premature ignition.

Premature Firing

Frühzündung; départ prématuré

The detonation of an explosive charge or the ignition of an electric blasting cap before the planned time. This can be a hazardous occurrence and is usually accidental.

Prequalification Test

Vorprüfung; test préliminaire

Brief test program conducted on an item or system to determine if it will meet only the most rigorous specified requirements.

Pre-splitting (Pre-shearing)

Vorspalten; tendage préliminaire

A → *Contour Blasting* method in which cracks for the final contour are created by firing a single row of holes prior to the blasting of the rest of the holes for the blast pattern.

Pressing of Rocket Propellant Charges

Pressen von Treibsätzen; moulage des propellants de roquette par pression

Rocket compositions of both double base and composite type are shaped into the desired form (e.g., star-shaped configurations) on extrusion or screw-type presses through a die or by casting and curing.

Press Molding of Explosives

Pressen von Sprengstoffen; moinlage d'explosifs pa pression

The purpose of compression by hydraulic presses is similar to that of casting, i.e. to attain a high loading density (→ *Brisance*) while at the same time imparting the desired shape to the charge.

Certain explosives (TNT, Tetryl, etc.) can be compacted by compression in the absence of any additives; sensitive explosives such as PETN (Nitropenta), RDX (Hexogen), or HMX (Octogen) have to be phlegmatized by the incorporation of wax. The wax reduces the impact sensitivity and, at the same time, acts as a binder.

Plastic binder materials: → *LX* and → *PBX*.

Pressure Cartridge

Druckgas-Patrone; cartouche génératrice de gaz

Pyrotechnic device in which propellant combination is used to produce pressurized gas for short duration.

Prills

denote the ammonium nitrate pellets obtained by cooling free falling droplets of the molten salt in so called prill towers. By special processing, they can be porous and are capable of absorbing a certain percentage of liquid hydrocarbons (→ *ANFO*). The ready made ANFO-explosive is also marketed under the name "Prills".

Primary Blast

Hauptsprengung; tir primaire

A blast that loosens rock ore from its original or natural location in the ground. A secondary blast may be used to reduce the rocks from the primary blast to smaller size for ease of handling.

Primary Explosive

Initialsprengstoff; explosif d'amorcage

A sensitive explosive which nearly always detonates by simple ignition from such means as spark, flame, impact and other primary heat sources of appropriate magnitude (→ *Initiating Explosives*).

Primer

A primary initiating device to produce a hot flame. A primary stimulus sensitive component generally is used to generate a brisant output for initiating detonating compositions. Infrequently used to initiate deflagrating compositions (→ *Squib;* → Detonator; → Initiator).

Primer Charge

Zündladung; charge d'amorçage

Secondary component in an → *Ignition Train*, which is ignited by an initiator, starts pressurization of a generator, and ignites the booster charge.

For the firing of industrial explosives, primers are prepared by inserting a blasting cap or an electric detonator in hole of a cartridge of a cap-sensitive explosive.

In military ammunition primers are charges used to initiate the main explosive charge of a weapon containing built in detonators.

Progressive Burning Powder

Progressiv-Pulver; poudre progressive

Gunpowder which burns at a progressively increasing rate, owing to the appropriate choice of the geometry of the powder grain and sometimes owing to a suitable grain surface coating. Examples are perforated powders (7-hole powder, 19-hole-powder, etc.).

Projectile Impact Sensitivity

Beschußempfindlichkeit; sensibilité à l'impact de projectiles

→ also *Armor Plate Impact Test* and → *Impact Sensitivity.*

The projectile impact sensitivity is the reaction of an explosive charge if hit by infantry projectiles. Impact safety is given if the charge does not fully explode at impact. The projectile impact sensitivity does not only depend on the type of explosive itself, but also on the nature of its confinement (metallic, plastic, thin-walled, or thick-walled). A single bullet impact by an ordinary or a hard steel cored projectile, or a machine gun burst, will create different reactions.

A test has been developed in Sweden: cylinders made of copper, brass, and aluminum (15 mm ∅) are brought to accurately adjusted and measured impact velocities (→ *Impact Sensitivity*).

Propellant

Treibstoff; produit propulsif; → Gunpowder

Explosive material with low rate of combustion. May be either solid or liquid. Will burn smoothly at uniform rate after ignition without depending on interaction with atmosphere. Single base propellant consists primarily of matrix of nitrocellulose. Double base propellant contains nitrocellulose and nitroglycerine. Composite propellant contains oxidizing agent in matrix of binder.

Propellant Types:

- a) Composite — Finely divided oxidizers dispersed in fuel matrix.
  - (1) Ammonium nitrate oxidizer
  - (2) Ammonium perchlorate oxidizer
  - (3) Nitramine (RDX or HMX) oxidizer
- b) Double-Base — Homogeneous colloidal propellant consisting of nitrocellulose dissolved in plasticizer comprised of nitroglycerine and inert materials
- c) Plastisol — PVC-composite or double-base propellant in which polymer is dissolved in plasticizier
- d) Composite Double-Base — Double base propellant containing dispersed phase of finely ground oxidizer and (usually) powdered fuel additive
- e) Single-Base — Colloid of nitrocellulose and inert plasticizers

Propellant Area Ratio

Klemmung; resserrement

In rocket technology, the ratio between the burning surface of the propellant and the smallest cross-section of the nozzle. It determines the resultant pressure in the combustion chamber of the rocket (other relevant keywords: → *Burning Rate,* → Gas Jet Velocity, → Rocket, → Solid Propellant Rocket, → Specific Impulse, → Thrust).

Propellant-actuated Power Devices

Any tool or special mechanized device or gas generator system which is actuated by a propellant or which releases and directs work through a propellant charge.

Propergol

In rocket technology, a collective term for all chemical propellants.

Propyleneglycol Dinitrate

methylnitroglycol; Propylenglykoldinitrat;
dinitrate de propylèneglycol propanediol dinitrate

$$\begin{array}{l} CH_3 \\ | \\ CH-O-NO_2 \\ | \\ CH_2-O-NO_2 \end{array}$$

colorless liquid
empirical formula: $C_3H_6N_2O_6$
molecular weight: 166.1
oxygen balance: -28.9%
nitrogen content: 16.87%
density (20 °C): 1.368 g/cm^3
lead block test: 540 cm^3/10 g

Propyleneglycol dinitrate is readily soluble in organic solvents, but is practically insoluble in water. It is obtained by nitration of propyleneglycol with mixed acid.

Propyl Nitrate

Propylnitrat; nitrate de propyle

$CH_3-CH_2-CH_2ONO_2$ and $\begin{array}{c} H_3C \\ \diagdown \\ C-O-NO_2 \\ \diagup | \\ H_3C H \end{array}$

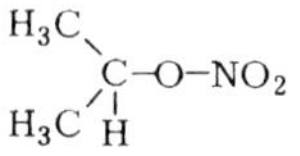

n-propyl nitrate, NPN

isopropyl nitrate, IPN

colorless liquid
empirical formula: $C_3H_7NO_3$
molecular weight: 105.1
energy of formation:
   n: −1911 kJ/kg
   iso: −2057 kJ/kg
enthalpy of formation:
   n: −2041 kJ/kg
   iso: −2184 kJ/kg
oxygen balance: −99.0 %
nitrogen content: 13.33 %
heat of explosion (H_2O liq.):
   n: 3272 kJ/kg
   iso: 3126 kJ/kg
density:
   n: 1.058 g/cm^3 (20 °C)
   iso: 1.036 g/cm^3 (20 °C)
impact sensitivity: up to 5 kp m = 49 N m no reaction

n-Propyl nitrate serves as a → *Monergol* in liquid propellant rockets.

Isopropylnitrate is used in → themobaric explosives together with Magnesium.

Pulsed Infusion Shotfiring

Stoßtränkungssprengen; Drucktränksprengen; tir d'imprégnation

This blasting technique combines the effect of an explosive charge in coal mine blasting with the effect of water pressure. The borehole is loaded with the explosive charge, after which water is pressed into the borehole with the aid of the so-called water infusion pipe, and the charge is ignited while maintaining the water pressure. The pressure shock in the water causes the coal to disintegrate into large lumps.

In addition, the water fog, which is produced at the same time, causes the dust to settle to the ground.

Pyrophoric

Materials that will ignite spontaneously. Exmples of pyrophoric substances in air white phosphorus, alcyl derivates of aluminium or zinc or finely devided metals, which are readily oxidizers.

Pyrotechnical Compositions

Feuerwerksätze; compositions pyrotechniques

Oxidizer – fuel mixtures, which give off bright or colored light (Bengal fireworks), evolve heat (thermites), produce fogs (also colored fogs), or give acoustic effects (howling, whistling, and banging).
Special black powder granules for pyrotechnics → *Black Powder.* The additives employed for colored light are:

>barium salts or boric acid for green;
>strontium salts for red;
>cupric oxide for blue;
>sodium salts for yellow.

Pyrotechnical Fuses

Feuerwerkszündschnüre; fusées pyrotechniques

Pyrotechnical fuses are ~ Safety' Fuses, which are specially adapted for pyrotechnical purposes by their diameter and their rigidity. They are cut into small (4~6 cm) segments.

Quantity – Distance Table

Sicherheitsabstands-Tabelle; tableau des distances des sécurité

A Table listing minimum recommended distances from explosive materials stores of various weights to some predetermined location.

Quick-Match

match cord; cambric; Stoppine

Quick-match serves to transfer ignition to pyrotechnic sets. It consists of between 2 and 16 spun cotton threads, which have been impregnated with black powder and dried. This impregnation is carried out by using an alcohol-water saturated black powder sludge, and the threads are drawn through this mixture and gauged by drawing them through an extruder die. The impregnation mass contains resin and gum arabic as binders. After the match cords have dried, they are cut into size; if they are to be used for larger fireworks, they receive an additional cover of paraffin-treated paper, and both ends are then tied. For additional safety, two Quickmatches are inserted into the paper sleeve.

Burning time is preset at between 30 to 100 s/m. A Quick match contained in paper tubes, is preset to a maximum of 40 m/s to avoid failure to ignition.

RDX

→ Hexogen

Recommended Firing Current

Soll-Zündimpulse; ampèrage recommande our le mise á feu

Current that must be applied to bridgewire circuit to cause operation of device within specified time.

Recommended Test Current

Maximum current that can be applied to bridgewire circuit for extended period of time without degrading prime material.

Reduced Sensitivity RS

This term describes improved properties of energetic materials. In the 1990s it was shown that with careful recrystallization techniques the sensitivity RDX can be reduced on a crystalline level. Nowadays, reduced sensitivity variants of RDX, assigned as RS-RDX or I-RDX (for insensitive), are provided by different manufacturer and tested, particularly in plastic bonded explosives PBX for Insensitive Munitions IM. Besides, RS-variants of HMX and other high explosives, and characterization and quality assessment techniques for Reduced Sensitivity are in the scope of actual research.

Regressive Burning

Degressiver Abbrand; brûlage regressive

Condition in which mass flow produced by the propellant grain decreases as web is consumed, due to a decreasing area, decreasing burn rate, or both (→ *Progressive Burning Powder*).

Relay

An explosive train component that requires explosive energy to reliably initiate the next element in the train. Specifically applied to small charges that are initiated by a delay element, and in turn, cause the functioning of a detonator.

Reliability

Zuverlässigkeit

Statistical evaluation of probability of device performing its design function.

Resonance

→ Erosive Burning.

Restricted Propellant

Propellant grain having portion of its surface area treated to control burning. → *Inhibited Propellant*

Restrictor:

Material applied to selected areas of propellant charge to prevent burning in these areas.

RID

Abbreviation for "Règlement Concernant le Transport International Ferroviaires des Marchandises Dangereuses". It contains the official regulations governing the haulage, admission, and packing for international railway traffic. → *ADR* are the corresponding regulations governing international motor traffic.

Table 26 shows the examination procedure as exemplified for the powder-form ammonium nitrate explosive Donarit 1 manufactured, in Germany.

Table 26. RID Test results of Donarit 1

Composition Components in %			External Apearance and Texture	Storage at 75 °C (167 °F) (closed weighting bottles)	Behavior on being heated in Wood's metal bath	Behavior when lit by a match	Behavior when lit with a 10-mm high, 5-mm wide gas flame	Behavior when thrown into a red-hot steel bowl	Behavior when heated inside a steel sheet box in a wood fire	Behavior when heated confined in a steel sleeve with escape diameter of:	Sensitivity under Fallhammer	Sensitivity under Fallhammer a 5-kp weitght falling from a height of cm: 15 20 30 40 50	Sensitivity in the friction tester
	Declared	Found											
ammonium nitrate	80	79.8	light-yellow fine-grained powder	weight loss after 2 days 0.2%; no nitrous gases	At 180 °C (365 °F) evolution of brown vapors; at 212 °C (414 °F) and 320 °C (608 °F) decomposition not accompanied by burning	ignition failed 5 times	ignition failed 5 times	catches fire and burns with a steady flame for 12/14/10 s	catches fire after 64–78 s; end of burning after 390–500 s; strongly hissing flame; the box bulges on all sides	2.0 mm⌀: explosion; t_1 = 16 s t_2 = 20 s 2.5 mm⌀: no explosion	no reaction decomposition explosion	6 4 3 1 0 0 0 0 0 0 0 2 3 5 6	at 36 kp pistil load no reaction
TNT	12	12.1											
nitroglycerine	6	5.9											
wood meal	2	2.2											

Rifle Bullet Impact Test

Beschussprobe

is a USA standard test procedure for explosives of military interest.

Approximately 0.5 pound of explosive is loaded in the same manner as it is loaded for actual use: that is, cast, pressed, or liquid in a 3-inch pipe nipple (2-inch inside diameter, 1/16-inch wall) closed on each end by a cap. The loaded item, in the standard test, contains a small air space which can, if desired, be filled by inserting a wax plug. The loaded item is subjected to the impact of caliber 30 bullet fired perpendicularly to the long axis of the pipe nipple from a distance of 90 feet.

Rocket

Rakete; roquette

Pressure vessel containing a propellant, which, on being ignited, produces hot gases, which, in turn, are expelled through a nozzle or nozzles to produce thrust.

Rocket Motor

Raketentriebwerk; moteur fusée; propulseur

The propulsion assembly of a rocket or → *Missile*. The driving force can be produced by burning liquid fuels in liquid oxidizers (liquid oxygen, nitric acid, or other oxidants such as liquid fluorine), by burning of solid propellants (→ *Solid Propellant Rockets*), by burning solid fuels in liquid oxidizers (→ *Hybrids*), or by catalytic decomposition of endothermal compounds (→ *Hydrazine*; → *Aerozin*; → *Aurol*).

Rocket Test Stand

Raketen-Prüfstand; banc d'essai

The test stand serves to determinate the thrusts and pressures which develop during the combustion process (→ *Thrust Determination*). Since we are interested in the combustion behavior at different temperatures, the test stands are mostly equipped with warm and cold chambers for conditioning prior to testing.

The design of some stands makes it possible to determine other thrust components (e.g., the side component in inclined nozzles) and torques.

Test stands may be designed for the engine to be tested in a vertical or in a horizontal position.

Modern test stands are also equipped for environmental testing (e.g., temperature changes, vibration, impact, and drop tests).

Rotational Firing

Delay blasting system used so that the detonating explosives will successfully displace the burden into the void created by previously detonated explosives in holes, which fired at an earlier delay period.

Round Robin Test

Ringversuch

Round Robin tests are testing procedures, developed by the joint effort of several institutes in different countries, with the purpose of obtaining comparable results. Such international tests are particularly useful if they are recognized as binding acceptance tests in the sales of munition from one country to another.

SAFE & ARM

Device for interrupting (safing) or aligning (arming) an initiation train of an explosive device, i.e., bomb or warhead.

Safety Diaphragm

Berstscheibe; diaphragme de securité

Diaphragm, usually metal, that will rupture in the event that excessive gas generator chamber pressure develops.

Safety Fuses

Schwarzpulverzündschnüre; mèches de sureté

Safety fuses are black powder cords with an external yarn winding adjusted to a definite combustion rate – usually 120 s/m. The purpose is to initiate the explosive charge by igniting the blasting cap of the primer. The fuse must be freshly cut in the plane perpendicular to its axis, and the plane of the cut must reach the ignition level of the cap. The length of the fuse depends on the safety period required. The structure of the fuse comprises a black powder core with one or two marking threads, the color of which indicates the identity of the manufacturer, two or three layers of yarn wound around it (jute, cotton, or some other yarn), a bitumen impregnation, and a plastic coating.

The black powder contains 65–74% potassium nitrate, and its grain size is 0.25–0.75 mm. A 1-meter length of fuse contains about 4–5 g of powder.

A special type of safety fuse is employed in Switzerland. The core is a pyrotechnical composition in meal form, which is sheathed in paper strips and has a large number of textile threads around it.

SAFEX INTERNATIONAL

SAFEX INTERNATIONAL, a non-profit making organisation for producers of explosives and pyrotechnics, was founded in 1954. The aim of SAFEX is to encourage the exchange of experience in the explosive industry. The information gained from accidents and incidents leads to a better understanding and can help members avoid similar events.

SAFEX INTERNATIONAL has more than 80 members in 40 countries from allover the world (2001). The organisation is strictly non-political; all information is for SAFEX members only. Every member is obliged to notify the secretary of any accident or incident within the plant. The secretary then sends out this information to all members; any further clarification can be requested from the secretary, who will in turn contact the member concerned.

Every third year a Congress is organised for the presentation of papers on common themes by the members. Admission is for SAFEX members only.

Sand Test

A performance test of an explosive, used in the USA. A known amount of the explosive is exploded in sand consisting of a single grain size (sieve) fraction; the magnitude determined is the amount of sand which passes a finer-meshed sieve following the fragmentation. The test descriptions follow:

(a) Sand test for solids.
A 0.4-g sample of explosive, pressed at 3000 psi into a No. 6 cap, is initiated by lead azide or mercury fulminate (or, if necessary, by lead azide and tetryl) in a sand test bomb containing 200 g of "on 30 mesh" Ottawa sand. The amount of azide of Tetryl that must be used to ensure that the sample crushes the maximum net weight of sand, is designated as its sensitivity to initiation, and the net weight of sand crushed, finer than 30 mesh, is termed the sand test value. The net weight of sand crushed is obtained by subtracting from the total amount crushed by the initiator when shot alone.

(b) Sand test for liquids.

The sand test for liquids is made in accordance with the procedure given for solids except that the following procedure for loading the test samples is substituted:

Cut the closed end of a No. 6 blasting cap and load one end of the cylinder with 0.20 g of lead azide and 0.25 g of tetryl, using a pressure of 3000 psi to consolidate each charge. With a pin, prick the powder train at one end of a piece of miner's black powder fuse, 8 or 9 inches long. Crimp a loaded cylinder to the pricked end, taking care that the end of the fuse is held firmly against the charge in the cap. Crimp near the mouth of the cap to avoid squeezing the charge. Transfer of 0.400 g of the test explosive to an aluminum cap, taking precautions with liquid explosives to insert the sample so that as little as possible adheres to the side walls of the cap; when a solid material is being tested, use material fine enough to pass through a No. 100 U.S. Standard Sieve. The caps used should have the following dimensions: length 2.00 inches, internal diameter 0.248 inch, wall thickness 0.025 inch. Press solid explosives, after insertion into the aluminum cap, by means of hand pressure to an apparent density of approximately 1.2 g per cubic centimeter. This is done by exerting hand pressure on a wooden plunger until the plunger has entered the cap to a depth of 3.93 centimeters. The dimensions of the interior of the cap are: height 5.00 cm, area of cross section 0.312 square centimeters. Insert the cylinder containing the fuse and explosive charge to Tetryl and lead azide into the aluminum cap containing the test explosive for the determination of sand crushed.

Scaled Distance

Abstandsberechnung

A factor relating similar blast effects from various size charges of the same explosive at various distances. Scaled distance referring to blasting effects is obtained by dividing the distance of concern by an exponential root of the mass of the explosive materials.

Screw Extruder

Schneckenpresse; extrudeuse à vis

These shaping machines, which are commonly employed in the plastics industry, were introduced at an early stage of the manufacture of explosives and gunpowders.

Many cartridging machines for gelatinous explosives utilize double screws as conveyors, but pressures are not allowed to build up to significant values.

Screw extruders were also used for filling ammunition with powder-for explosives, since these can be compacted by the application of pressure.

Continuously charged, continuos operation horizontal screw extruders are employed, in particular, to impart the desired profile to → *Double Base Propellants* (e.g., shaping of tubes or special profiles for rocket propellants).

Secondary Blasting

Knäppern; pétardage de blocs

Blasting to reduce the size of boulders resulting from a primary blast (→ *Mud Cap*).

Secondary Explosives

Sekundär-Sprengstoffe; explosifs secondaires

Explosives in which the detonation is initiated by the detonation impact of an initial (primary) explosive. Accordingly, this definition includes all explosives used to obtain blasting effects.

Materials such as → *Ammonium Nitrate* or → *Ammonium Perchlorate* are classified as tertiary explosives, which are less sensitive to detonation impact.

Seismic Explosives

Seismische Sprengstoffe; explosifs sismiques

Seismic explosives produce the pressure impact during seismic measurements, which are carried out in prospecting for geological deposits, particularly oil horizons. Such explosives must detonate even under high hydrostatic pressures.

For practical reasons, the shape of the cartridges must differ from the conventional; cartridges which can be coupled, canisters for effecting explosions in coastal areas, and canned blasting agents, which resist the water pressure in the boreholes even for several days ("sleeper charges").

Seismo-Gelit

Trade name of a sensitized gelatinous special explosive, distributed in Germany and exported by Orica Gemany GmbH. It is used for seismic prospecting. The explosive can be supplied in sort plastic cartridges and in rigid plastic tubes threaded for coupling.

density of cartridge:	1.6 g/cm^3
weight strength:	90 %
detonation velocity at cartridge density, unconfined:	6000 m/s

Seismograph

An instrument, useful in blasting operations, which records ground vibration. Particle velocity, displacement, or acceleration are generally measured and recorded in three mutually perpendicular directions.

Seismometer, Falling Pin

An instrument used to indicate relative intensity of ground vibration. It consists of a level glass plate on which a series of 1//4″ diameter steel pins (6″ to 15″ lengths) stand upright inside of metal tubes. The use of the falling pin seismometer is based on the theory that the length of the pin toppled depends upon the amount of ground vibration present.

Seismoplast 1

Trade name of a plastic, water-resistant cap-sensitive explosive used for seismic exploration under extreme conditions such as pressures up to 500 bar and temperatures as low as −40 °C. Manufactured and distributed by Orica Germany GmbH.

density:	1.54 g/cm^3
weight strength:	75 %
detonation velocity, unconfined:	7250 m/sec.

Semiconductor Bridge (SCB) Igniter

A solid-state device that when driven with a low-energy current pulse, produces a plasma discharge that ignites energetic materials pressed against the SCB. A micro convective process transfers the plasma's energy into the explosive causing ignition in a few tens of microseconds. The SCB consists of a small polysilicon volume formed on a silicon substrate; typical bridge dimensions are 100 μm long by

360 μm wide by 2 μm thick. While components using the SCB igniter function at one-tenth the energy of conventional bridgewire devices, they have been shown to be explosively safe meeting both no-fire and electrostatic discharge (ESD) requirements. SCB chips are processed using standard semiconductor fabrication techniques; consequently, circuitry on the same die can be incorporated to provide a smart igniter that can be used in architectures to sense environments, provide precise delays, or respond to coded signals, to name a few possibilities. SCB devices can be used for commercial, military, and government applications that range from sophisticated intelligent devices to rock blasting. R. W. Bickes, Jr. and A. C. Schwarz at Sandia National Laboratories patented the SCB in 1987; Sandia patents for other semiconductor bridges and devices have also been issued.

Semigelatin Dynamites

Semigelatin dynamites are so named because of their consistency. These so-called semigelatins contain ammonium nitrate and wood meal as their main components, and also 10–14 % of a weakly gelatinized nitroglycerine.

Semtex

Trade name of a plastic explosive (→ *Plastic Explosives*) from the Czech firm Explosia, Pardubice-Semtin.

Semtex consists of → *Pentaerythritol Tetranitrate* and styrene-butadiene copolymer as a plasticizer.

detonation rate:	5000 m/s
oxygen balance:	−44.0 %
critical diameter:	15 mm

Sensitivity

Empfindlichkeit; sensibilité

The sensitivity of an explosive to heat, mechanical stress, shock, impact, friction impact and detonation impact (initiability) determine its handling safety and its application potential.

All explosives are, intrinsically, sensitive to impact and shock. The introduction of additives – oil, paraffin etc. – may diminish the sensitivity to mechanical stress.

Testing methods ensuring uniform evaluation were developed accordingly. Some of them are included in the Railroad Traffic Regulations

(→ *RID*), since certain sensitivity limits are clearly specified by law for the transportation of explosives within individual countries, as well as for international traffic. For details: → *Friction Sensitivity*; → *Impact Sensitivity* → *Heat Sensitivity*.

For the behavior or explosives at elevated temperatures, → *Stability*.

Series Electric Blasting Cap Circuit

Zündkreis in Serienschaltung; circuit de détonateurs, électrique couplés en série.

An electric blasting cap circuit that provides for one continuos path for the current through all caps in the circuit.

Series in Parallel Electric Blasting Cap Circuit

Zündkreis in Serien-Parallel-Schaltung; circuit combiné paralléle et en série

A combination of series and parallel where several series of caps are placed in parallel.

Also → *Parallel Connection*.

Setback

Rückstoß; recul

The relative rearward movement of component parts in a projectile, missile, or fuze undergoing forward acceleration during launching.

set forward

Relative forward movement of component parts which occurs in a projectile, missile, or bomb in flight when impact occurs. The effect is caused inertia and is opposite in direction to setback.

Shaped Charges

Hollow Charges; Hohlladung; charge creuse
(→ also *Munroe Effect*)

A shaped charge is an explosive charge with a hollow space facing the target.

A rotationally symmetric shaped charge is an explosive charge with an axis of symmetry which acts preferentially in the direction of the rotational axis. Shaped charges lined in a rotationally symmetric manner can pierce steel sheets eight times as thick as the diameter of the charge.

Liners for shaped charges are made of inert material, usually a metal. The lining acts as an energy carrier, since the energy of the explosive charge is concentrated on a small cross-section of the target.

The detonation of the explosive charge causes the lining material to collapse and to converge in the axis of symmetry of the charge. During this process the colliding metal mass separates into a large mass portion moving slowly and a smaller mass portion moving forward at very high speed. Only the fast moving portion with its high kinetic energy produces the perforation effect in the target: It forms a jet of very small diameter and correspondingly high density of energy. The slow moving portion is left as a conglomerated molten slug after detonation.

The main parameters to characterise a lined shaped charge are the detonation velocity, the density of the explosive, the geometry of the detonation wave, the shape of the lining, the lining material and its wall thickness.

Cutting charge

A plane-symmetrical shaped charge (cutting charge) is an explosive charge with a hollow space, which acts longitudinally in the plane of symmetry ("roof-shaped" charges).

Plane charge

In a plane charge the opening angle of the conical liner is larger than 100°. When the explosive is detonated, the lining no longer converges into a jet in the axis of symmetry, so that no jet or slug can be formed out of the collapse point; rather, the lining is turned inside out. The resulting sting is much thicker and much shorter with a weaker perforating power, but a larger perforation diameter than that made by a shaped charge.

Projectile-forming charge; self-forging fragment;
EFP (explosively formed projectile)

In a projectile-forming charge, the geometry of the lining is such that all its elements have approximately the same velocity. The strength of the material is chosen so that it can easily absorb the residual differences in the velocities. In this way a projectile with a greater kinetic energy is obtained, which consist, roughly speaking, of the entire mass of the lining, and which can also be employed against distant targets.

The shaped-charge effect was first described in 1883. Shortly before the Second World War, *Thomanek* found that the piercing power of the shaped charge could be significantly increased by lining the hollow space.

The first theoretical treatment of the subject was by *Trinks* in 1943/44 in a report submitted by the Research Department of the German Army Weapons Command.

The first non-classified study on the subject was that by *Birkhoff Mac Dougall, Pugh* and *Taylor*: Explosives with Lined Cavities, J. Appl. Physics. Vol. 19, p. 563 (1948).

For the first non-classified study concerning an interpretation of jet extension and the attendant increase in the duration of the effect, see *Pugh, Eichelberger* and *Rostoker*: Theory of Jet Formation by Charges with Lined Conical Cavities. J. Appl. Physics, Vol. 23, pp. 532–536 (1952).

Sheathed Explosives

ummantelte Sprengstoffe; explosifs gainés

Permitted explosives which are enveloped in a special cooling "sheath"; they are now obsolete.

High-safety explosives, such as sheathed explosives, whose structure is nevertheless homogeneous, are known as "explosives equivalent to sheathed" (Eq. S.); → *Permissibles*.

Shelf Life

Lebensdauer; durée de vie

The length of time of storage during which an explosive material, generator, rocket motor, or component retains adequate performance characteristics under specified environmental conditions.

Shock Wave

Stoßwelle; onde de choc

Intense compression wave produced by detonation of explosive. → *Detonation, Shock Wave Theory.*

Shot Anchor

A device that anchors explosive charges in the borehole so that the charge will not be blown out by the detonation of other charges.

Shot Firer

Sprengmeister; boutefeu

That qualified person in charge of and responsible for the loading and firing of a blast (same as a → *Blaster*).

Shunt

A short-circuiting device provided on the free ends of the leg wires of electric blasting caps to protect them from accidental initiation by extraneous electricity.

Silver Acetylide

silver carbide; Silberkarbid; Acetylensilber; acétylure dargent

C_2Ag_2

molecular weight: 239.8
oxygen balance: −26.7 %
deflagration point: 200 °C = 392 °F

Silver carbide is very sensitive to impact. It is prepared by bubbling acetylene through a slightly acidic or neutral silver nitrate solution.

Silver Azide

Silberazid; azoture dargent

AgN_3

molecular weight: 149.9
nitrogen content: 28.03 %
volume of detonation gases: 224 l/kg
density: 5.1 g/cm^3
melting point: 251 °C = 484 °F
lead block test: 115 cm^3/10 g
deflagration point: 273 °C = 523 °F

Silver azide is sensitive to light, insoluble in water, and soluble in ammonia, from which it can be recrystallized. It is prepared from sodium azide and solutions of silver salts (depending on the working conditions) as a cheesy, amorphous precipitate.

It gives a very satisfactory initiating effect which is superior to that of lead azide. Nevertheless, its practical use is limited, because of its high sensitivity to friction, and because its particular texture makes the dosing difficult.

Silvered Vessel Test

In this testing procedure the propellant sample (about 50 g) is heated in an insulating *Dewar* vessel, and the rise in temperature produced by the heat of decomposition of the powder is determined. The powder sample is heated to 80 °C (176 °F); the time is determined in which the powder reaches 82 °C (180 °F) by its own heat development on decomposition.

Frey's variant of the silvered vessel test has been in use in the Germany. In its variant, different amounts of heat are supplied to the electric heating elements mounted inside the Dewar flask, and the temperature differences between the interior of the Dewar vessel and the furnace are measured by thermocouples. A calibration curve is plotted from the values thus obtained, and the heat of decomposition of the propellant is read off the curve. In this way, the decomposition temperature at a constant storage temperature can be determined as a function of the storage time, and the heat of decomposition of the propellants can thus be compared with each other. If the measurements are performed at different storage temperatures, the temperature coefficient of the decomposition rate can be calculated. (→ also *Differential Thermal Analysis*.)

Silver Fulminate

Knallsilber; fulminate d'argent

$CNOAg$

white, crystalline powder
molecular weight: 149.9
oxygen balance: −10.7 %
nitrogen content: 9.34 %

Silver fulminate is prepared by the reaction employed in the preparation of → *Mercury Fulminate*, i.e., by reacting a solution of silver in nitric acid with alcohol. Like mercury fulminate, it is also toxic.

Silver fulminate is much more sensitive than mercury fulminate. Since its detonation development distance is very short, its initiation effect is superior to that of mercury fulminate, but the compound is too sensitive to be used commercially.

An altogether different product, known as *Berthollet's* detonating silver (which is not a fulminate), is obtained when a solution of freshly precipitated silver oxide in ammonia is allowed to evaporate. Its probable formula is Ag_3N. It is highly sensitive and explodes even during the evaporation of the ammoniacal solution.

Single Base Powders

Nitrocellulose-Pulver; poudre à simple base

Such powders mainly consist of nitrocellulose and stabilizers as well as other additives such as dinitrotoluene in some formulations. Nitrocellulose is gelatinized with the aid of solvents, mostly etheralcohol mixtures, and additives are incorporated and gelatinized by prolonged kneading. The mixture is shaped into tubes, perforated tubes, flakes, etc., by extrusion and cutting, and the solvents are removed by evaporation, displacement by warm water, vacuum drying, etc., and the material is surface-treated. The purpose of the surface treatment is to let phlegmatization agents diffuse into the material, thus retarding the combustion rate in the surface layers and to attain a progressive burning rate (→ *Progressive Burning Powders*).

SINCO® Ignition booster and gas mixture for motor vehicle safety

SINCO® was developed by the Dynamit Nobel GmbH Company as an environmentally compatible and particularly stable class of substances for rapid gas evolution. It involves a pyrotechnic gas mixture based on nitrogen-rich fuels and oxygen carriers as reactants. A solid combustion residue consisting essentially of alkali carbonates, non-toxic gas products consisting of nitrogen, water vapour, carbon dioxide and oxygen together with heat are liberated during the reaction between fuels and the oxygen carriers.

The gas mixtures themselves are free from heavy metals and have high toxicological compatibility. In the acute oral toxicity test carried out according to the EU Directive, the LD_{50} value was greater than 2500 mg/kg.

In addition the pyrotechnic mixtures are characterised by high thermal stability. This is also necessary in order to guarantee a constant reaction characteristic over a long period of time and even after thermal stressing.

Stable reaction of the mixtures is possible only with tamping. This property reduces the potential risk that may occur in the event of improper handling or possible misuse.

Because of its properties, SINCO® is suitable for personal protection in passive safety systems in motor vehicles. In addition to the use of SINCO® in pressure elements for seat-belt tensioners or lock tensioners, the gas mixture is also suitable for driver and passenger gas generators. In this case the mixture also performs the function of a booster charge in the igniter elements of the gas generators in addition to the main task of gas evolution.

The proportion of solids formed, which can be controlled via the composition of the mixture, promotes the process of tablet ignition in the gas generator combustion chamber.

SINOXID Primer Composition

SINOXID is the trademark used for the traditional primer compositions of DYNAMIT NOBEL AG. The formulation was developed by Rathsburg and Herz and patented as tetracenetricinate primer composition in 1928.

The term SINOXID is made up of "sine" and "oxide" and means "without rust". It underlines the fact that this composition is not susceptible to corrosion as against mercury fulminante or potassium chlorate mixtures. SINOXID compositions consist of the following components: Lead tricinate, → *tetracene*, → *Barium Nitrate*, lead dioxide, antimony trisulfide and calcium silicide. These components meet all requirements currently applied in ammunition technology. SINOXID compositions feature very good chemical stability and storage life, they are abrasion-, erosion- and corrosion-free and ignite propellants with precision.

SINTOX Primer Composition

SINTOX is the international registered trademark for newly developed primer compositions of DYNAMIT NOBEL AG. They are required if the ambient air in closed firing ranges must not be polluted with combustion products containing lead, antimony or barium.

→ *Diazodinitrophenol* or the newly developed strontium diazodinitroresorcinate are used as initial explosives. Special types of → *Zinc Peroxide* are used as oxidizers. Additionally, the primer compositions may contain substances like titanium. → *Tetracene* may also be required as a sensitizer.

For the primer composition, the residual content of lead, barium or antimony compounds is smaller than 0.01 %. Zinc is emitted as non-toxic zinc oxide.

In terms of corrosion and erosion, SINTOX primer compositions behave like → *SINOXID Primer Compositions*. There is no adverse effect on hit accuracy.

Skid Test

This test is intend to simulate a bare explosive charge accidently hitting a rigid surface at an oblique angle during handling. An uncased hemispherical charge, 254 mm in diameter, is dropped in free fall onto a rigid target. In a second version, the charge swings down in a harness on the end of a cable and strikes the target at a pre-determined angle*).

Slurries

Sprengschlamm; bouillies

Slurries consist of saturated aqueous solutions of ammonium nitrate (saturated ammonium nitrate solution at 20 °C = 68 °F contains about 65% NH_4NO_3) and other nitrates, which also contain additional amounts of undissolved nitrates in suspension and fuels which take up the excess oxygen of the nitrate; the structure of the nitrate solution can be significantly effected by added thickeners (e.g., → *Guar Gum*) and cross linking agents (e.g., borax). The most important fuel is aluminum powder; Water soluble fuels such as glycol can also be employed; the nitrates may also include nitrates of organic amines, e.g. → *Methylamine Nitrate* (MAN).

Slurries may contain sensitizing additives (e.g. TNT; PETN, etc.); sensitization can also be achieved by introducing finely dispersed air bubbles, e.g. by introducing air-filled → *Microballoons* (in this form the bubbles will not be compressed by static pressure). Sensitized slurries can be cap-sensitive and may detonate even when the diameter of the bore-hole is small; → *Emulsion Slurries*. Sensitized explosive slurries in the form of cartridges can be utilized in boreholes of conventional and large diameters. In addition, explosive slurries may be pumped into large boreholes.

Addition of rock salt, which reduces the detonation temperature (→ *Permissibles*), may impart a certain degree of safety against firedamp to the slurry explosives.

Slurrit

Trade name for slurry *blasting agents* distributed in Norway by DYNO; → *Slurries* Slurrit 5 is cartridged in large-hole dimensions. Slurrit 110 and 310 are mixed on site and pumped into the borehole by special trucks. Non cap sensitive, best initiation by cast booster.

density: 1.25 g/cm^3

---

* Gibbs & Popolato, LASL Explosive property Data, University of California Press, 1980.

Small Arms Ammunition Primers

Anzündhütchen; amorces

Small percussion-sensitive explosive charges, encased in a cap and used to ignite propellant powder (→ *Percussion Cap*).

Snakehole

Sohlenbohrloch; trou de fond

A borehole drilled in a slightly downward direction from the horizontal into the floor of a quarry. Also, a hole driven under a boulder.

Sodatol

A pourable 50:50 mixture of sodium nitrate with → TNT

Sodium Chlorate

Natriumchlorat; chlorate de sodium

$NaClO_3$

molecular weight: 106.4
oxygen balance: +45.1%
density: 2.48 g/cm^3
melting point: 248 °C = 478 °F

Sodium chlorate, though containing more oxygen than potassium chlorate, has the disadvantage of being more hygroscopic. Like all other chlorates, it must not be used in contact with ammonium salts and ammonium nitrate explosives.

Its practical significance in explosives is very limited.

Sodium Nitrate

Natronsalpeter, Natriumnitrat; nitrate de soidum; SN

$NaNO_3$

colorless crystals
molecular weight: 85.0
energy of formation: −1301 kcal/kg −5443 kJ/kg
enthalpy of formation: −1315 kcal/kg −5503 kJ/kg
oxygen balance: +47.1%
nitrogen content: 16.48%
density: 2.265 g/cm^3
melting point: 317 °C = 603 °F

The salt is hygroscopic, readily soluble in water, less so in ethanol, methanol, and glycerin. It is used in industrial explosives and in 8-black blasting powder as an oxidizer.

Specifications

net content (by nitrogen determination in Lunges nitrometer): not below	98.5 %
moisture: not more than	0.2 %
insolubles in water: not more than	0.05 %
NH_4-, Fe-, Al-, Ca-, Mg- and K-salts:	none
NaCl: not more than	0.02 %
Na_2SO_4: not more than	0.2 %
reaction:	neutral
Abel test 80 °C (176 °F): not under	30 min

Sodium Perchlorate

Natriumperchlorat; perchlorate de sodium

$NaClO_4$

colorless crystals
molecular weight: 122.4
oxygen balance: +52.3 %
density: 2.5 g/cm^3
melting point: 482 °C = 900 °F (anhydrous product)

Sodium perchlorate is hygroscopic and is readily soluble in water and alcohol. Standard product contains 1 mole of crystal water.

Soil Grain Powder

Weichkornpulver; poudre à grain souple

Type of → *Black Powder* for firework manufacture. Soft grain powder is not compacted in hydraulic presses.

Solid Propellant Rockets

Feststoff-Raketen; roquettes à propergol solide

These rockets operate on homogeneous solid propellants. Following ignition, the propellant charge burns, and it is not possible to interrupt or to control the combustion process (for certain possibilities in this respect → *Hybrids*). The course and the rate of the combustion process may be modified by suitable shaping of the charge (front of cigarette burner, internal burner, all-side burner, and charges with special configurations), by varying its composition and grain size, and by incorporating special accelerating or retarding additives. The propellant charge must be carefully examined for cracks, since in their presence the combustion will not proceed uniformly. If case-bonded charges are employed, adequate cohesion between the wall of the combustion chamber and the propellant charge (→ *Case Bonding*) must be ensured.

The advantages of solid rockets are the short time needed for the actuation, long storage life, and a simple design.

The burning process in the rocket motor is influenced by:

the thermodynamic performance values of the propellant (→ *Thermodynamic Calculation of Decomposition Reactions*), the burning characteristics of the propellant grain (→ *Burning Rate*), depending on its shape, and by the pressure influence on the burning rate. The pressure exponent can be zero in the case of modern propellants ("Plateau", "Mesa", → *Burning Rate*). The pressure function of the burning rate cannot be described by a universal equation, but within smaller pressure ranges the equation of Saint-Robert or *Vieille* equation is applicable:

$$r = a\, p^{\alpha} \tag{1}$$

r: rate of burning normal to the burning surface
p: pressure
α: pressure exponent
a: constant

(→ also *Burning Rate, Charbonnier equation*.) At any time during the reaction, equilibrium must exist between the gas produced

$$r \cdot f_T \cdot \rho \tag{2}$$

f_T: burning surface
ρ: density of propellant

and the gas discharged through the nozzle

$$p \cdot f_m \cdot C_D \tag{3}$$

f_m: nozzle cross section
C_D: mass flow coefficient

The ratio f_m/f_T = nozzle cross section to the burning surface at any time is called → *Propellant Area Ratio* ("Klemmung") *K*; the equations (2) and (3) are considered to be equal: equation (1) can be rearranged to

$$p = \frac{a}{C_D}\, \rho K^{\frac{1}{1-\alpha}} \tag{4}$$

Equation (4) allows plotting of the gas pressure-time diagram, if a, C_D and α are known and the course of the propellant area ratio *K* with the burning time can be assumed. Modification of the pressuretime diagram may be caused by the pressure falling off along the propellant grain (*Bernoulli's* equation), by → *Erosive Burning*, by the igniter system, and by irregular combustion of the remaining propellant. The thrust-time diagram (→ *Thrust, Thrust Determination*) can be derived from the pressure-time diagram obtained from equation (4).

Spacing

Bohrlochabstand; distance entre trous

The distance between boreholes measured parallel to the free face toward which rock is expected to move.

Spark Detonators

Spaltzünder; amorce électrique à l'étincelle

Spark detonators, like bridgewire detonators, were employed in the past to produce electric initiation of explosive charges. The priming charge itself, containing current conducting additives, served as a current conductor through the priming pill itself. Relatively high voltages were required to produce the ignition, so that such devices were safe from stray currents.

Spark detonators have now been substituted by → *Bridgewire Detonators*. If there is danger of stray currents, low-sensitivity primer types are employed, which can be actuated only by a strong current pulse.

Special Effects

These are special arrangements for the simulation of dangerous events in the military, western, and science fiction scene in motion pictures and television programmes.

Frequently, specially designed fireworks are used for creating these effects, e.g. tiny detonators (→ *Bullet Hit Squib*) for the simulation of bullet impacts.*)

Specific Energy

spezifische Energie; force

The specific energy of an explosive is defined as its working performance per kg, calculated theoretically from the equation of state for ideal gases:

---

* Clark, Frank F., Special Effects in Motion Pictures, Society of Motion Picture and Television Engineers, Inc., 862 Scarsdale Avenue, Scarsdale, N.Y. 10583, Second Printing 1979.

$f = pV = nRT$

where p is the pressure, V is the volume, n is the number of moles of the explosion gases per kg (*also Volume of Explosive Gases*), R is the ideal gas constant, and T is the absolute temperature of the explosion. If we put the volume equal to unity, i.e., if the loading density is unity, the specific energy becomes

$f = p$

i.e., is equal to the pressure which would be exerted by the compressed explosion gases in their confinement, if the latter were indestructible. This is why the term "specific pressure" is also frequently used, and why the magnitude f is often quoted in atmospheres.

Nevertheless, strictly speaking, f is an energy per unit mass and for this reason is reported in meter-tons per kg. The value of T will have this dimension if R is taken as $0.8479 \cdot 10^{-3}$ mt·K·mol.

In accordance with recent standardisation regulations, the energy data are also reported in joules. For more details on the calculation → *Thermodynamic Calculation* of *Decomposition Reactions*; also → *Strength*.

Specific Impulse

spezifischer Impuls; impulse spécifique

The specific impulse of a propellant or a pair of reacting liquids in rocket motors is the most important characteristics of the performance. It is the → *Thrust*×time (*i.e.*, the impulse) per unit weight of propellant:

$$I_s = \frac{F \cdot t}{W}$$

I_s specific impulse
F: thrust
T: time
W: weight of propellant

It is measured in kilopond-seconds (or Newton-seconds) per kilogram of propellant*). It can be evaluated by the equation

$I_s = \sqrt{2J(H_c - H_e)}$ N s/kg

J: mechanical heat equivalent

---

* Since the numerical values of kp and kg are the same, the apparent dimension of the specific impulse is simply the second. For this reason, all impulse data can be directly compared with each other, even if other unit systems are employed.

H_c: enthalpy of the reaction products in the rocket chamber (at chamber pressure and chamber temperature)   Unit: kcal/kg

H_e: enthalpy of the reaction products at the nozzle exit.   kcal/kg

The equation can be solved with the aid of computers, considering various equilibrium reactions; → *Thermodynamic Calculation of Decomposition Reactions*.

The relation of the specific impulse to the temperature of the reaction gas in the rocket chamber is

$$I_s = k_1 \sqrt{T_c \cdot N} = k_2 \sqrt{\frac{T_c}{M}}$$

T_c: flame temperature in the chamber
N: number of moles per weight unit
M: average molecular weight of the flame gases
$k_1; k_2$:   constants

The value for the specific impulse is high if the reaction heat is high and produces a high flame temperature, and if the average molecular weight of the reaction products is low. Data concerning specific impulses are only comparable if they refer to the same working pressure in the combustion chamber; a frequently employed standard value is 1000 lbs/sq. in. approx. 70 bar in test chambers.

For more details on this subject, see: *Barrére, Jaumotte, Fraeijs de Veubeke Vandenkerckhove*: Raketenantriebe. Elsevier, Amsterdam, 1961. Also: *Dadieu, Damm, Schmidt*: Raketentreibstoffe Springer, Wien 1968; also → *Gas Jet Velocity* and → *Thermodynamic Calculation of Decomposition Reactions*.

Springing

Vorkesseln; agrandissement par explosion

The practice of enlarging the bottom of a blast hole by the use of a relatively small charge of explosive material. Typically used in order that a larger charge of explosive material can be loaded in a subsequent blast in the same borehole.

Squib

Anzünder; allumeur (électrique)

Used in a general sense to mean any of the various small size pyrotechnic or explosive devices. Specifically, a small explosive device, similar in appearance to a detonator, but loaded with low explosive, so that its output is primarily heat (flash). Usually electrically initiated and provided to initiate the action of pyrotechnic devices and rocket propellants. An *electric squib* is a tube containing a flammable material and a-mall charge of powder compressed around a fine resistance wire connected to electrical leads or terminal (→ *Initiator,* → *Bullet Hit Squib*).

Stability

Stabilität; stabilité

A distinction must be made between chemical and physical stability. While physical stability is important, particularly in the evaluation of solid propellants, the chemical stability is of prime importance in the estimation of the course of decomposition of nitrate esters. The nitrate esters which are processed for use as propellants – unlike nitro compounds, which are relatively stable under these conditions – undergo a steady decomposition, which is due to imperfect purification of the starting materials and to the effect of other parameters such as temperature and air humidity. The rate of this decomposition is autocatalyzed by the acidic decomposition products and may in certain cases produce spontaneous ignition. In order to reduce the decomposition rate as much as possible, suitable stabilizers are added to the powders, which are capable of accepting the acid cleavage products with formation of the corresponding nitro compounds (→ *Stabilizers*). The stability is controlled by means of several tests (→ *Hot Storage Tests*).

A distinction must be made between tests of short duration, in which the possible decomposition reactions are accelerated by a considerable rise in temperature, and the so-called service-life tests or surveillance tests, which take place over several months and may sometimes take more than a year. Short-duration tests alone do not suffice for a reliable estimate of the stability, at least where imperfectly known products are concerned.

An estimation of the probable → *Shelf Life* of aged propellants can be made by chromatography*). If e.g. diphenylamine is used as a stabilizer, the transformation into the nitro derivatives up to hexanitrodiphenylamine can be analysed; if this stage is reached, decomposition of the powder occurs.

---

* Volk, F., Determination of the Shelf Life of Solid Propellants, Propellants and Explosives **1**, 59–65 (1976).
  Volk, F., Determination of the Lifetime of Gun-propellants using Thin layer Chromatography, Propellants and Explosives **1**, 90–97 (1976).

Stabilizers

Stabilisatoren; stabilisateurs

Stabilizers are generally defined as compounds which, when added in small amounts to other chemical compounds or mixtures, impart stability to the latter.

In propellant chemistry, especially so in the case of nitrocellulose-containing powders, the stabilizers employed are compounds which, owing to their chemical structure, prevent the acid-catalyzed decomposition of nitrocellulose, nitroglycerine, and similar nitrate esters.

They exert their stabilizing effect by binding the decomposition products, such as the free acid and nitrous gases; the stabilizers themselves are converted into relatively stable compounds at the same time. Neither stabilizers nor their secondary products should give a chemical reaction (saponification) with nitroglycerine or nitrocellulose. Compounds used as stabilizers are mostly substitution products of urea and diphenylamine. Readily oxidizable compounds – higher alcohols, camphor, unsaturated hydrocarbons (vaselines) – may also be employed. For such compounds to be effective, their homogeneous incorporation into the powder must be easy, they must not be too volatile, and must not be leached out by water. Many stabilizers also display plasticizing (gelatinizing) properties; accordingly, they have both a stabilizing effect and – in the manufacture of powders – a gelatinizing (softening) effect.

Pure stabilizers include:
- diphenylamine
- Akardite I (*asym*-diphenylurea)

Stabilizers with a gelatinizing effect include:
- Centralite I (*sym*-diethyldiphenylurea)
- Centralite II (*sym*-dimethyldiphenylurea)
- Centralite III (methylethyldiphenylurea)
- Akardite II (methyldiphenylurea)
- Akardite III (ethyldiphenylurea)

Substituted urethanes:
- ethyl- and methylphenylurethanes
- diphenylurethane

Pure gelatinizers, without a stabilizing effect, include:
- dibutyl phthalate
- diamyl phthalate
- camphor

For formulas and properties see appropriate keywords.

Stemming

Besatz; bourrage

In mining, "stemming" refers to the inert material used to plug up a borehole into which the explosive charge has been loaded. The "classical" stemming materials are mud or clay noodles. Stemming brings about more economical utilization of the explosive charge, provided the explosive columns employed are short and the detonation is effected at the mouth of the borehole. Stemming is mandatory if there is any danger of firedamp. The strongest stemming is not necessarily the best; if the stemming is too strong, deflagration may take place. In coal mining water stemming cartridges proved to be the best; they are plastic tubes filled with water or water gel and closed at both ends, which are easily inserted into the borehole, do not stem too strongly, and make a significant contribution to the settling of dust and fumes.
→ also *Confinement*.

Storage

Lagerung; magasinage

The safe keeping of explosive materials, usually in specially designed structures called *Magazines*.

Stray Current Protection

Streustromsicherheit; protection contre les courants vagabonds

The increasingly large consumption of electric current has resulted in intensified stray currents. The stray current safety of an electric primer is the maximum current intensity at which the glowing wire just fails to attain the ignition temperature of the charge in the primer. To improve protection against stray currents, the "A" bridgewire detonators, which were formerly used in Germany have now been replaced by the less sensitive "U" detonators (→ *Bridgewire Detonators*).

Strength

Arbeitsvermögen; force

Also → *Bulk Strength, Weight Strength.*

The performance potential of an explosive cannot be described by a single parameter. It is determined by the amount of gas liberated per unit weight, the energy evolved in the process (→ *Heat of Explosion*), and by the propagation rate of the explosive (detonation velocity → *Detonation*). If an explosive is to be detonated in a borehole, the relevant parameter is its "strength"; here the criterion of the performance is not so much a high detonation rate as a high gas yield and a high heat of explosion. If, on the other hand, a strong disintegration effect in the nearest vicinity of the detonation is required, the most important parameters are the detonation rate and the density (→ *Brisance*).

A number of conventional tests and calculation methods exist for determining the comparative performance of different explosives. The determinations of the detonation rate and density require no conventions, since they are both specific physical parameter.

Lead block test and ballistic mortar test

Practical tests for comparative strength determination are the lead block test and the declination of a ballistic mortar. In both cases relatively small amounts of the explosive (of the order of 10 g) are initiated by a blasting cap. In the lead block test, the magnitude measured is the volume of the pear-shaped bulge made in the block borehole by the sample introduced into it; in the ballistic mortar test the magnitude which is measured is the deflection angle; this angle is taken as a measure of the recoil force of a heavy steel weight suspended as a pendulum bob, after the exploding cartridge has fired a steel projectile out of a hole made in the bob. The performance of the explosive being tested is reported as the percentage of that of → *Blasting Gelatin*, which is conventionally taken as 100 % (For further details → *Ballistic Mortar*). In both cases the explosive is enclosed in a confined space, so that, for all practical purposes, the parameter measured is the work of decomposition of an explosive in a borehole. The disadvantage of both methods is that the quantity of the sample used in the test (exactly or approximately 10 g) is quite small, and for this reason accurate comparative data can be obtained only with more sensitive explosives; less sensitive materials require a longer detonation development distance (→ *Detonation*), within which a considerable proportion of the 10-g sample does not fully react. Practical methods for determining the performance of explosives requiring much larger samples (up to 500 g) include the following.

Jumping mortar test

Two halves with finely ground surfaces fitting exactly onto one another form a mortar with a borehole. One of the halves is embedded in the ground at a 45° angle, while the other half is projected like a shot, when the explosive charge is detonated in the hole; the distance to which it has been thrown is then determined. A disadvantage of the method is that when high-brisance explosives are tested, the surfaces must be reground after each shot. The method gives excellent results with weaker → *Permitted Explosives*.

Vessel mortar test

This test is also based on the determination of the range distance of a heavy projectile. The explosive is suspended in a thick-walled vessel, and an accurately fitting cap of the vessels is projected. This apparatus is stronger, and the weight of the charge may be made as large as 500 g.

Large lead block test

The device consists of a lead block with linear dimensions three times as large as normal. The block has been used to obtain information about slurries; the method is too expensive for practical work, since more than one ton of lead must be cast for each shot.

The crater method

This method is based on the comparison of the sizes (volumes) of the funnels produced in the ground by underground explosions. It is used for explosives with a large critical diameter only if no other method is available, since it is inaccurate and the scatter is large.

→ *Aquarium Test*

The sample is exploded under water (in a natural water reservoir or in a man-made pool), and the pressure of the resulting impact wave is measured with the aid of lead or copper membranes.

→ *Specific Energy*

For calculations of performance parameters of explosives → *Thermodynamic Calculation* of *Decomposition Reactions*. As far as the strength of propellants and explosives is concerned, the most relevant thermodynamically calculable parameter is the → *Specific Energy*. This is the amount of energy which is released when the gases in the body of the explosive (assumed to be compressed in their initial state) are allowed to expand at the explosion temperature while performing useful work. In order to illustrate the working performance obtainable from explosive materials, this magnitude is conventionally reported in meter-tons per kilogram; in this book, it is also given in joules (J).

The calculated values of the specific energy agree well with the performance data obtained by conventional tests. This is particularly true of the tests in which larger samples are employed, but the apparatus required for such tests is not always available, and the tests themselves are relatively expensive.

The following empirical formula relating the specific energy to the relative weight strength is valid in most cases:

weight strength (%) = $0.0746 \times$ spec. energy (in mt/kg)

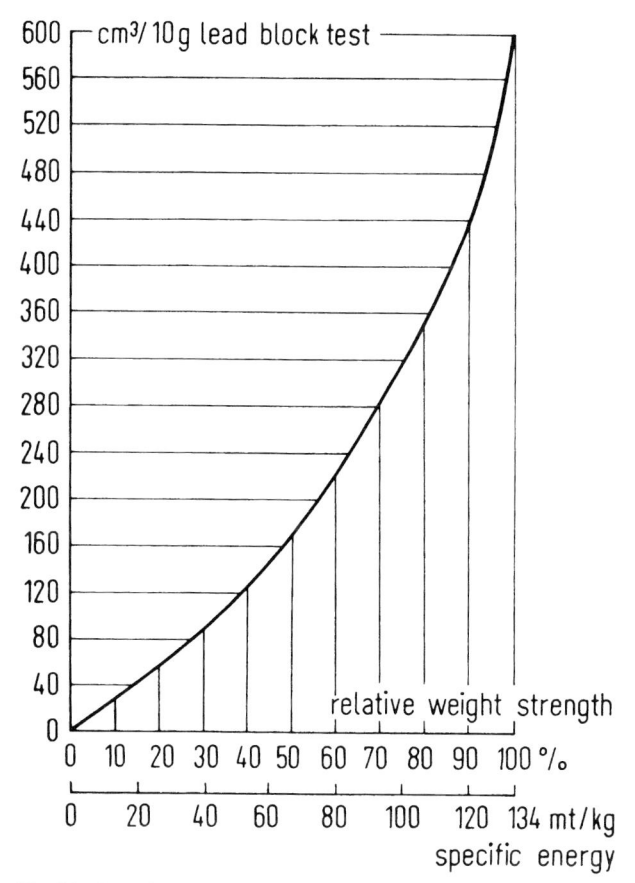

Fig. 22. Specific energy and relative weight strength in relation to lead block test values.

The relationship between the size of the lead block excavation and the specific energy is not linear. The true relationship may be seen in Fig. 22 (representation of experimental results).

The relationship between weight strength and the coefficient d'utilisation pratique (c.u.p) used in France (→ *Lead Block Test*) can be given by the empirical formula weight strength (%) = 0.645×(%) c.u.p. and (%) c.u.p. = 1.55×(%) weight strength.

Strontium Nitrate

Strontiumnitrat; nitrate de strontium

$Sr(NO_3)_2$

colorless crystals
molecular weight: 211.7
energy of formation: −968.3 kJ/mole
enthalpy of formation: −4622 kJ/kg
oxygen balance: +37.8 %
nitrogen content: 13.23 %
density: 2,99 g/cm^3
melting point: 570 °C

Strontium nitrate is used in pyrotechnics as a flame-coloring oxidizer for red-colored fireworks. Recently, strontium nitrate is used as oxidizer in gas generating propellants; e.g. for airbacks.

Styphnic Acid

trinitroresorcinol; 2,4,6-trinitro-1,3-dihydroxybenzene; Trinitroresorcin; trinitrorésorcinol; acide styphnique; Trizin; TNR

yellow-brown to red-brown crystals
empirical formula: $C_6H_3N_3O_8$
molecular weight: 245.1
energy of formation: −493.1 kcal/kg = −2063.1 kJ/kg
enthalpy of formation: −510.0 kcal/kg = −2133.8 kJ/kg
oxygen balance: −35.9 %
nitrogen content: 17.15 %
volume of explosion gases: 814 l/kg
heat of explosion
   (H_2O liq.): 706 kcal/kg = 2952 kJ/kg
   (H_2O gas): 679 kcal/kg = 2843 kJ/kg
specific energy: 89 mt/kg = 874 kJ/kg
density: 1.83 g/cm^3
melting point: 176 °C = 349 °F
lead block test: 284 cm^3/10 g
deflagration point: 223 °C = 433 °F
impact sensitivity: 0.75 kp m = 7.4 N m
friction sensitivity: at 36 kg = 353 N
   pistil load no reaction
critical diameter of steel sleeve test: 14 mm

Styphnic acid is prepared by dissolving resorcinol in concentrated sulfuric acid and nitrating the resulting solution with concentrated nitric acid. It is a relatively weak explosive. Its lead salt (→ *Lead Styphnate*) is used as an initiating explosive.

Substainer Charge

Component (optional) of ignition system (train) that maintaines operating pressure until thermal equilibrium is obtained.

Sulfur

Schwefel; soufre

S

atomic weight: 32.07
melting point: 113 °C = 235 °F
boiling point: 445 °C = 833 °F
density: 2.07 g/cm^3

Sulfur is used with charcoal as a fuel component in → *Black Powder*. Sublimated sulfur is not completely soluble in carbon sulfide and contains traces of sulfuric acid; the use of sublimated sulfur for black powder production is therefore not permitted.

Supercord 40 and Supercord 100

Trade names of → *Detonating Cords* containing 40 and 100 g PETN/m distributed in Germany and exported by Orica Germany GmbH. It is covered with red-colored plastic. It is used for the initiations of ANFO blasting agents and for → *Contour Blasting*.

Surface Treatment

Oberflächenbehandlung; traitement de surface

When gunpowder burns in the chamber of a weapon, the internal-ballistic energy of the powder charge is best exploited if the gas pressure is kept constant almost up to the emergence of the projectile from the barrel, despite the fact that the gas volume keeps growing larger during that period, owing to the movement of the projectile. It follows that gas evolution from the powder charge should be slow at first, while towards the end of the combustion process, it must be quicker ("progressive burning"). This is achieved mainly by imparting a suitable shape to the powder granule (in a seven-hole powder, the surface area increases during combustion, and the combustion is therefore progressive); progressive combustion is also enhanced to a considerable extent by surface treatment, i.e., by allowing phlegmatizing, combustionretarding substances (such as Centralit, dibutyl phthalate, camphor, dinitrotoluene, etc.) to soak into the powder. A careful surface treatment is an excellent way of keeping the maximum pressure peak of the combustion curve low.

Susan Test*)

The Susan Sensitivity Test is a projectile impact test. The explosive to be tested is loaded into a projectile shown in Fig. 23 and thrown against a steel target. The reaction on impact is recorded by measuring the shock wave pressure by a gauge 10 ft (3.1 m) away. The percentage of the reaction (0 = no reaction; 40 = fast burning reactions; 70 = fully reacted TNT; 100 = detonation) is plotted against the projectile velocity (0 to 1600 ft/s = 488 m/s). → *Plastic Explosives* with rubberlike binders give better results than cast RDX/TNT mixtures.

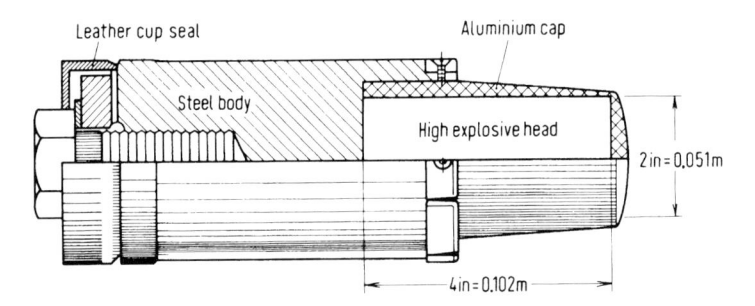

Fig. 23. The Susan projectile.

---

* Information, results, and figure obtained from Brigitta M. Dobratz, Properties of Chemical Explosives and Explosive Simulants, publication UCEL-51319, University of California, Livermore.

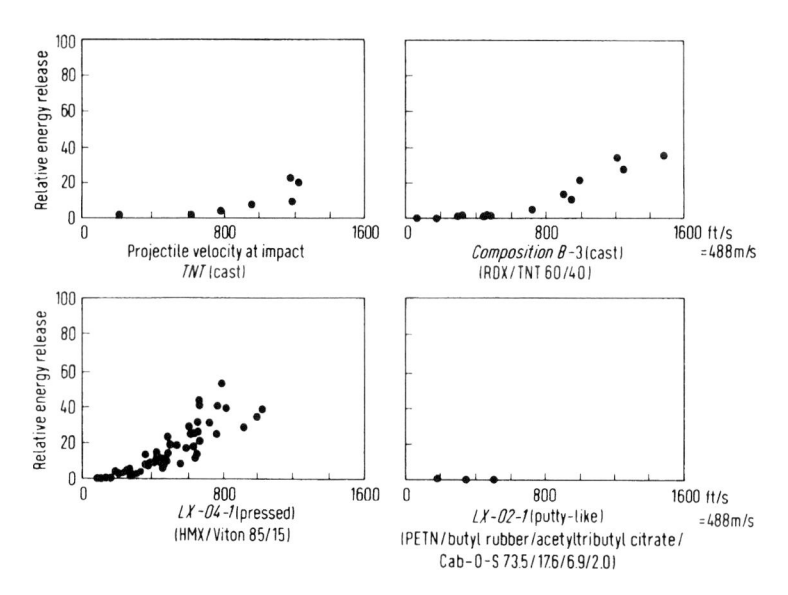

Fig. 24. Test results.

Sympathetic Propagation

→ Detonation, Sympathetic Detonation

Tacot

tetranitrodibenzo-1,3a,4,6a-tetrazapentalene;
tétranitrodibenzo-tétraza-pentaléne

orange red crystals
empirical formula: $C_{12}H_4N_8O_8$
molecular weight: 388.1
oxygen balance: -74.2%
nitrogen content: 28.87%
melting point (under decomposition): 378 °C = 712 °F
density: 1.85 g/cm^3

heat of detonation, experimental (H_2O liq.)*):
  980 kcal/kg = 4103 kJ/kg
detonation velocity, confined:
  7250 m/s = 23800 ft/s at ρ = 1.64 g/cm^3
impact sensitivity: 7 kp m = 69 N m
(Quoted from the prospectus leaflet of DU PONT.)

The compound is prepared by direct nitration of dibenzotetrazapentalene in sulfuric acid solution.

Tacot is insoluble in water and in most organic solvents; its solubility in acetone is only 0.01 %. It is soluble in 95 % nitric acid, and is sparingly soluble in nitrobenzene and dimethylformamide. It does not react with steel or with nonferrous metals.

The explosive is of interest because of its exceptionally high stability to high temperatures; it remains serviceable:

after	10 minutes	at 660 °F = 350 °C
after	4 hours	at 620 °F = 325 °C
after	10 hours	at 600 °F = 315 °C
after	2 weeks	at 540 °F = 280 °C
after	4 weeks	at 530 °F = 275 °C

Taliani Test

An improved version of the manometric test developed by *Obermüller* in 1904. The method was considerably modified, first by *Goujan* and, very recently, by *Brissaud*. In all modifications of the method, the test tube containing the sample preheated to the desired temperature is evacuated, and the increase in pressure produced by the gaseous decomposition products is measured with a mercury manometer. The test is usually terminated when the pressure has attained 100 mm Hg. The test temperature are:

for nitrocellulose	135 °C = 275 °F
for propellants	110 °C = 230 °F.

The sample must be thoroughly dried before the test; the result would otherwise also include all other components which increase the pressure on being heated, such as water and organic solvents. Since the result is also affected by the nitroglycerine content of the propellant sample, the test can only be used in order to compare propellants of the same kind with one another. This, in addition to the high testing temperature, makes the applicability of the *Taliani* test for propellants questionable. Another disadvantage is the necessity for thorough dry-

---

* Value quoted from *Brigitta M. Dobratz*, Properties of Chemical Explosives and Explosive Simulants, University of California, Livermore.

ing, since in the course of the drying operation the test sample is altered in an undesirable manner, and the experimental stability data may show better values than its true stability. The latter disadvantage does not apply to nitrocellulose testing.

Tamping

Verdämmen; bourrage

Synonymous with → *Stemming*

Tamping Pole

Ladestock; bourroir

A wooden or plastic pole used to compact explosive charges for stemming.

No. 8 Test Cap

Prüfkapsel → Zündkapsel; detonateur d'epreuve

Defined by the Institute of Makers of Explosives (USA):

A No. 8 test blasting cap is one containing 2 grams of a mixture of 80 % mercury fulminate and 20 % potassium chlorate, or a cap of equivalent strength.

In comparison: the European test cap: 0.6 g PETN; the commercial No. 8 cap: 0.75 Tetryl (German).

Test Galleries

Versuchsstrecken; Sprengstoffprüfstrecken; galeries d'essai

→ *Permissibles.*

Tetramethylammonium Nitrate

Tetramethylammoniumnitrat; nitrate de tétraméthylammonium

$(CH_3)_4N\ NO_3$

colorless crystals empirical formula: $C_4H_{12}N_2O_3$
molecular weight: 136.2
energy of formation: -562.3 kcal/kg = -2352.7 kJ/kg
enthalpy of formation: -599.3 kcal/kg = -2507.3 kJ/kg

oxygen balance: −129.2 %
nitrogen content: 20.57 %

During the Second World War, this compound was utilized as a fuel component in fusible ammonium nitrate mixtures. It can be homogeneously incorporated into the melt.

Tetramethylolcyclopentanone Tetranitrate

Nitropentanon; tétranitrate de tétraméthylolcyclopentanone; Fivonite

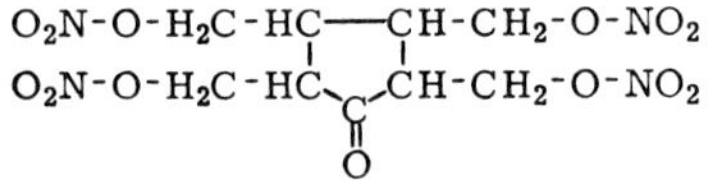

$$\begin{array}{c} O_2N-O-H_2C-HC\!\!-\!\!-\!\!CH-CH_2-O-NO_2 \\ |\quad\quad\quad | \\ O_2N-O-H_2C-HC\!\!\diagdown\!\!\diagup\!\!CH-CH_2-O-NO_2 \\ \underset{O}{\overset{\|}{C}} \end{array}$$

colorless crystals
empirical formula: $C_9H_{12}N_4O_{13}$
molecular weight: 384.2
energy of formation: −398.2 kcal/kg = −1666 kJ/kg
enthalpy of formation: −420.6 kcal/kg = −1760 kJ/kg
oxygen balance: −45.8 %
nitrogen content: 14.59 %
density: 1.59 g/cm^3
melting point: 74 °C = 165 °F
lead block test: 387 cm^3/10 g
detonation velocity, confined:
  7040 m/s = 23 100 ft/s at ρ = 1.55 g/cm^3

Condensation of formaldehyde with cyclopentanone yields a compound with four –CH$_2$OH groups, which can be nitrated to the tetranitrate. Analogous derivatives of hexanone, hexanol, and pentanol can be prepared in the same manner, but in the case of pentanol and hexanone the fifth hydroxyl group also becomes esterified:

tetramethylolcyclohexanol pentanitrate "Sixolite";
tetramethylolcyclohexanone tetranitrate "Sixonite";
tetramethylolcyclopentanol pentanitrate "Fivolite".

Table 27. *Thermochemical data*

Compound	Empirical Formula	Molecular Weight	Oxygen Balance %	Energy of Formation kcal/kg	Energy of Formation kJ/kg	Enthalpy of Formation kcal/kg	Enthalpy of Formation kJ/kg
Sixolite	$C_{10}H_{15}N_5O_{15}$	445.3	−44.9	−334	−1397	−357	−1494
Sixonite	$C_{10}H_{14}N_4O_{13}$	398.2	−56.3	−402	−1682	−425	−1778
Fivolite	$C_9H_{13}N_5O_{15}$	431.2	−35.3	−325	−1360	−348	−1456

2,3,4,6-Tetranitroaniline

Tetranitroanilin; tétranitroaniline; TNA

```
      NH₂
O₂N⌬NO₂
    │ NO₂
    NO₂
```

        pale yellow crystals
        empirical formula: $C_6H_3N_5O_8$
        molecular weight: 273.1
        energy of formation: −25.5 kcal/kg = −107 kJ/kg
        enthalpy of formation: −42.8 kcal/kg = −179 kJ/kg
        oxygen balance: −32.2 %
        nitrogen content: 25.65 %
        volume of explosion gases: 813 l/kg
        heat of explosion
           (H_2O liq.): 1046 kcal/kg = 4378 kJ/kg
           (H_2O gas): 1023 kcal/kg = 4280 kJ/kg
        specific energy: 123 mt/kg = 1204 kJ/kg
        density: 1.867 g/cm^3
        melting point: 216 °C = 421 °F (decomposition)
        deflagration point: 220−230 °C = 428−446 °F
        impact sensitivity: 0.6 kp m = 6 N m

Tetranitroaniline is soluble in water, hot glacial acetic acid, and hot acetone, and is sparingly soluble in alcohol, benzene, ligroin and chloroform.

It is prepared by nitration of 3-nitroaniline or aniline with a H_2SO_4 HNO_3 mixture; the yield is moderate.

Tetranitrocarbazole

Tetranitrocarbazol; tétranitrocarbazol; TNC

```
O₂N⌬   ⌬NO₂
   │ NH │
   NO₂  NO₂
```

        yellow crystals
        gross formula: $C_{12}H_5N_5O_8$
        molecular weight: 347.2
        energy of formation: +28.3 kcal/kg = +118.5 kJ/kg
        enthalpy of formation: +13.0 kcal/kg = +54.4 kJ/kg
        oxygen balance: −85.2 %
        nitrogen content: 20.17 %

melting point: 296 °C = 565 °F
heat of explosion (H_2O liq.): 821 kcal/kg = 3433 kJ/kg

Tetranitrocarbazole is insoluble in water, ether, alcohol, and carbon tetrachloride, and is readily soluble in benzene. It is not hygroscopic.

It is prepared by the nitration of carbazole; preparation begins with sulfuric acid treatment until the compound becomes fully soluble in water, after which the sulfonic acid derivative is directly converted to the nitro compound by adding mixed acid without previous isolation.

Tetranitromethane

Tetranitromethan; tétranitrométhane; TNM

$$\begin{array}{c} O_2N \quad NO_2 \\ \diagdown C \diagup \\ \diagup \quad \diagdown \\ O_2N \quad NO_2 \end{array}$$

colorless liquid with a pungent smell
empirical formula: CN_4O_8
molecular weight: 196.0
energy of formation: +65.0 kcal/kg = +272.1 kJ/kg
enthalpy of formation: +46.9 kcal/kg = +196.4 kJ/kg
oxygen balance: +49.0 %
nitrogen content: 28.59 %
volume of explosion gases: 685 l/kg
heat of explosion*): 526 kcal/kg = 2200 kJ/kg
specific energy: 69.1 mt/kg = 677 kJ/kg
density: 1.6377 g/cm^3
solidification point: 13.75 °C = 56.75 °F
boiling point: 126 °C = 259 °F vapor pressure

Pressure millibar	Temperature °C	°F
12	20	68
57	50	122
420	100	212
1010	126	259 (boiling point)

detonation velocity, confined:
6360 m/s = 20 900 ft/s at ρ = 1.637 g/cm^3

---

* The presence of small amounts of impurities may easily increase the experimental value to above 1000 kcal/kg.

Tetranitromethane is insoluble in water, but soluble in alcohol and ether. The volatile compound strongly attacks the lungs. The oxygen-rich derivative is not explosive by itself, but forms highly brisant mixtures with hydrocarbons such as toluene.

Tetranitromethane is formed as a by-product during nitration of aromatic hydrocarbons with concentrated acids at high temperatures, following opening of the ring. It can also be prepared by reacting acetylene with nitric acid in the presence of mercury nitrate as a catalyst. According to a more recent method, tetranitromethane is prepared by introducing a slow stream of ketene into cooled 100% nitric acid. When the reaction mixture is poured into ice water, tetranitromethane separates out.

Mixtures of tetranitromethane with organic fuels are very sensitive to impact and friction and may react spontaneously by detonation or fast deflagration.

Tetranitronaphthalene

Tetranitronaphthalin; tétranitronaphthaléne; TNN

yellow crystals
empirical formula: $C_{10}H_4N_4O_8$
molecular weight: 308.2
energy of formation: +23.7 kcal/kg = +99.2 kJ/kg
enthalpy of formation: +8.4 kcal/kg = +35.3 kJ/kg
oxygen balance: −72.7%
nitrogen balance: 18.18%
melting point:
    softening of the isomer mixture
    begins at 190 °C = 374 °F

Tetranitronaphthalene is a mixture of isomers, which is obtained by continued nitration of dinitronaphthalenes.

The tetrasubstituted compound can only be attained with difficulty. The crude product is impure and has an irregular appearance. It can be purified with glacial acetic acid.

Tetracene

tetrazolyl guanyltetrazene hydrate; Tetrazen; tétrazéne*)

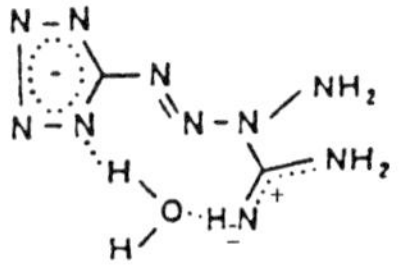

[structural formula]

feathery, colorless to pale yellow crystals
empirical formula: $C_2H_8N_{10}O$
molecular weight: 188.2
energy of formation: +270.2 kcal/kg = +1130 kJ/kg
enthalpy of formation: +240.2 kcal/kg = +1005 kJ/kg
oxygen balance: −59.5%
nitrogen content: 74.43%
density: 1.7 g/cm^3
lead block test: 155 cm^3/10 g
deflagration point: ca. 140 °C = 294 °F
impact sensitivity: 0.1 kp m = 1 N m

Tetrazene is classified as an initiating explosive, but its own initiation effect is weak.

It is practically insoluble in water, alcohol, ether, benzene, and carbon tetrachloride. It is prepared by the reaction between aminoguanidine salts and sodium nitrite.

Tetrazene is an effective primer which decomposes without leaving any residue behind. It is introduced as an additive to erosion-free primers based on lead trinitroresorcinate in order to enhance the response. Its sensitivity to impact and to friction are about equal to those of mercury fulminate.

Specifications

moisture: not more than	0.3%
reaction, water extract to universal indicator paper:	no acid indication
mechanical impurities:	none
pouring density; about	0.3 g/cm^3
deflagration point: not below	138 °C = 280 °F

---

* The structural formula found in the earlier literature: HN-C(=NH)-NH-NH-N=N-C(=NH)-NH-NH-NO was corrected in 1954 by *Patinkin* (Chem. Zentr. 1955, p. 8377).

Tetryl

trinitro-2,4,6-phenylmethylnitramine; Tetryl; tétryl;
Tetranitromethylanilin; pyronite; tetralita

[Structural formula: benzene ring with NO$_2$ groups at 2, 4, 6 positions and N(CH$_3$)-NO$_2$ group]

pale yellow crystals
empirical formula: $C_7H_5N_5O_8$
molecular weight: 287.1
energy of formation: +35.3 kcal/kg = +147.6 kJ/kg
enthalpy of formation: +16.7 kcal/kg
= +69.7 kJ/kg
oxygen balance: −47.4%
nitrogen content: 24.39%
volume of explosion gases: 861 l/kg
heat of explosion
 (H$_2$O gas): 996 kcal/kg = 4166 kJ/kg } calculated*)
 (H$_2$O liq.): { 1021 kcal/kg = 4271 kJ/kg
   { 1140 kcal/kg = 4773 kJ/kg experimental**)
specific energy: 123 mt/kg = 1208 kJ/kg
density: 1.73 g/cm^3
melting point: 129.5 °C = 265 °F
heat of fusion: 19.1 kcal/kg = 80 kJ/kg
lead block test: 410 cm^3/10 g
detonation velocity, confined:
 7570 m/s = 24 800 ft/s at ρ = 1.71 g/cm^3
deflagration point: 185 °C = 365 °F
impact sensitivity: 0.3 kp m = 3 N m
friction sensitivity: 36 kp = 353 N
pistil load reaction
critical diameter of steel sleeve test: 6 mm

Tetryl is poisonous; it is practically insoluble in water, sparingly soluble in alcohol, ether, and benzene, and is more readily soluble in acetone.

It is obtained by dissolving mono- and dimethylaniline in sulfuric acid and pouring the solution into nitric acid, with cooling.

Tetryl is a highly brisant, very powerful explosive, with a satisfactory initiating power which is used in the manufacture of primary and

---

 * computed by the "ICT-Thermodynamic-Code".
 ** value quoted from *Brigitta M. Dobratz*, Properties of Chemical Explosives and Explosive Simulants, University of California, Livermore.

secondary charges for blasting caps. Owing to its relatively high melting point, it is employed pressed rather than cast.

Specifications

melting point: not less than	128.5 °C = 270 °F
volatiles, incl. moisture: not more than	0.10 %
benzene insolubles, not more than	0.07 %
ash content, not more than	0.03 %
acidity, as HNO_3, not more than	0.005 %
alkalinity	none

Tetrytol

A pourable mixture of 70 % Tetryl and 30 % TNT.

Thermobaric Explosives

*TBX, Single- event FAE, explosifs thermobarique,
Thermobare Sprengstoffe*

Type of → *FAE* with solid fuel, mainly using aluminium or magnesium. Because their reaction with atmospheric oxygen only produces solid oxides the blast wave is primarily generated by heat of combustion ("thermobaric") instead of expanding explosion gases.

The advantage to classical FAE is the shorter delay between distribution of the fuel cloud and ignition of the fuel/air mixture. In this third and fourth generation of fuel air explosives there is no more need for an atomization charge which makes them a real "single event FAE".

The peak pressure of TBX is obviously lower than in typical high explosives, but with up to 200 milliseconds it lasts over 200 times longer. In the 90's Russia already brought a thermobaric warhead into service, the Anti-Tank Rocket RPG-7 (RPO-A) "Satan's pipe" against snipers in urban areas. A thermobaric mixture developed at NSWC -Indian Head in 2001 contains high-insensitive explosive on → HMX- Aluminium- HTPB -basis called PBXIH-135 (a Type of SFAE – solid fuel air explosive). PBXIH-135 is used in the laser-guided bomb BLU 118/B.

Thermite

An incendiary composition consisting of 2.75 parts of black iron oxide (ferrosoferric oxide) and 1.0 part of granular aluminum.

Thermodynamic Calculation of Decomposition Reactions

Important characteristics of explosives and propellants may be calculated from the chemical formula and the → *Energy of Formation* (*enthalpy* in the case of rocket propellants) of the explosive propellant components under consideration. The chemical formula of mixtures may be obtained by the percentual addition of the atomic numbers of the components given below, for example:

composition:
    8% nitroglycerine
    30% nitroglycol
    1.5% nitrocellulose
    53.5% ammonium nitrate
    2% DNT
    5% wood dust;

The number of atoms of C, H, O and N per kilogram are calculated from Table 28.

	C	H	O	N
nitroglycerine 13.21 C; 22.02 H; 39.62 O; 13.21 N; 8% thereof:	1.057	1.762	3.170	1.057
nitroglycol 13.15 C; 26.30 H; 39.45 O; 13.15 N; 30% thereof:	3.945	7.890	11.835	3.945
nitrocellulose (12.5% N) 22.15 C; 27.98 H; 36.30 O; 8.92 N; 1.5% thereof:	0.332	0.420	0.545	0.134
ammonium nitrate 49.97 H, 37.48 O; 24.99 N; 53.5% thereof:	–	26.73	20.052	13.37
dinitrotoluene 38.43 C; 32.94 H; 21.96 O; 10.98 N; 2% thereof:	0.769	0.659	0.439	0.220
wood meal 41.7 C; 60.4 H; 27.0 O; 53.5% thereof:	2.085	3.02	1.35	–
	8.19	40.48	37.39	18.73

As a result, the formula for one kilogram of the explosive can be written:

$C_{8.19}H_{40.48}O_{37.39}N_{18.73}$.

The same calculation has to be made for gun and rocket propellants as the first step.

In decomposition processes (detonation in the case of high explosives or burning processes in the case of gunpowders and rocket propellants), the kilogram $C_a H_b O_c N_d$ is converted into one kilogram

$n_1\ CO_2 + n_2\ H_2O + n_3\ N_2 + n_4\ CO + n_5\ H_2 + n_6\ NO$.

In the case of industrial explosive with a positive → *Oxygen Balance*, the occurrence of free oxygen O_2, and in the case of explosive with a very negative oxygen balance, e.g., TNT, the occurrence of elementary carbon C have to be considered. If alkali metal salts such as $NaNO_3$ are included, the carbonates of these are taken as reaction products, e.g., Na_2CO_3. The alkaline earth components, e.g., $CaNO_3$ are assumed to form the oxides, e.g., CaO; chlorine will be converted into HCl; sulfur into SO_2.

Exact calculations on burning processes in rocket motors must include dissociation phenomena; this is done on computer facilities (at leading national institutes*), and the relevant industrial laboratories in this field are nowadays equipped with computers and programs. The following explanations are based on simplifying assumptions.

1. Conventional Performance Data of Industrial Explosives.

The explosion of an industrial explosive is considered as an isochoric process, i.e. theoretically it is assumed that the explosion occurs confined in undestroyable adiabatic environment. Most formulations have a positive oxygen balance; conventionally it is assumed, that only CO_2, H_2O, N_2 and surplus O_2 are formed. The reaction equation of the example above is then

$C_{8.19} H_{40.48} O_{37.39} N_{18.73} =$

$8.19\ CO_2 + \dfrac{40.48}{2} H_2O + \dfrac{18.73}{2} N_2 + \dfrac{1}{2}(37.39 - 2 \times 8.19 - \dfrac{40.48}{2})O_2 =$

$8.19\ CO_2 + 20.24\ H_2O + 9.37\ N_2 + 0.39\ O_2$

The real composition of the explosion gases is slightly different; CO and traces of NO are also formed.

1.1 Heat of explosion.

Table 33 also lists the enthalpies and energies of formation of the explosives and their components.

---

* The data for the heat of explosion, the volume of explosion gases and specific energy given in this book for the individual explosives have been calculated with the aid of the "ICT-Code" in the Fraunhofer Institut für CHEMISCHE TECHNOLOGIE, D-76318 Pfinztal, including consideration of the dissociation phenomena. Therefore, the values have been changed in comparison to the figures listed in the first edition of this book (computed without dissociation).

Thermodynamic Calculation of Decomposition Reactions

In the case of isochoric explosion, the value for the energy of formation referring to constant volume has to be employed. The heat of explosion is the difference of the energies between formation of the explosive components and the reaction products, given in

Table 29. Energy of formation of the example composition:

Component	Energy of Formation kcal/kg	Thereof: %	
nitroglycerine	− 368.0	8	= − 29.44
nitroglycol	− 358.2	30	= −107.46
nitrocellulose (12,5 % N)	− 605.6	1.5	= − 9.08
ammonium nitrate	−1058	53.5	= −566.03
DNT	− 70	2	= − 1.40
wood dust	−1090	5	= − 54.5
			−767.91

Table 30. Energy of formation of the reaction products:

Component	Energy of Formation kcal/mol	Mole Number	Portion of Component
CO_2	−94.05	8.19	− 770.27
H_2O (vapor)*	−57.50	20.24	−1163.80
			−1934.1

The heat of explosion is obtained by the difference
−767.91 −(−1934.1) = +1934.1 −767.91 = 1167
say **1167** kcal/kg or **4886** kJ/kg (H_2O gas).

1.2 Volume of explosion gases.

The number of moles of the gaseous reaction products are multiplied by 22.4 l, the volume of 1 mole ideal gas at 0 °C (32 °F) and 1 atmosphere. The number of moles in the example composition:

---

* Conventionally, the computed figure for the heat of explosion for industrial explosives is given based on H_2O vapor as the reaction product. The values for the individual explosive chemicals (now calculated by the "ICT-Code" are based on H_2O liquid. They are directly comparable with results obtained by calorimetric measurements.

CO$_2$:   8.19
H$_2$O: 20.24
N$_2$:   9.37
O$_2$:   0.39

sum: 38.19   ×22.4 = 855 l/kg *Volume of Explosion Gases.*

1.3 Explosion temperature.

The heat of explosion raises the reaction products to the explosion temperature. Table 31 gives the internal energies of the reaction products in relation to the temperature. The best way to calculate the explosion temperature is to assume two temperature values and to sum up the internal energies for the reaction product multiplied by their corresponding mole number. Two calorific values are obtained, of which one may be slightly higher than the calculated heat of explosion and the other slightly lower. The explosion temperature is found by interpolation between these two values.

For the example composition at: 3600 K and 3700 K. Table 31 gives for

	3600 K	Mole Number	Product kcal	3700 K	Product kcal
CO$_2$	38.76	8.19	317.4	40.10	328.4
H$_2$O	30.50	20.24	617.3	31.63	640.2
N$_2$	20.74	9.37	194.3	21.45	201.0
O$_2$	22.37	0.39	8.7	23.15	9.0
		38.19	1138		1178

The interpolated temperature value for 1167 kcal/kg is 3670 K.

For industrial nitroglycerine-ammonium nitrate explosives, the following estimated temperature values can be recommended:

Table 32

Heat of Explosion Found kcal/kg	Lower value K	Upper Value K
900	2900	3000
950	3000	3100
1000	3100	3200
1050	3300	3400
1100	3400	3500
1150	3500	3600
1200	3700	3800

1.4 Specific energy.

The concept of specific energy can be explained as follows. When we imagine the reaction of an explosive to proceed without volume expansion and without heat evolution, it is possible to calculate a theoretical thermodynamic value of the pressure, which is different from the shock wave pressure (→ *Detonation*); if this pressure is now multiplied by the volume of the explosive, we obtain an energy value, the "specific energy", which is the best theoretically calculable parameter for the comparison of the → *Strengths* of explosives. This value for explosives is conventionally given in meter-tons per kg.

The specific energy results from the equation

$f = n\, RT_{ex}$.
f: specific energy
n: number of gaseous moles
T_{ex}: detonation temperature in degrees Kelvin
R: universal gas constant (for ideal gases).

If f is wanted in meter-ton units, R*) has the value 8.478×10^{-4}.

The values for the considered example composition are

$n = 38.19$
$T_{ex} = 3670$ K
$f = 38.19 \times 8.478 \times 10^{-4}\, 3670 = $ **118.8** mt/kg.

For the significance of specific energy as a performance value, → *Strength*.

1.5 Energy level.

Because higher loading densities involve higher energy concentration, the concept "energy level" was created; it means the specific energy per unit volume instead of unit weight. The energy level is

$I = \rho \cdot f$
I: energy level mt/l
ρ: density in g/cm^3
f: specific energy mt/kg

Since the example composition will have a gelatinous consistency, ρ may be assumed as 1.5 g/cm^3. The energy level is then

$I = 1.5 \times 118.8 = $ **178.2** mt/l.

1.6 Oxygen balance.

→ oxygen balance

---

* For the values of R in different dimensions, see the conversion tables on the back fly leaf.

2. Explosive and Porpellant Composition with a Negative Oxygen Balance

Calculation of gunpowders.

The decomposition reactions of both detonation and powder combustion are assumed to be isochoric, i.e., the volume is considered to be constant, as above for the explosion of industrial explosives.

The first step is also the sum formula

$C_a H_b O_c N_d$

as described above, but now

$c < 2a + \frac{1}{2} b$

The mol numbers n_1, n_2, etc. cannot be directly assumed as in case the of positive balanced compositions. More different reaction products must be considered, e.g.,

$C_a H_b O_c N_d = n_1 CO_2 + n_2 H_2 O + n_3 N_2 + n_4 CO + n_5 H_2 + n_6 NO$;

CH_4 and elementary carbon may also be formed; if the initial composition contains Cl-, Na-, Ca-, K-, and S-atoms (e.g., in black powder), the formation of HCl, Na_2O, Na_2CO_3, K_2O, K_2CO_3, CaO, SO_2 must be included. Further, the occurrence of dissociated atoms and radicals must be assumed.

The mole numbers n_1, n_2, etc., must meet a set of conditions: first, the stoichiometric relations

$a = n_1 + 2 n_5$ \hfill (1)

(carbon containing moles)

$h = 2 n_2 + 2 n_5$ \hfill (2)

(hydrogen containing moles)

$c = 2 n_1 + n_2 + n_4 + n_6$ \hfill (3)

(oxygen containing moles)

$d = 2 n_3 + n_6$ \hfill (4)

(nitrogen containing moles);

second, the mutual equilibrium reactions of the decomposition products: the water gas reaction

$H_2O + CO = CO_2 + H_2$;

the equilibrium is influenced by temperature and is described by the equation

$K_1 = \dfrac{[CO]\ [H_2O]}{[H_2\ [CO_2]}$ \hfill (5)

K_1: equilibrium constant

$[CO_2]$, $[H_2]$, $[H_2O]$ and $[CO]$: the partial pressures of the four gases.

The total mole number n is not altered by the water gas reaction; K_1, is therefore independent of the total pressure p, but depends on the temperature ($\rightarrow$ Table 36). Equation (5) can be written as

$$K_1 = \frac{n_2 \cdot n_4}{n_1 \cdot n_5} \tag{5a}$$

The reaction for NO formation must also be considered

$1/2\ N_2 + CO_2 = CO + NO$

with the equilibrium equation

$$K_2 = \frac{[CO] \cdot [NO]}{[N_2]^{1/2} \cdot [CO_2]} = \frac{\frac{p}{n} n_4 \cdot \frac{p}{n} n_6}{\left(\frac{p}{n}\right)^{1/2} \cdot n_3^{1/2} \cdot \frac{p}{n} n_1} \text{ or}$$

$$K_2 = \sqrt{\frac{p}{n}} \frac{n_4 \cdot n_6}{\sqrt{n_3} \cdot n_1} \tag{6}$$

K_2: equilibrium constant,
p: total pressure; $p/n \cdot n_1$, etc., the partial pressures
n: total number of moles

Because NO formation involves an alteration of n, the equilibrium constant K_2 depends not only on the temperature, but also on the total pressure p.

For the calculation of the six unknown mol numbers $n_1 \ldots n_6$, there are six equations. Alteration of the mole numbers cause alteration of the values for the reaction heat, the reaction temperature, the reaction pressure, and hence the constants K_1 and K_2. Calculations without the aid of a computer must assume various reaction temperatures to solve the equation system, until the values for the reaction heat as a difference of the energies of formation and the internal energy of the reaction products are the same (as shown above for the detonation of industrial explosives). This is a long trial and error calculation; therefore the use of computer programs is much more convenient.

For low caloric propellants and for highly negative balanced explosives, such as TNT, the formation of element carbon must be assumed (*Boudouard* equilibrium):

$CO_2 + C = 2\ CO$

with the equilibrium equation

$$K_3 = \frac{[CO]^2}{[CO_2]} \tag{7}$$

Explosion fumes with a dark color indicate the formation of carbon. The calculation becomes more complicated, if dissociation processes are taken into consideration (high caloric gunpowders; rocket propellants).

In the computer operation, the unknown mole numbers are varied by stepwise "iteration" calculations, until all equation conditions are satisfied. The following results are obtained.

> heat of explosion
> temperature of explosion
> average molecular weight of the gaseous explosion products
> total mole number
> specific energy (force)
> composition of the explosion gases
> specific heat ratio c_p/c_v
> covolume
> etc.

The basis data for internal ballistic calculations are thus obtained (also → Ballistic Bomb; → Burning Rate). As an example for the calculation of a double base gunpowder*):

composition

nitrocellulose (13.25 % N):	57.23 %
nitroglycerine:	40.04 %
potassium nitrate:	1.49 %
Centralite I:	0.74 %
ethanol (solvent rest):	0.50 %

Enthalpy of formation of the composition: -2149.6 kJ/kg $= -513.8$ kcal/kg

The results (loading density assumed: 210 kg/m³)

sum formula for
1 kg powder:
$C_{18.14}H_{24.88}O_{37.41}N_{10.91}K_{0.015}$

temperature of explosion:	3841 K
pressure:	305.9 MPa = 3059 bar
average molecular weight of gases:	27.28 g/mol
total mole number:	36.66 mol/kg
specific energy (force):	1.17×10^6 Nm/kg = 1170 kJ/kg
kappa $\kappa = c_p/c_v$):	1.210
covolume:	9.37×10^{-4} m³/kg

---

* Calculated by the ICT-Fortran programm.

The composition of the reaction products (mol %):

    28.63 % H_2O
    28.39 % CO
    21.07 % CO_2
      4.13 % H_2
    14.63 % N_2
      0.21 % O_2
      0.48 % NO
      0.37 % KOH
      1.50 % OH ⎫
      0.42 % H ⎬ dissociated atoms and radicals
      0.09 % O ⎭
      0.02 % K

2. Rocket Propellants

Raketentreibstoffe; propellants de roquette

The calculation of the performance data of rocket propellants is carried out in the same manner as shown above for gunpowders, but the burning process in the rocket chamber proceeds at constant pressure instead of constant volume. For the evaluation of the heat of reaction, the difference of the enthalpies of formation instead of the energies must now be used; for the internal heat capacities, the corresponding enthalpy values are listed in Table 38 below (instead of the energy values in Table 31); they are based on the average specific heats c_p at constant pressure instead of the c_v values. The first step is to calculate the reaction temperature T_c and the composition of the reaction gases (mole numbers n_1, etc.). The second step is to evaluate the same for the gas state at the nozzle exit (transition from the chamber pressure p_c to the nozzle exit pressure p_e e.g., the atmospheric pressure). The basic assumption is, that this transition is "isentropic", i.e., that the entropy values of the state under chamber pressure and at the exit are the same. This means that the thermodynamic transition gives the maximum possible output of kinetic energy (acceleration of the rocket mass).

The calculation method begins with the assumption of the temperature of the exit gases, e.g., $T_e = 500$ K. The transition from the thermodynamical state in the chamber into the state at the nozzle exit is assumed to be instantaneous, i.e. the composition of the gases remains unchanged ("frozen" equilibria). The entropy of the exit gases at the assumed temperature T_e is assumed to be the same as the entropy of the gases in the chamber (known by calculation); the assumed value Ta is raised until both entropy values are equal. Since both states are known, the corresponding enthalpy values can be calculated. Their difference is the source of the kinetic energy of the rocket (half the mass × square root of the velocity); the → *Specific Impulse* (mass × velocity) can be calculated according to the following equation (the same as shown on p. 293):

$I_s = \sqrt{2J(H_c-H_e)}$ Newton seconds per kilogramm

or

$I_{sfroz} = 92.507 \sqrt{H_c-H_e}$

J: mechanical heat equivalent

		unit:
I_{sfroz}:	specific impulse for frozen equilibrium	Ns/kg
H_c:	enthalpy of the burnt gases in the chamber	kcal/kg
H_e:	enthalpy of the gases at the nozzle exit	kcal/kg
T_c:	gas temperature in the chamber	K
T_e:	gas temperature at the nozzle exit with frozen equilibrium	K

The computer program also allows calculations for shifting (not frozen) equilibria by stepwise iteration operations.

The following parameters can also be calculated with computer facilities:

> chamber temperature (adiabatic flame temperature);
> temperature of exit gas with frozen equilibrium;
> composition of exit gas with shifting equilibrium;
> temperature of the burnt gases in the chamber and at the nozzle exit;
> average molecular weight, of the burnt gases in the chamber and at the nozzle exit;
> total mole number of the burnt gases in the chamber and at the nozzle exit;
> specific impulse at frozen and with shifting equilibrium;
> ratio of specific heats c_p/c_v;

An example for the calculation of a double bases rocket propellant:

Composition:
nitrocellulose	52.15 % (13.25 % N)
nitroglycerine	43.54 %
diethylphthalate	3.29 %
Centralit I	1.02 %

sum formula for 1 kg propellant: $C_{19.25}H_{25.96}O_{36.99}N_{10.76}$
chamber pressure: 7.0 MPa (1015 p.s.i.)

Results of the computer calculation*):

temperature in the rocket chamber:	3075 K
temperature at the nozzle exit: (frozen equilibrium)	1392 K
temperature at the nozzle exit: (equilibrium flow)	1491 K
average molecular weight:	26.33 g/mol
total mole number:	37.98 mol/kg
kappa ($\kappa = c_p/c_v$):	1.216
specific impulse for frozen equilibrium:	2397 Ns/kg
specific impulse for shifting equilibrium:	2436 Ns/kg

For more detailed calculations (e.g. if the presence of free carbon must be assumed), the reader is referred to *M. A. Cook:* The Science of High Explosives, Chapman & Hall, London 1958 and, by the same author: The Science of Industrial Explosives, copyright 1974 by IRECO CHEMICALS, Salt Lake City, USA. They contain basic data on heat capacities and equilibria constants concerned, as well as computing programs for hand and machine calculations.

The data for the following Tables have been taken from this book. The data for enthalpies and energies of formation are taken from H. Bathelt, F. Volk, M. Weindel, ICT-Database of Thermochemical Values, Sixth update (2001), FRAUNHOFER-INSTITUT FÜR CHEMISCHE TECHNOLOGIE, D-76318 Pfinztal-Berghausen.

---

* Calculated by the "ICT-Thermodynamic-Code"

Table 33. Enthalpy and energy of formation of explosive and propellant components and their number of atoms per kg. Reference Temperature: 298.15 K = 25 °C = 77 °F; reference state of carbon: as graphite.

(1): primary explosive   (7): stabilizer   (13): thickener
(2): secondary explosive (8): gelatinizer   (14): anticaking agent
(3): tertiary explosive   (9): burning moderator   (15): anti acid
(4): propellant component (10): polymer binder   (16): component of permitted explosives
(5): oxidizer   (11): prepolymer   (17): slurry component
(6): fuel   (12): hardener

Component	Empirical Formula	Enthalpy of Formation kcal/kg	Enthalpy of Formation kJ/kg	Energy of Formation kcal/kg	Energy of Formation kJ/kg	Number of Atoms per kilogram				Uses
						C	H	O	N	
Arkadit I	$C_{13}H_{12}ON_2$	−138.2	−578	−117.3	−490	61.25	56.54	4.71	9.42	(7)
Arkadit II	$C_{14}H_{14}ON_2$	−112.7	−472	−90.5	−379	61.86	61.86	4.42	8.84	(7); (8)
Arkadit III	$C_{15}H_{16}ON_2$	−151.9	−636	−128.5	−538	62.42	66.58	4.16	8.32	(7); (8)
aluminum	Al	−	−	−	−	−	−	−	−	Al: 37.08 (6); (17)
ammonium chloride	H_4NCl	−1405	−5878	−1372	−5739	−	74.77	−	18.69	Cl: 18.69 (16)
Ammoniumdinitramide	$H_4N_4O_4$	−288.58	−1207.4	−259.96	−1086.6	−	32.24	32.24	32.24	(4); (5)
ammonium nitrate	$H_4O_3N_2$	−1092	−4567	−1058	−4428	−	49.97	37.48	24.99	(3); (5); (16); (17)
ammonium oxalate · H_2O	$C_2H \cdot O_5N_2$	−2397	−10031	−2362	−9883	14.07	70.36	35.18	14.07	(6); (16)
ammonium perchlorate	H_4O_4NCl	−602	−2517	−577	−2412	−	34.04	34.04	8.51	Cl: 8.51 (3); (5)
barium nitrate	O_5N_2Ba	−907.3	−3796	−898.2	−3758	−	−	22.96	7.65	Ba: 3.83 (5)
calcium carbonate	CO_3Ca	−2882	−12059	−2873	−12022	10.00	−	30.00	−	Ca: 10.00 (15)

325   Thermodynamic Calculation of Decomposition Reactions

calcium nitrate	O_6N_2Ca	-1367	-5718	-1352	-5657	—	—	—	36.56	12.19	Ca: 6.09	(5); (17)
calcium stearate	$C_{36}H_{70}O_4Ca$	-1092	-4567	-1055	-4416	59.30	115.31	6.59	—	Ca: 1.65	(6); (14)
camphor	$C_{10}H_{16}O$	-513	-2146	-480	-2008	65.69	105.10	6.57	—	(8)
Centralite I	$C_{17}H_{20}O N_2$	-93.5	-391.5	-68.2	-285.6	63.34	74.52	3.73	7.45	(7); (8)
Centralite II	$C_{15}H_{16}O N_2$	-60.8	-254	-37.3	-156	62.42	66.58	4.16	8.32	(7); (8)
Centralite III	$C_{15}H_{18}O N_2$	-119.1	-499	-94.7	-396	62.90	70.76	3.93	7.86	(7); (8)
coal (pit coal dust)		-16	-67	0	0	72.08	49.90	5.24	—	(6)
diamyl phthalate	$C_{13}H_{26}O_4$	-721	-3017	-692	-2895	58.75	84.85	13.06	—	(8)
dibutyl phthalate	$C_{13}H_{22}O_4$	-723	-3027	-696	-1913	57.47	79.02	14.37	—	(8)
diethyleneglycol dinitrate	$C_4H_8O_7N_2$	-532	-2227	-507	-2120	20.40	40.79	35.69	10.20	(4)
dimethylhydrazine	$C_2H_8N_2$	$+198$	$+828$	$+247$	$+1035$	33.28	133.11	—	33.28	(4)
2.4-dinitrotoluene	$C_7H_6O_4N_2$	-89.5	-374.5	-70.0	-292.8	38.43	32.94	21.96	10.98	(4); (6)
2.6-dinitrotoluene		-57.7	-241.2	-38.1	-159.5					(4); (6)
diphenylamine	$C_{12}H_{11}N$	-183.6	-768.2	$+204.6$	$+856.0$	70.92	65.01	—	5.91	(7)
diphenylurethane	$C_{15}H_{15}O_2N$	-278.1	-1164	-256.0	-1071	62.16	62.16	8.29	4.14	(7); (8)
Ethriol trinitrate	$C_5H_{11}O_9N_3$	-426	-1783	-401	-1678	22.30	40.88	33.44	11.15	(4)
ethylenediamine dinitrate	$C_2H_{13}O_6N_4$	-839.2	-3511	-807.4	-3378	10.75	53.73	32.24	21.49	(2)
ferrocene	$C_{10}H_{10}Fe$	$+199$	$+833$	$+215$	$+899$	53.76	53.76	—	—	Fe: 5.38	(4); (10)
glycidyl azide polymer	$C_3H_5O N_3$	$+340,1$	$+1423$	$+366,9$	$+1535$	30.28	50,46	10,09	30,28	(9)
glycol	$C_2H_6O_2$	-1752	-7336	-1714	-7177	32.22	96.66	32.22	—	(6)
guanidine nitrate	$CH_6O_3N_4$	-758	-3170	-726.1	-3038	8.19	49.14	24.57	32.76	(4)
guar gum		-1671	-7000	-1648	-6900	37.26	55.89	31.05	—	(13)
hexanitrodiphenylamine	$C_{12}H_5O_{12}N_7$	$+22.5$	$+94.3$	$+38.7$	$+162$	27.32	11.38	27.32	15.94	(2)

Table 33. (continued)

Component	Empirical Formula	Enthalpy of Formation kcal/kg	Enthalpy of Formation kJ/kg	Energy of Formation kcal/kg	Energy of Formation kJ/kg	Number of Atoms per kilogram C	H	O	N		Uses
Hexanitrohexaaza-isowurtzitane (CL20)	$C_6H_6N_{12}O_{12}$	+ 299.0	+ 920.5	+ 240.3	+ 1005.3	13.69	13.69	27.39	27.39		(2); (4)
Hexogen (RDX)	$C_3H_6O_6N_6$	+ 72	+ 301.4	+ 96.0	+ 401.8	13.50	27.01	27.01	17.01		(2)
hydrazine	H_4N_2	+ 377.5	+ 1580	+ 433.1	+ 1812	–	124.80	–	62.40		(6)
hydrazine nitrate	$H_5O_3N_3$	– 620.7	– 2597	– 586.4	– 2453	–	52.60	31.56	31.56		(2); (4)
iron acetylacetonate	$C_{15}H_{21}O_6Fe$	– 859	– 3593	– 836	– 3498	42.47	59.46	16.99	–		(9)
kerosene		– 540	– 2260	– 500	– 2100	71.90	135.42	–	–		(6)
lead acetylsalicylate · H_2O	$C_{18}H_{16}O_9Pb$	– 823	– 3444	– 810	– 3391	30.85	27.42	15.42	–	Fb: 1.71 Pb: 3.43	(9)
lead azide	N_6Pb	+ 391.4	+ 1638	+ 397.5	+ 1663	–	–	–	20.60	Pb: 2.03	(1)
lead ethylhexanoate	$C_{16}H_{30}O_4Pb$	– 724	– 3027	– 703	– 2940	32.41	60.78	8.10	–	Fb: 3.02	(9)
lead nitrate	O_6N_2Pb	– 326.1	– 1364	– 318.9	– 1334	–	–	18.11	6.04	Pb: 2.13	(5)
lead styphnate	$C_6H_3O_9N_3Pb$	– 427.1	– 1787	– 417.6	– 1747	12.81	6.41	19.22	6.41	Mg: 11.86	(1)
magnesium carbonate	O_3CMg	–3106	–12994	–3095	–12950	11.86	–	35.60	–		(14); (15)
mannitol hexanitrate	$C_6H_8O_{18}N_6$	– 357.2	– 1494	– 336.2	– 1407	13.27	17.70	39.82	13.27		(2)
methylamine nitrate (MAN)	$CH_6O_3N_2$	– 896	– 3748	– 862	– 3604	10.63	63.78	31.89	21.26		(17)
metriol trinitrate	$C_5H_9O_9N_3$	– 398.2	– 1666	– 373.8	– 1564	19.60	35.27	35.27	11.76		(4)
Nitroaminoguanidine	$C_1H_5N_5O_2$	+ 44.3	+ 185.5	+ 74.2	+ 310.2	8.40	41.99	16.79	41.99		(4)
nitrocellulose, 13.3 % N		– 577.4	– 2416	– 556.1	– 2327	21.10	25.83	36.66	9.50		(4)
nitrocellulose, 13.0 % N		– 596.1	– 2494	– 574.6	– 2404	21.55	26.64	36.52	9.28		(4)
nitrocellulose, 12.5 % N		– 627.2	– 2624	– 605.6	– 2534	22.15	27.98	36.30	8.92		(4)

nitrocellulose, 12.0 % N		− 658.4	− 2755	− 636.5	− 2663	22.74	29.33	36.08	8.57	(4)
nitrocellulose, 11.5 % N		− 689.6	− 2885	− 667.4	− 2793	23.33	30.68	35.86	8.21	(4)
nitrocellulose, 11.0 % N		− 720.7	− 3015	− 698.4	− 2922	23.94	32.17	35.65	7.84	(4)
nitrodiphenylamine	$C_{12}H_{10}O_2N_2$	+ 88.13	+ 369	+ 107.5	+ 450	56.01	46.68	9.34	9.34	(4)
nitroglycerine	$C_3H_5O_9N_3$	− 392.0	− 1633	− 369.7	− 1540	13.21	22.02	39.62	13.21	(2); (4)
nitroglycol	$C_2H_4O_6N_2$	− 381.6	− 1596	− 358.2	− 1499	13.15	26.30	39.45	13.15	(2)
nitroguanidine (picrite)	$CH_4O_2N_4$	− 213.3	− 893.0	− 184.9	− 773	9.61	38.42	19.21	38.42	(4)
nitromethane	CH_3O_2N	− 442.8	− 1853	− 413.7	− 1731	16.39	49.17	32.70	16.39	
Nitrotriazolone (NTO)	$C_2H_2N_4O_3$	− 185.14	− 774.60	− 164.69	− 689.10	15.37	15.37	23.06	30.75	(2)
nitrourea	$CH_3O_3N_3$	− 642.5	− 2688	− 617.2	− 2582	9.52	28.55	28.55	28.55	
Octogen (HMX)	$C_4H_8O_8N_8$	+ 60.5	+ 253.3	+ 84.5	+ 353.6	13.50	27.01	27.01	27.01	
paraffin (solid)		− 540	− 2261	− 500	− 2094	71.0	148	−	−	(2)
PETN (Nitropenta)	$C_5H_8O_{12}N_4$	− 407.4	− 1705	− 385.0	− 1611	15.81	25.30	37.95	12.65	(6)
petroleum		− 440	− 1842	− 400	− 1675	70.5	140	−	−	(2)
picric acid	$C_6H_3O_7N_3$	− 259.3	− 1085	− 242.4	− 1015	26.20	13.10	30.55	13.10	(2)
polybutadiene, carboxy terminated	$(C_4H_6)_{100}C_2H_2O_4$	− 140	− 586	− 107	− 448	73.10	109.65	0.73	−	(6); (10)
polyisobutylene	$(CH_2)_n$	− 374	− 1570	− 331	− 1386	71.29	142.58	−	−	(10)
polypropyleneglycol	$(C_3H_5O)_n$-H_2O	− 888.1	− 3718	− 852.9	− 3571	51.19	103.37	17.56	−	(11)
polyvinyl nitrate	$(C_2H_3O_3N)_n$	− 275.4	− 1152	− 252.1	− 1055	22.46	33.68	33.68	11.23	(4) (values for % N = 15.73)
potassium nitrate	O_3NK	−1169	− 4891	−1157	− 4841	−	−29.67	9.89	K: 9.89	(5); (16)
n-propyl nitrate	$C_3H_7O_3N$	− 487.8	− 2041	− 436.8	− 1911	28.55	66.63	28.55	9.52	(4)
sodium hydrogen carbonate	CHO_3Na	−2705	−11318	−2691	−11259	11.90	11.90	35.71	Na: 11.90	(15)

Table 33. (continued)

Component	Empirical Formula	Enthalpy of Formation kcal/kg	Enthalpy of Formation kJ/kg	Energy of Formation kcal/kg	Energy of Formation kJ/kg	Number of Atoms per kilogram				Uses
						C	H	O	N	
sodium nitrate	O_3NNa	−1315	−5506	−1301	−5447	–	−35.29	11.76	Na. 11.76	(5); (16)
tetranitromethane	CN_4O_8	+ 46.9	+ 196.4	+ 65.0	+ 272.1	5.10	–	40.81	20.40	
Tetryl	$C_7H_5O_8N_5$	+ 16.7	+ 67.9	+ 35.3	+ 147.6	24.40	17.40	27.86	17.40	(2)
TNT (trinitrotoluene)	$C_5H_5O_6N_3$	− 70.6	− 295.3	− 52.4	− 219.0	30.82	22.01	26.40	13.20	(2)
toluene diisocyanate (TDI)	$C_9H_6O_2N_2$	− 179.7	− 752	− 162.7	− 681	51.71	34.47	11.49	11.49	(12)
triaminoguanidine nitrate	$CH_9N_7O_3$	− 68.8	− 287.9	− 35.2	− 147.2	5.98	53.86	17.95	41.89	(4)
triaminotrinitrobenzene	$C_6H_6N_6O_6$	− 129.4	− 541.3	− 108.7	− 455.0	23.25	23.25	23.25	23.25	(2); (4)
trimethylamine nitrate	$C_3H_{10}O_3N_2$	− 673.1	− 2816	− 636.7	− 2664	24.57	81.90	24.57	16.38	(17)
1,3,3-trinitroazetidine (TNAZ)	$C_3H_4N_4O_6$	+ 45.29	+ 189.50	+ 66.84	+ 279.77	15.62	20.82	31.23	20.82	(2)
trinitrobenzene	$C_5H_3O_5N_3$	− 48.8	− 204	− 32.1	− 135	28.15	14.08	28.15	14.08	(2)
trinitrochlorobenzene	$C_6H_2O_6N_3Cl$	+ 25.9	+ 108	+ 40.4	+ 169	24.24	8.08	24.24	12.12	Cl: 4.04
trinitropyridine	$C_5H_2O_6N_4$	+ 88,0	+ 368.5	+ 104.6	+ 437.9	23.35	9.34	28.03	18.68	(2)
trinitropyridine-1-oxide	$C_5H_2O_7N_4$	+ 102.5	+ 428.9	+ 119.2	+ 499.1	21.73	8.69	30.42	17.38	(2)
trinitroresorcinol (styphnic acid)	$C_6H_3O_8N_3$	− 510.0	− 2134	− 493.1	− 2063	24.48	12.24	32.64	12.24	(2) lead salt
urea	CH_4ON_2	−1326	−5548	−1291	−5403	16.65	66.60	16.65	33.30	(5); (17)
urea nitrate	$CH_5O_4N_3$	−1093	−4573	−1064	−4452	8.12	40.62	32.49	24.37	
water	H_2O	−3792	−15880	−3743	−15670	–	111.01	55.51	–	(17)
wood dust		−1116	−4672	−1090	−4564	41.7	60.4	27.4	–	(6)

Thermodynamic Calculation of Decomposition Reactions

Table 34. Enthalpy and energy of formation of gaseous reaction products.

Product	Formula	Molecular Weight	Enthalpy of Formation kcal/mol	Enthalpy of Formation kJ/mol	Energy of Formation kcal/mol	Energy of Formation kJ/mol
carbon monoxide	CO	28.01	−26.42	−110.6	−26.72	−111.9
carbon dioxide	CO	44.01	−94.05	−393.8	−94.05	−393.8
water (vapor)	H_2O	18.02	−57.80	−242.0	−57.50	−240.8
water (liquid)			−68.32	−286.1	−67.43	−282.3
nitrogen monoxide	NO	30.01	+21.57	+ 90.3	+21.57	+ 90.3
nitrogen dioxide (gas)	NO_2	46.01	+ 7.93	+ 33.2	+ 8.23	+ 34.5
nitrogen	N_2	28.02	± 0	± 0	± 0	± 0
oxygen	O_2	32.00	± 0	± 0	± 0	± 0
hydrogen chloride	HCl	36.47	−22.06	− 92.4	−22.06	− 92.4

Table 35. Moles per kilogram of inert Components

Component	Formula	mol/kg
aluminum oxide	Al_2O_3	9.808
barium sulfate	$BaSO_4$	4.284
calcium carbonate	$CaCO_3$	9.991
guhr (silicic acid)	SiO_2	16.65
iron oxide	Fe_2O_3	6.262
magnesium sulfate	$MgSO_4$	11.858
potassium chloride	KCl	13.413
sodium chloride	NaCl	17.11
talc	$Mg_3(SiO_{10})(OH)_2$	2.636 (21 atoms)
water (slurry component)*)	H_2O	55.509

Heat of evaporation of H_2O:
555.5 kcal/kg = 2325.9 kJ/kg = 10.01 kcal/mol = 41.91 kJ/mol

Thermodynamic Calculation of Decomposition Reactions

Table 36. Data of solid explosion reaction products.

Product	Formula	Molecular Weight	Energy of Formation kcal/mol	Molar Heat of Fusion kcal/mol	Number of Atoms in Molecule	Molar Heat of Sublimation kcal/mol
sodium chloride	NaCl	58.44	− 97.98	2	6.73	50.3
potassium chloride	KCl	74.56	−104.03	2	6.28	48.1
magnesium chloride	$MgCl_2$	95.23	−152.68	3	10.3	
sodium carbonate	Na_2CO_3	105.99	−269.4	6	8.0	
potassium carbonate	K_2CO_3	138.21	−274.0	6	7.8	
calcium carbonate	$CaCO_3$	100.09	−287.6	5		
magnesium carbonate	$MgCO_3$	84.33	−261.0	5		
barium carbonate	$BaCO_3$	197.37	−289.8	5		
lead carbonate	$PbCO_3$	267.22	−166.2	5		
aluminum oxide	Al_2O_3	101.96	−399.6	5	28.3	115.7 (at 2480 K)
iron oxide	Fe_2O_3	159.70	−196.1	5	−	
lead	Pb	207.21	± 0	1	1.21	46.34
mercury	Hg	200.61	± 0	1	−	14.0

Table 37. Molar internal energies of reaction products $\bar{c}_v(T-T_0)$; $T_0 = 25\ °C \approx 300$ K.

Temperature K	CO kcal/mol	CO kJ/mol	CO_2 kcal/mol	CO_2 kJ/mol	H_2O kcal/mol	H_2O kJ/mol	H_2 kcal/mol	H_2 kJ/mol	O_2 kcal/mol	O_2 kJ/mol	N_2 kcal/mol	N_2 kJ/mol	NO kcal/mol	NO kJ/mol	C*) kcal/mol	C*) kJ/mol
1000	3.65	15.28	6.34	26.55	4.64	19.43	3.41	14.28	4.33	18.13	3.59	15.03	3.77	15.78	3.31	13.86
1200	4.86	20.35	8.59	35.97	6.26	26.21	4.47	18.72	5.62	23.53	4.78	20.01	5.02	21.02	4.33	18.13
1400	6.11	25.35	10.91	45.68	7.99	33.45	5.58	23.36	6.93	29.02	6.01	25.16	6.30	26.38	5.32	22.27
1500	6.75	28.26	12.10	50.66	8.89	37.22	6.14	25.71	7.58	31.74	6.64	27.80	6.95	29.10	5.88	24.62
1600	7.39	30.94	13.31	55.73	9.81	41.07	6.72	28.14	8.22	34.42	7.28	30.48	7.61	31.86	6.38	28.71
1700	8.05	33.71	14.51	60.75	10.75	45.01	7.31	30.61	8.67	36.30	7.92	33.16	8.28	34.67	6.92	28.97
1800	8.69	36.39	15.77	66.03	11.52	48.23	8.13	34.04	9.61	40.24	8.57	35.88	9.02	37.77	7.48	31.32
1900	9.34	39.11	16.94	70.92	12.43	52.04	8.64	36.17	10.17	42.58	9.20	38.52	9.66	40.44	7.95	33.27
2000	19.99	41.83	18.13	75.91	13.38	56.02	9.17	38.39	10.77	45.09	9.84	41.20	10.31	43.16	8.46	35.42
2100	10.64	44.55	19.33	80.93	14.35	60.08	9.74	40.78	11.39	47.68	10.49	43.92	10.97	45.93	8.94	37.43
2200	11.31	47.35	20.56	86.08	15.34	64.22	10.32	43.20	12.04	50.41	11.15	46.68	11.64	48.73	9.53	39.90
2300	11.98	50.16	21.81	91.31	16.36	68.49	10.92	45.72	12.71	53.21	11.81	49.44	12.31	51.54	10.10	42.28
2400	12.65	52.97	23.07	96.59	17.39	72.81	11.54	48.31	13.39	56.06	12.48	52.25	12.99	54.38	10.68	44.71
2500	13.33	55.81	24.34	101.91	18.43	77.16	12.17	50.95	14.09	58.99	13.15	55.05	13.67	57.23	11.28	47.22
2600	14.02	58.70	25.61	107.22	19.49	81.60	12.81	53.63	14.80	61.96	13.83	57.90	14.36	60.12	11.88	49.74
2700	14.70	61.55	26.90	112.63	20.56	86.08	13.47	56.39	15.52	64.98	14.51	60.75	15.06	63.05	12.50	52.32
2800	15.39	64.44	28.20	113.07	21.63	90.56	14.13	49.16	16.25	68.03	15.20	63.64	15.75	65.94	13.12	54.93
2900	16.08	67.33	29.50	123.51	22.72	95.12	14.80	61.96	17.00	71.17	15.88	66.48	16.45	68.87	13.75	57.57
3000	16.78	70.26	30.81	123.00	23.81	99.69	15.47	64.77	17.75	74.31	16.57	69.37	17.15	71.80	14.30	59.87
3100	17.47	73.15	32.12	134.48	24.91	104.29	16.16	67.66	18.50	77.45	17.26	72.26	17.85	74.73	15.04	62.97
3200	18.17	76.08	33.44	140.01	26.02	108.94	16.84	70.50	19.26	80.64	17.96	75.19	18.56	77.71	15.69	65.69
3300	18.87	79.01	34.77	145.58	27.13	113.59	17.54	73.43	20.03	83.86	18.65	78.08	19.26	80.64	16.35	68.45
3400	19.57	81.94	36.10	151.15	28.25	118.28	18.23	76.32	20.80	87.08	19.35	81.01	19.97	83.61	17.01	71.22
3500	20.27	84.87	37.43	156.71	29.37	122.97	18.93	79.25	21.58	90.35	20.05	83.94	20.68	86.58	17.67	73.98
3600	20.97	87.86	38.76	162.28	30.50	127.70	19.64	82.23	22.37	93.66	20.74	86.83	21.39	89.55	18.34	76.78

Thermodynamic Calculation of Decomposition Reactions

Temperature K	CO kcal/mol	CO kJ/mol	CO_2 kcal/mol	CO_2 kJ/mol	H_2O kcal/mol	H_2O kJ/mol	H_2 kcal/mol	H_2 kJ/mol	O_2 kcal/mol	O_2 kJ/mol	N_2 kcal/mol	N_2 kJ/mol	NO kcal/mol	NO kJ/mol	C*) kcal/mol	C*) kJ/mol
3700	21.67	90.73	40.10	167.89	31.63	132.43	20.35	85.20	23.15	96.92	21.45	89.81	22.11	92.51	19.01	79.59
3800	22.38	93.71	41.44	173.50	32.76	137.16	21.06	88.17	23.94	100.23	22.15	92.74	22.82	95.54	19.69	82.44
3900	23.08	96.64	42.78	179.11	33.89	141.89	21.78	91.19	24.73	103.54	22.85	95.67	23.53	98.52	20.36	85.24
4000	23.79	99.61	44.13	184.77	35.03	146.67	22.49	94.16	25.53	106.89	23.55	98.60	24.25	101.53	21.04	88.09
4100	24.50	102.58	45.47	190.38	36.17	151.44	23.21	97.18	26.33	110.24	24.26	101.57	24.97	104.54	21.73	90.98
4200	25.20	105.51	46.82	196.03	37.32	156.25	23.94	100.23	27.13	113.59	24.96	104.50	25.68	107.52	22.41	93.83
4300	25.91	108.49	48.17	201.68	38.46	161.03	24.66	103.25	27.93	116.94	25.67	107.48	26.40	110.53	23.10	96.71
4400	26.62	111.46	49.52	207.34	39.61	165.84	25.39	106.30	28.74	120.33	26.37	110.41	27.12	113.55	23.79	99.60
4500	27.33	114.43	50.88	213.03	40.76	170.66	26.12	109.36	29.54	123.68	27.08	113.38	27.84	116.56	24.49	102.53
4600	28.04	117.40	52.23	218.68	41.91	175.47	26.85	112.42	30.35	127.07	27.79	116.35	28.56	119.58	25.17	105.38
4700	28.75	120.38	53.59	224.38	43.07	180.33	27.58	115.47	31.16	130.46	28.50	119.32	29.28	122.59	25.86	108.27
4800	29.46	123.35	54.95	230.07	44.22	185.14	28.31	118.53	31.98	133.90	29.21	122.30	30.00	125.61	26.56	111.20
4900	30.17	126.32	56.31	235.76	45.38	190.00	29.04	121.59	32.79	137.29	29.91	125.23	30.72	128.62	27.26	114.13
5000	30.88	129.29	57.67	241.46	46.54	194.86	29.78	124.68	33.60	140.68	30.62	128.20	31.44	131.63	27.95	117.02
5100	31.60	132.31	59.03	247.15	47.70	199.71	30.52	127.78	34.42	144.11	31.33	131.17	32.16	134.65	28.65	119.95
5200	32.31	135.28	60.39	252.85	48.86	204.57	31.25	130.84	35.24	147.54	32.04	134.15	32.89	137.71	29.35	122.88
5300	33.02	138.25	61.76	258.59	50.02	209.43	31.99	133.94	36.06	150.98	32.76	137.12	33.61	140.72	30.05	125.81

* If carbon separation occurs.

Table 38. Molar internal enthalpies of reaction products $\bar{c}_v (T-T_0)$; $T_0 = 25\,°C \approx 300$ K.

Temperature K	CO kcal/mol	CO kJ/mol	CO_2 kcal/mol	CO_2 kJ/mol	H_2O kcal/mol	H_2O kJ/mol	H_2 kcal/mol	H_2 kJ/mol	O_2 kcal/mol	O_2 kJ/mol	N_2 kcal/mol	N_2 kJ/mol	NO kcal/mol	NO kJ/mol
700	4.99	20.89	7.68	32.16	5.98	25.04	4.75	19.89	5.22	21.85	4.93	20.64	5.11	21.39
900	6.60	27.63	10.33	43.25	8.00	33.49	6.21	26.00	6.91	28.93	6.52	27.29	6.75	28.30
1100	8.24	34.50	13.04	54.59	10.12	42.37	7.71	32.28	8.62	36.09	8.14	34.00	8.43	35.29
1200	9.08	38.01	14.43	60.41	11.22	46.98	8.47	35.46	9.49	39.73	8.97	37.55	9.28	38.86
1300	9.92	41.53	15.84	66.32	12.34	51.67	9.25	38.73	10.36	43.38	9.81	41.07	10.14	42.46
1400	10.78	45.14	17.25	72.23	13.48	56.44	10.04	42.03	11.25	47.10	10.65	44.59	11.01	46.10
1500	11.62	48.66	18.70	78.30	14.45	60.50	11.06	46.31	12.54	52.50	11.50	48.15	11.95	50.03
1600	12.47	52.21	20.07	84.03	15.56	65.15	11.77	49.28	13.30	55.69	12.33	51.63	12.79	53.55
1700	13.32	55.77	21.46	89.86	16.71	69.96	12.50	52.33	14.10	59.04	13.17	55.14	13.64	57.11
1800	14.16	59.29	22.85	95.67	17.87	74.82	13.26	55.52	14.91	62.43	14.00	58.62	14.49	60.67
1900	15.03	62.93	24.28	101.66	19.06	79.80	14.04	58.79	15.76	65.99	14.87	62.26	15.36	64.31
2000	15.81	66.20	25.73	107.73	20.29	84.91	14.84	62.14	16.63	69.63	15.73	65.86	16.23	67.96
2100	16.77	70.22	27.19	113.84	21.51	90.06	15.66	65.57	17.51	73.31	16.60	69.50	17.11	71.64
2200	17.65	73.90	28.66	120.00	22.79	95.42	16.49	69.04	18.41	77.08	17.47	73.15	17.99	75.32
2300	18.54	77.63	30.16	126.28	24.01	100.53	17.33	72.56	19.32	80.89	18.35	76.83	18.88	79.05
2400	19.42	81.31	31.62	132.39	25.28	105.85	18.19	76.16	20.24	84.74	19.23	80.52	19.78	82.82
2500	20.30	85.00	33.11	138.63	26.54	111.12	19.04	79.72	21.16	88.60	20.11	84.20	20.66	86.50
2600	21.19	88.72	34.61	144.91	27.83	116.52	19.91	83.36	22.11	92.60	20.99	87.90	21.56	90.27
2700	22.09	92.49	36.12	151.23	29.12	121.93	20.78	87.01	23.06	96.55	21.88	91.61	22.46	94.04
2800	22.98	96.22	37.63	157.56	30.42	127.37	21.67	90.73	24.01	100.53	22.77	96.34	23.36	97.81
2900	23.88	99.99	39.15	163.92	31.73	132.85	22.55	94.42	24.97	104.55	23.67	99.11	24.27	101.62
3000	24.78	103.75	40.68	170.35	33.04	138.34	23.48	98.31	25.94	108.61	24.56	102.83	25.17	105.39
3100	25.68	107.52	42.21	176.73	34.36	143.87	24.34	101.91	26.91	112.67	25.46	106.60	26.08	109.20

Thermodynamic Calculation of Decomposition Reactions

Temperature K	CO kcal/mol	CO kJ/mol	CO_2 kcal/mol	CO_2 kJ/mol	H_2O kcal/mol	H_2O kJ/mol	H_2 kcal/mol	H_2 kJ/mol	O_2 kcal/mol	O_2 kJ/mol	N_2 kcal/mol	N_2 kJ/mol	NO kcal/mol	NO kJ/mol
3200	26.57	111.25	43.73	183.10	35.67	149.35	25.21	105.64	27.88	116.73	26.35	110.33	26.98	112.97
3300	27.47	115.02	45.26	189.50	37.00	154.92	26.14	109.45	28.87	120.88	27.24	114.05	27.89	116.78
3400	28.37	118.79	46.80	195.95	38.33	160.49	27.05	113.26	29.85	124.98	28.15	117.86	28.81	120.63
3500	29.28	122.60	48.34	202.40	39.66	166.06	27.96	117.07	30.84	129.13	29.05	121.63	29.72	124.44
3600	30.18	126.36	49.88	208.85	40.99	171.63	28.88	120.02	31.83	133.27	29.95	125.40	30.63	128.25
3700	31.09	130.17	51.43	215.34	42.33	177.24	29.79	124.73	32.83	137.46	30.85	129.17	31.55	132.10
3800	32.00	133.98	52.97	221.79	43.67	182.85	30.71	128.58	33.83	141.65	31.76	132.98	32.47	135.95
3900	32.89	137.71	54.51	228.23	45.01	188.46	31.69	132.69	34.82	145.79	32.65	136.71	33.37	139.72
4000	33.80	141.52	56.06	234.72	46.35	194.07	32.55	136.29	35.82	149.98	33.56	140.52	34.29	143.57
4100	34.71	145.33	57.61	241.21	47.70	199.72	33.48	140.18	36.83	154.21	34.46	144.28	35.21	147.42
4200	35.62	149.14	59.17	247.74	49.05	205.37	34.41	144.07	37.83	158.39	35.37	148.09	36.13	151.28
4300	36.53	152.95	60.72	254.23	50.40	211.02	35.34	147.97	38.85	162.66	36.28	151.90	37.05	155.13
4400	37.44	156.76	62.28	260.77	51.76	216.72	36.27	151.86	39.85	166.85	37.19	155.71	37.97	158.98
4500	38.35	160.57	63.84	267.30	53.11	222.37	37.20	155.76	40.78	170.75	38.10	159.52	38.89	162.83
4600	39.26	164.38	65.40	273.83	54.47	228.07	38.13	159.65	41.88	175.35	39.00	163.29	39.81	166.58
4700	40.17	168.19	66.96	280.36	55.84	233.80	39.07	163.59	42.89	179.58	39.91	167.10	40.73	170.54
4800	41.08	172.00	68.51	286.85	57.18	239.41	40.00	167.48	43.90	183.81	40.81	170.87	41.64	174.35
4900	41.99	175.81	70.07	293.38	58.54	245.11	40.93	171.37	44.92	188.08	41.72	174.68	42.57	178.24
5000	42.90	179.62	71.64	299.96	59.90	250.80	41.87	175.31	45.94	192.35	42.63	178.49	43.49	182.09

Values for carbon C can be taken from table 37.

Thermodynamic Calculation of Decomposition Reactions

Table 39. Equilibrium constants.

Temperature K	$K_1^*) = \dfrac{[CO]\cdot[H_2O]}{[CO_2]\,[H_2]}$	$K_2 = \dfrac{[CO]\cdot[NO]}{[N_2]^{1/2}\,[CO_2]}$	$K_3^{**}) = \dfrac{[CO]^2}{[CO_2]}$
1000	0.6929	1.791×10^{-16}	2.216×10^{-3}
1200	1.3632	2.784×10^{-13}	5.513×10^{-2}
1400	2.1548	5.238×10^{-11}	5.346×10^{-1}
1500	2.5667	4.240×10^{-10}	1.317
1600	2.9802	2.638×10^{-9}	2.885
1700	3.3835	1.321×10^{-8}	5.744
1800	3.7803	5.520×10^{-8}	10.56
1900	4.1615	1.982×10^{-7}	18.15
2000	4.5270	6.254×10^{-7}	29.48
2100	4.8760	1.767×10^{-6}	45.61
2200	5.2046	4.536×10^{-6}	67.67
2300	5.5154	1.072×10^{-5}	96.83
2400	5.8070	2.356×10^{-5}	134.2
2500	6.0851	4.858×10^{-5}	181.0
2600	6.3413	9.467×10^{-5}	238.1
2700	6.5819	1.755×10^{-4}	306.5
2800	6.8075	3.110×10^{-4}	387.0
2900	7.0147	5.295×10^{-4}	480.2
3000	7.2127	8.696×10^{-4}	586.8
3100	7.3932	1.383×10^{-4}	706.9
3200	7.5607	2.134×10^{-3}	841.0
3300	7.7143	3.207×10^{-3}	989.1
3400	7.8607	4.704×10^{-3}	1151
3500	7.9910	6.746×10^{-3}	1327
3600	8.1144	9.480×10^{-3}	1517
3700	8.2266	1.307×10^{-2}	1720
3800	8.3310	1.772×10^{-2}	1936
3900	8.4258	2.364×10^{-2}	2164
4000	8.5124	3.108×10^{-2}	2406
4100	8.5926	4.030×10^{-2}	2656
4200	8.6634	5.160×10^{-2}	2919
4300	8.7296	6.530×10^{-2}	3191
4400	8.7900	8.173×10^{-2}	3474
4500	8.8442	1.013×10^{-1}	3765
4600	8.8888	1.243×10^{-1}	4064
4700	8.9304	1.511×10^{-1}	4370
4800	8.9698	1.823×10^{-1}	4684
4900	9.0001	2.181×10^{-1}	5003
5000	9.0312	2.591×10^{-1}	5329
5100	9.0524	3.056×10^{-1}	5659
5200	9.0736	3.581×10^{-1}	5993
5300	9.0872	4.171×10^{-1}	6331

* Revised values published in JANAF Thermochemical Tables, The Dow Chemical Company, Midland, Michigan, 1965. The values for K_2 and K_3 are taken from Cook, The Science of High Explosives, Chapman & Hall, London 1958.

** Applies only in the presence of elementary carbon.

Thrust

Schub; poussée

In rocket technology, the recoil force produced by rearward gas discharge. It is expressed in tons, kiloponds, or newtons, and is one of the most important characteristic rocket parameters. The initial weight of a rocket must remain within a certain relation to the thrust. The launching thrust chosen is usually higher than the cruising thrust; this can be achieved by the use of → *Boosters.*

Thrust Determination

Schubmessung; mesurage de la poussée

The determination of the thrust of a rocket motor involves recording the time diagram of the force (tons, kp, or newtons) during combustion. The force is allowed to act on a support, with a pick-up element thrust cell interposed between them. The measurement is carried out by the aid of a strain gauge element (variation of resistance with pressure) or of a piezo-quartz element, and the results are recorded on an oscillograph connected in a compensation circuit. Modern measuring and computation techniques yield the total thrust time (impulse) immediately.

The same technique is applied for the determination of the pressure in the combustion chamber. The pressure cell must be attached to the previously prepared measuring points on the combustion chamber.

→ *Solid Propellant Rockets* and → *Specific Impulse.*

TNT

2,4,6-trinitrotoluene; Trinitrotoluol;
trinitrotoluene; Trotyl; tolite

$$\text{O}_2\text{N} - \underset{\underset{\text{NO}_2}{|}}{\overset{\overset{\text{CH}_3}{|}}{\text{C}_6\text{H}_2}} - \text{NO}_2$$

pale yellow crystals; if granulated; flakes
empirical formula: $C_7H_5N_3O_6$
molecular weight: 227.1
energy of formation: -52.4 kcal/kg = -219.0 kJ/kg
enthalpy of formation: -70.6 kcal/kg = -295.3 kJ/kg
oxygen balance: $-73,9\%$
nitrogen content: 18.50%
volume of explosion gases: 825 l/kg

heat of explosion
(H₂O gas): 871 kcal/kg = 3646 kJ/kg ⎫ calculated*)
(H₂O liq.): { 900 kcal/kg = 3766 kJ/kg ⎭
{ 1090 kcal/kg = 4564 kJ/kg experimental;**)
heat of detonation
specific energy: 92.6 mt/kg = 908 kJ/kg
density, crystals: 1.654 g/cm^3
density, molten: 1.47 g/cm^3
solidification point: 80.8 °C = 177.4 °F
heat of fusion: 23.1 kcal/kg = 96.6 kJ/kg
specific heat at 20 °C = 68 °F:
 0.331 kcal/kg = 1.38 kJ/kg
vapor pressure:

Pressure millibar	Temperature °C	°F
0.057	81	178 (melting point)
0.14	100	212
4	150	302
14	200	392
86.5	250	482 (beginning decomposition)

lead block test: 300 cm^3/10 g
detonation velocity, confined:
 6900 m/s = 22 600 ft/s at ρ = 1.60 g/cm^3
deflagration point: 300 °C = 570 °F
impact sensitivity: 1.5 kp m = 15 N m
friction sensitivity: up to 353 N
 no reaction
critical diameter of steel sleeve test: 5 mm

TNT is almost insoluble in water, sparingly soluble in alcohol, and soluble in benzene, toluene and acetone.

It is produced by nitration of toluene with mixed nitric and sulfuric acid in several steps. The trinitration step needs high concentrated mixed acids with free SO$_3$. There are batchwise and continous nitration methods. TNT for military use must be free from any isomer other than the 2,4,6 (the specifications). This can be done by recrystallization in organic solvents (alcohol; benzene) or in 62% nitric acid. The non-symmetrical isomers can be destroyed by washing with an aqueous

---
 * computed by the "ICT-Thermodynamic-Code".
 ** value quoted from *Brigitta M. Dobratz*, Properties of Chemical Explosives and Explosive Simulants, University of California, Livermore.

sodium sulfite solution; this processing, however, brings large quantities of red colored waste waters.

The purity grade of the product is determined by its solidification point. The minimum value for military purposes is 80.2 °C; the value for the pure 2,4,6-isomer is 80.8 °C; owing to the nitric acid recrystallization procedure, products with solidification points 80.6 and 80.7 °C are currently available.

TNT is still the most important explosive for blasting charges of all weapons. It is very stable, neutral, and does not attack metals; it can be charged by casting as well by pressing; it is insensitive and needs no phlegmatizers. It can be applied pure and mixed with ammonium nitrate (→ *Amatols*), aluminum powder (→ *Tritonal*), with RDX (→ *Composition B*), and combinations (→ *Torpex*, → *HBX*, → *Trialenes*). Furthermore, TNT is an important component of industrial explosives.

Cast charges of TNT are insensitive to blasting caps and need a booster for safe initiation. Pressed TNT is cap-sensitive.

Specifications

Appearance:	pale yellow flakes or crystals
solidification point, depending on quality requirement:	
not less than	80.6 °C = 177.1 °F
(the point for TNT as a component in industrial explosives can be lower)	80.4 °C = 176.7 °F 80.2 °C = 176.4 °F
volatiles: not more than	0.1%
tetranitromethane:	none
acidity as H_2SO_4: not more than	0.005%
alkalinity as Na_2CO_3: not more than	0.001%
benzene – insolubles: not more than	0.05%
ash content: not more than	0.01%

Other specification characteristics may also be included in the list, e.g. referring to the behavior or pressed specimens at 70 °C = 158 °F (→ *Exudation*).

Table 40. Data of the non-symmetrical TNT Isomers

TNT Isomer	Melting Point °C	°F	Heat of Fusion kcal/kg	kJ/kg	Beginning of Decomposition °C	°F
2,3,4-	112	234	25.8	108	282	540
2,3,5-	97	207	20.3	85	283	542
2,3,6-	108	226	24.9	104	280	536
2,4,5-	104	219	26.3	110	262	504
3,4,5-	132	270	21.2	89	288	550

TNT Isomer	Energy of Formation kcal/kg	Enthalpy of Formation kJ/kg	kcal/kg	kJ/kg
2,3,4-	+34.2	+143	+15.9	+ 67
2,3,5-	− 5.9	− 25	−24.2	−101
2,3,6-	+ 0.7	+ 3	−17.6	− 74
2,4,5-	+ 2.0	+ 8	−16.3	− 68
3,4,5-	+13.0	+ 54	− 5.3	− 22

Toe

Fuß-Vorgabe; distance entre le trou et la surface libre din massit

The perpendicular distance from blast hole to the free face measured at the floor elevation of the quarry pit.

Torpex

Castable mixtures of RDX (Hexogen), TNT, and aluminum powder, e.g. a 41:41:18 mixture. It contains 1% added wax. Other phlegmatized mixtures of similar compositions are → *DBX* and → *HBX*.

> density: 1.81 g/cm^3
> detonation velocity, confined:
> 7600 m/s = 24 900 ft/s at ρ = 1.81 g/cm^3

Tracers

Leuchtspur; compositions lumineuses

Tracers are slowly burning pyrotechnical compositions, used in tracer bullets, signalling cartridges, tracer rockets, and light-track shells. The colored flame is due to the presence of added salts such as sodium, barium, strontium, and copper salts. The signalling formulations also comprise smoke generators, including colored smoke generators and staining formulations which mark ground and water surfaces with organic dyes.

Transmission of Detonation

→ *Detonation; Sympathetic Detonation*

Transport Regulations

Transportvorschriften; règlement de transport

→ *ADR* (Road); → *RID* (Rail); → *IATA DGR, ICAO TI* (Air); → *IMDG Code* (Maritime).

Trauzl Test

(→ Lead Block Test)

Trauzl, an officer in the pioneer corps of the Austrian army, proposed the lead block method for the determination of the strength of explosive materials. The first international standardization of his method was enacted in 1904.

Trialenes

Mixtures of TNT, Hexogen, and aluminum powder in the following proportions: 80:10:10, 70:15:15, 60:20:20, 50:10:40, and 50:25:25. They were used as fillings for bombs and torpedo warheads in the Second World War.

Triaminoguanidine Nitrate

Triaminoguanidinnitrat; nitrate de triaminoguanidine; TAGN

$$H_2N-N=C{\begin{matrix}NH-NH_2 \cdot HNO_3 \\ NH-NH_2\end{matrix}}$$

colorless crystals
empirical formula: $CH_9N_7O_3$
molecular weight: 167.1
energy of formation: -35.2 kcal/kg = -147.2 kJ/kg
enthalpy of formation:
    -68.8 kcal/kg = -287.9 kJ/kg
oxygen balance: -33.5%

nitrogen content: 58.68 %
volume of explosion gases: 1163 l/kg
heat of explosion
  (H_2O liq.) 950 kcal/kg = 3974 kJ/kg
  (H_2O gas): 835 kcal/kg = 3492 kJ/kg
density: 1.5 g/cm^3
melting point: 216 °C = 420 °F (decomposition)
lead block test: 350 cm^3/10 g
detonation velocity, confined:
  5300 m/s at ρ = 0.95 g/cm^3
deflagration point: 227 °C = 440 °F
impact sensitivity:
  0.4 kp m = 4 N m
friction sensitivity: over 12 kp = 120 N
  pistil load crackling

This compound is prepared by reacting one mole of guanidine dinitrate with 3 moles of hydrazine hydrate at 100 °C for four hours. The reaction is accompanied by the liberation of ammonia.

The product is distinguished by high contents of hydrogen and nitrogen. TAGN serves as an ingredient for LOVA gun propellants with high force but comparable low compustion temperature. It is chemically not stable in connection to nitrate esters and some transition metal compounds, e.g. copper.

1,3,5-Triamino-2,4,6-trinitrobenzene

Triaminotrinitrobenzol; triaminotrinitrobenzène TATB

bright yellow crystals
empirical formula: $C_6H_6N_6O_6$
molecular weight: 258.1
energy of formation:
  −108.7 kcal/kg = −455.0 kJ/kg
enthalpy of formation:
  −129.4 kcal/kg = 541.3 kJ/kg
oxygen balance: −55.8 %
nitrogen content: 32.6 %
heat of explosion (H_2O liq.):
  732 kcal/kg = 3062 kJ/kg
density: 1.93 g/cm^3

melting point: 350 °C = 600 °F (decomp.)
lead block test: 175 cm^3/10 g
detonation velocity, confined:
 7350 m/s at ρ = 1.80 g/cm^3
deflagration point: 384 °C
impact sensitivity: 5 kp m = 50 N m
friction sensitivity: at 35 kp = 353 N
 pistil load no reaction

TATB is obtained by nitration of trichlorobenzene and conversion of the trinitrotrichlorobenzene to TATB.

It resists heat up to 300 °C (570 °F) and is very insensitive to friction and impact; the → *Critical Diameter* is high. Therefore the lead block test value listed above may be too low in comparison with its other performance data.

Direct contact with some transistion metals (e.g. copper) must be avoided.

1,3,5-Triazido-2,4,6-trinitrobenzene

Triazidotrinitrobenzol; triazidotrinitrobenzene; TATNB

green-yellow crystals
empirical formula: $C_6N_{12}O_6$
molecular weight: 336.2
oxygen balance: −28.6 %
nitrogen content: 50.0 %
volume of explosion gases: 755 l/kg
specific energy: 170 mt/kg = 1666 kJ/kg
density: 1.805 g/cm^3
melting point: 131 °C = 268 °F (decomp.)
lead block test: 470 cm^3/10 g
impact sensitivity: 0.5 kp m = 5 N m

The product can be obtained by treating 2,4,6-trichloro- 1,3,5-trinitrobenzene with an alkali metal azide in alcoholic solution. It is a lead-free → *initiating* and powerful explosive and does not produce toxic fumes (→ *Lead-free Priming Compositions*). The product undergoes a slow conversion into hexanitrosobenzene, thus losing its initiating power. This reaction reaches

at 20 °C = 68 °F after 3 years: 2.7%
at 35 °C = 95 °F after 1 year: 9.5%
at 50 °C = 122 °F after 10 days: 2.6%
at 50 °C = 122 °F after 6 years: 50%

TATNB can be "dead pressed", like mercury fulminate.

Tricycloacetone Peroxide

Acetonperoxid; peroxyde de tricycloacétone

colorless crystals
empirical formula: $C_9H_{18}O_6$
molecular weight: 222.1
oxygen balance: −151.3%
melting point: 91 °C = 196 °F
lead block test: 250 cm^3/10 g
impact sensitivity: 0.03 kp m = 0.3 N m
friction sensitivity: reaction at 0.01 kp = 0.1 N pistil load

This compound is formed from acetone in sulfuric acid solution when acted upon by 45% hydrogen peroxide. It displays the properties of primary explosives. It is not used in practice because of its tendency to sublimation and the relatively high friction sensitivity.

Triethyleneglycol Dinitrate

triglycol dinitrate, Triglykoldinitrat; dinitrate de triéthyleneglycol; TEGN

pale yellow liquid
empirical formula: $C_6H_{12}N_2O_8$

Triethyleneglycol Dinitrate

molecular weight: 240.1
energy of formation: −598.9 kcal/kg = −2505.8 kJ/kg
enthalpy of formation: −626.0 kcal/kg = −2619.2 kJ/kg
oxygen balance: −66.7%
nitrogen content: 11.67%
volume of explosion gases: 1065 l/kg
heat of explosion
  (H_2O liq.): 793 kcal/kg = 3317 k/kg
specific energy:
  91.7 mt/kg = 899 kJ/kg
density: 1.335 g/cm^3
lead block test: 320 cm^3/10 g
deflagration point: 195 °C = 383 °F
impact sensitivity: 1.3 kp m = 12.7 N m

Triglycol dinitrate is less volatile than → *Diethylenglycol Dinitrate*. It gelatinizes nitrocellulose just as well as diglycol dinitrate, i.e., better than nitroglycerine.

Its chemical stability is better than that of nitroglycerine or nitrocellulose, and at least as good as that of diglycol dinitrate.

Triglycol dinitrate is prepared by nitration of triglycol with mixed acid. The solubility of triglycol dinitrate in the waste acid is very high (∼9%). It is particularly suited for the production of low caloric → *Double Base Propellants*.

Specifications

density 20/4:	1.1230–1.1234 g/cm^3
boiling analysis,	
start: not below	280 °C = 536 °F
90% distilled: not over	295 °C = 563 °F
moisture: not more than	0.5%
chlorides: not more than	traces
acid, as H_2SO_4: not more than	0.02%
saponification value, as Na_2O: not more than	0.05%
reducing matter ($AgNO_3$-NH_3-test):	none

Trimethylamine Nitrate

Trimethylaminnitrat; nitrate de trimèthylamine

$$\begin{array}{c} H_3C \\ H_3C-N \cdot HNO_3 \\ H_3C \end{array}$$

colorless crystals
empirical formula: $C_3H_{10}N_2O_3$
molecular weight: 122.1
energy of formation: -636.7 kcal/kg = -2664.1 kJ/kg
enthalpy of formation: -637.1 kcal/kg = -2816.2 kJ/kg
oxygen balance: -104.8%
nitrogen content: 22.95%
volume of explosion gases: 1244 l/kg
explosion heat
  (H_2O liq.): 511 kcal/kg = 2140 kJ/kg
specific energy: 70.5 mt/kg = 691 J/kg

This salt, as well as other methylamine nitrates, has been proposed as a component of castable charges and of → *Slurries*.

Trimethyleneglycol Dinitrate

Trimethylenglykoldinitrat; dinitrate de trimethléneglycol

$$\begin{array}{l} CH_2-O-NO_2 \\ | \\ CH_2 \\ | \\ CH_2-O-NO_2 \end{array}$$

colorless oil
empirical formula: $C_3H_6N_2O_6$
molecular weight: 166.1
oxygen balance: -28.9%
nitrogen content: 16.87%
density: 1.393 g/cm³
lead block test: 540 cm³/10 g
deflagration point: 225 °C – 437 °F
(decomposition begins at 185 °C = 365 °F)
impact sensitivity: up to 2 kp m = 20 N m
  no reaction

Trimethyleneglycol dinitrate is less volatile than nitroglycol, but more so than nitroglycerine. Its solubilities in various solvents are similar to those of nitroglycerin. Like nitroglycerine, it forms satisfactory gels with nitrocelluloses. It causes headaches. Trimethyleneglycol dinitrate is prepared by nitration of trimethylene glycol with nitric acid or mixed acid at 0–10 °C. It is less impact-sensitive than nitroglycerine and is much more stable to store. It is now not longer used.

Trinitroaniline

Picramid: TNA

$$\underset{NO_2}{\underset{|}{O_2N-C_6H_2(NH_2)-NO_2}}$$

orange red crystals
empirical formula: $C_6H_4N_4O_6$
molecular weight: 228.1
energy of formation: -69.8 kcal/kg = -292.2 kJ/kg
enthalpy of formation: -88.0 kcal/kg = -368.1 kJ/kg
oxygen balance: -56.1%
nitrogen content: 24.56%
volume of explosion gases: 838 l/kg
heat of explosion
   (H_2O liq.): 858 kcal/kg = 3589 kJ/kg
   (H_2O gas): 834 kcal/kg = 3488 kJ/kg
density: 1.762 g/cm^3
melting point: 188 °C = 370 °F
lead block test: 310 cm^3/10 g
detonation velocity, confined:
   7300 m/s at ρ = 1.72 g/cm^3
deflagration point: 346 °C = 654 °F
impact sensitivity: 1.5 kp m = 15 N m
friction sensitivity: at 36 kp = 353 N
   pistil load no reaction
critical diameter of steel sleeve test: 3.5 mm

Trinitroaniline is prepared by reacting trinitrochlorobenzene with ammonia or by nitration of 4-nitroaniline.

Trinitroanisole

methyl picrate: 2,4,6-trinitrophenylmethylether;
Pikrinsäuremethylether Methoxytrinitrobenzol

$$\underset{NO_2}{\underset{|}{O_2N-C_6H_2(OCH_3)-NO_2}}$$

pale yellow crystals
empirical formula: $C_7H_5N_3O_7$
molecular weight: 243.1
energy of formation: -131.0 kcal/kg = -548.2 kJ/kg
enthalpy of formation: -150.6 kcal/kg = -630.1 kJ/kg

oxygen balance: −62.5 %
nitrogen content: 17.29 %
volume of explosion gases: 844 l/kg
heat of explosion
   (H_2O liq.): 903 kcal/kg = 3777 kJ/kg
   (H_2O gas): 874 kcal/kg = 3656 kJ/kg
specific energy: 99.1 mt/kg = 972 kJ/kg
density, crystals: 1.61 g/cm^3
density, molten: 1.408 g/cm^3
heat of fusion: 19.3 kcal/kg = 80.8 kJ/kg
melting point: 68 °C = 155 °F
lead block test: 295 g/cm^3
detonation velocity, confined:
   6800 m/s = 22 300 ft/s at ρ = 1.57 g/cm^3
deflagration point: 285 °C = 545 °F
impact sensitivity: 2.0 kp m = 20 N m
friction sensitivity: at 36 kp = 353 N
   pistil load no reaction
critical diameter of steel sleeve test: 12 mm

Trinitroanisole is insoluble in water, but is soluble in hot alcohol and ether. It is toxic.

It is prepared by treating dinitrochlorobenzene with methyl alcohol and alkali and nitration of the dinitroanisole thus obtained. Recrystallization from methanol yields the pure, pale yellow product.

It is very low sensitive explosives. Its performance is intermediate between that of TNT and picric acid. It has found use as a bomb filling component. It produces skin eczemas and is not safe physiologically. Owing to this and its very low melting point, the compound is only of limited practical importance.

1,3,3-Trinitroazetidine

Trinitroazetidin, TNAZ

```
        NO₂
         \
          N ── CH₂
          │    │
    H₂C ──C── NO₂
          │
          NO₂
```

empirical formula: $C_3H_4N_4O_6$
molecular weight: 192.09
energy of formation: +66.84 kcal/kg = +279.77 kJ/kg
enthalpy of formation: +45.29 kcal/kg = +189.50 kJ/kg
oxygen balance: −16.66 %
nitrogen content: 29.2 %

heat of explosion
(H2O liq.): 1516 kcal/kg = 6343 kJ/kg
(H2O gas): 1440 kcal/kg = 6024 kJ/kg
specific energy: 138.5 mt/kg = 1358 kJ/kg
density: 1.84 g/cm3
melting point.: 101 C

Several synthetic routes for trinitroazetidine have been described, e.g. from epichlorhydrin and tert. butylamine to give 1-tert.-butylazetidine and subsequent stepwise nitration to yield TNAZ.

Trinitroazetidine's performance data as an explosive lies between → Hexogen and → Octogen, but it is considerably less sensitive and therefore attractive for → LOVA (Low Vulnerability Ammunition) applications.

1,3,5-Trinitrobenzene

Trinitrobenzol; trinitrobenzène; TNB

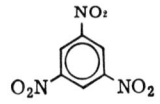

pale green-yellow crystals
empirical formula: $C_6H_3N_3O_6$
molecular weight: 213.1
energy of formation: -32.1 kcal/kg = -134.5 kJ/kg
enthalpy of formation: -48.8 kcal/kg = -204.2 kJ/kg
oxygen balance: -56.3%
nitrogen content: 19.72%
volume of explosion gases: 805 l/kg
heat of explosion
(H$_2$O liq.): 947 kcal/kg = 3964 kJ/kg
(H$_2$O gas): 926 kcal/kg = 3876 kJ/kg
specific energy: 107 mt/kg = 1050 kJ/kg
density: 1.76 g/cm^3
solidification point: 123.2 °C = 253.7 °F
heat of fusion: 16.0 kcal/kg = 67.2 kJ/kg
vapor pressure:

Pressure millibar	Temperature °C	°F	
0.5	122	252	melting point
2	150	302	
14	200	392	
133	270	518	

lead block test: 325 cm^3/10 g
detonation velocity, confined:
   7300 m/s = 23900 ft/s at ρ = 1.71 g/cm^3
impact sensitivity: 0.75 kp m = 7.4 N m
friction sensitivity: up to 36 kp = 353 N
   pistil load no reaction

Trinitrobenzene is insoluble in water, sparingly soluble in hot alcohol, and is readily soluble in acetone, ether, and benzene.

Trinitrobenzene is formed by of decarboxylation of trinitrobenzoic acid. It can also be prepared from trinitrochlorobenzene by reduction with copper in alcohol. Further nitration of dinitrobenzene also yields trinitrobenzene, but the reaction must be carried out under very severe conditions (high SO$_3$-concentration in the mixed acid, high nitration temperature), and the yields are low.

All the above syntheses are difficult and uneconomical. For this reason, no practical application has been found, despite the fact that its strength and detonation velocity are superior to those of TNT, and that it is very stable.

Specifications

solidification point: not below	121 °C = 250 °F
moisture, volatile matter: not more than	0.1 %
glow residue: not more than	0.2 %
insoluble in benzene: not more than	0.2 %
HNO$_3$: not more than	traces
sulfate, as N$_2$SO$_4$: not more than	0.02 %
acid, as H$_2$SO$_4$: not more than	0.005 %
alkali:	none
Abel test 80 °C = 176 °F: not under	30 min

Trinitrobenzoic Acid

Trinitrobenzosäure; acide trinitrobenzoique

$$\underset{NO_2}{\underset{|}{O_2N}}\text{—}\underset{}{C_6H_2}\text{—}\underset{}{\overset{COOH}{|}}\text{—}NO_2$$

yellow needles
empirical formula: C$_7$H$_3$N$_3$O$_8$
molecular weight: 257.1
energy of formation: −358.4 kcal/kg = −1500 kJ/kg

enthalpy of formation: -374.6 kcal/kg = -1567 kJ/kg
oxygen balance: -46.7%
nitrogen content: 16.35%
volume of explosion gases: 809 l/kg
heat of explosion
  (H_2O liq.): 719 kcal/kg = 3008 kJ/kg
  (H_2O gas): 700 kcal/kg = 2929 kJ/kg
specific energy: 88.8 mt/kg = 871 kJ/kg
lead block test: 283 cm^3/10 g
impact sensitivity: 1 kp m = 10 N m
friction sensitivity: at 36 kp = 353 N
  pistil load no reaction
critical diameter of steel sleeve test: 2 mm

Trinitrobenzoic acid is sparingly soluble in water, and soluble in alcohol and ether. It is prepared by oxidation of TNT with nitric acid, or with a solution of $KClO_3$ in nitric acid, or with a chromic acid mixture.

The crude product is purified by dissolving it in a dilute sodium carbonate solution and reprecipitating with sulfuric acid. If trinitrobenzoic acid is exposed to water vapor for a long period of time, → *Trinitrobenzene* is formed as a result of the liberation of CO_2.

Trinitrochlorobenzene

picryl chloride; Trinitrochlorbenzol; trinitrochlorobenzene; chlorure de picryle

pale yellow needles
empirical formula: $C_6H_2N_3O_6Cl$
molecular weight: 247.6
energy of formation: +40.4 kcal/kg = +169 kJ/kg
enthalpy of formation: +25.9 kcal/kg = 108.2 kJ/kg
oxygen balance: -45.3%
nitrogen content: 16.98%
density: 1.797 g/cm^3
solidification point: 83 °C = 181 °F
heat of fusion: 17.5 kcal/kg = 73.3 kJ/kg
vapor pressure:

Pressure millibar	Temperature °C	°F	
0.05	83	181	melting point
0.2	100	212	
2.0	150	302	
12.5	200	392	
100	270	518	

lead block test: 315 cm^3/10 g
detonation velocity, confined:
   7200 m/s = 23600 ft/s at ρ = 1.74 g/cm^3
deflagration point: 395–400 °C
impact sensitivity: 1.6 kp m = 16 N m
friction sensitivity: up to 36 kp = 353 N
   pistil load no reaction

Trinitrochlorobenzene is sparingly soluble in alcohol and benzene, somewhat more soluble in ether, and insoluble in water.

It is prepared by nitration of dinitrochlorobenzene. Manufacture is difficult, and highly concentrated acid must be employed.

Trinitrochlorobenzene is just as insensitive as TNT, its brisance is somewhat higher, and its density and heat stability are excellent.

2,4,6-Trinitrocresol

Trinitrometakresol; trinitrométacrésol

$$\underset{\underset{NO_2}{\big|}}{\overset{\overset{OH}{\big|}}{O_2N-\bigcirc-NO_2}}-CH_3$$

yellow needles
empirical formula: $C_7H_5N_3O_7$
molecular weight: 243.1
energy of formation: −229.7 kcal/kg = −961.2 kJ/kg
enthalpy of formation: −248.0 kcal/kg = −1038 kJ/kg
oxygen balance: −62.5 %
nitrogen content: 17.95 %
volume of explosion gases: 844 l/kg
heat of explosion
  (H_2O liq.): 805 kcal/kg = 3370 kJ/kg
  (H_2O gas): 776 kcal/kg = 3248 kJ/kg
specific energy: 87.8 mt/kg = 861 kJ/kg

density: 1.68 g/cm^3
melting point: 107 °C = 225 °F
heat of fusion: 25.5 kcal/kg = 107 kJ/kg
lead block test: 285 cm^3/10 g
detonation velocity, confined:
  6850 m/s = 22500 ft/s at ρ = 1.62 g/cm^3
deflagration point: 210 °C = 410 °F
impact sensitivity: 1.2 kp m = 12 N m
friction sensitivity: up to 36 kp = 353 N
  pistil load no reaction

Trinitrocresol is readily soluble in alcohol, ether, and acetone, and is sparingly soluble in water.

It is prepared by nitration of m-cresoldisulfonic acid. During the First World War, 60:40 mixtures of trinitrocresol and picric acid were used (under the name of Kresylith) as grenade fillings, since they melt at a temperature as low as 85 °C (185 °F).

Trinitromethane

Nitroform

$$HC\begin{matrix}NO_2\\-NO_2\\NO_2\end{matrix}$$

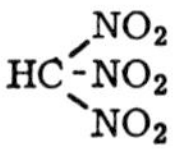

oil with a pungent smell
empirical formula: CHN_3O_6
molecular weight: 151.0
oxygen balance: +37.1 %
nitrogen content: 27.83 %
density: 1.59 g/cm^3
melting point: 22 °C = 72 °F
boiling point at 23 millibar pressure: 48 °C = 118 °F

Nitroform is obtained when acetylene is introduced into nitric acid; it may also be prepared from tetranitromethane.

Nitroform cannot be used on its own either as an oxidant or as an explosive. It is possible, however, to add nitroform to formaldehyde and to prepare explosives from the resulting trinitroethanol product (→ *Bi-trinitroethylamine* and *Bi-trinitroethylurea*).

Its hydrazin salt HNF is proposed as halogen free oxidizer for propellant formulations.

Trinitronaphthalene

Trinitronaphthalin; trinitronaphthalène; Naphtit; Trinal

brownish crystals
empirical formula: $C_{10}H_5N_3O_6$
molecular weight: 263.2
oxygen balance: -100.3%
nitrogen content: 15.97%
volume of explosion gases: 723 l/kg
heat of explosion
  (H_2O liq.): 842 kcal/kg = 3521 kJ/kg
  (H_2O gas): 819 kcal/kg = 3425 kJ/kg
specific energy: 76 mt/kg = 746 kJ/kg
melting point: 115 °C = 239 °F (beginning softening of the isomer mixture)
detonation velocity: 6000 m/s = 19700 ft/s
deflagration point: 350 °C = 660 °F
impact sensitivity: 2 kp m = 19 N m

Trinitronaphthalene is soluble in glacial acetic acid, and is sparingly soluble in alcohol and ether. It is prepared by dissolving mononitronaphthalene in concentrated sulfuric acid and adding mixed acid. The product thus obtained – a mixture of α-(1,3,5-), β-(1,3,8), and γ-(1,4,5)-isomers – melts above 115 °C (239 °F).

Trinitronaphthalene is difficult to detonate. It was used, in mixture with other compounds, as grenade filling, especially in France and Belgium. It is of no technological interest at present.

Trinitrophenoxethyl nitrate

Trinitrophenylglykolethernitrat; nitrate de trinitrophénoxethyle

yellowish-white crystals
empirical formula: $C_8H_6N_4O_{10}$
molecular weight: 318.2
energy of formation: -189.8 kcal/kg = -794.0 kJ/kg

enthalpy of formation: −208.4 kcal/kg = −871.9 kJ/kg
oxygen balance: −45.3%
nitrogen content: 17.61%
volume of explosion gases: 878 l/kg
heat of explosion
  (H_2O liq.): 935 kcal/kg = 3911 kJ/kg
  (H_2O gas): 906 kcal/kg = 3792 kJ/kg
specific energy: 115 mt/kg =1131 kJ/kg
density: 1.68 g/cm^3
melting point: 104.5 °C = 219.6 °F
lead block test: 350 cm^3/10 g
detonation velocity, confined:
  7600 m/s = 25 000 ft/s at ρ = 1.65 g/cm^3
deflagration point: >300 °C = >570 °F
impact sensitivity: 0.8 kp m = 7.9 N m

The compound is insoluble in water, but soluble in acetone and toluene. It is very stable and gelatinizes nitrocellulose on heating.

It is prepared by nitration of the corresponding dinitro compound.

2,4,6-Trinitrophenylnitraminoethyl Nitrate

Trinitrophenylethanolnitraminnitrat; nitrate de trinitrophenylnitramineéthyl; Pentryl

yellowish crystals
empirical formula: $C_8H_6N_6O_{11}$
molecular weight: 362.2
oxygen balance: −35.4%
nitrogen content: 23.19%
density: 1.75 g/cm^3
melting point: 128 °C = 262 °F
lead block test: 450 cm^3/10 g
deflagration point: 235 °C = 455 °F
impact sensitivity: 0.4 kp m = 4 N m

The compound is soluble in water, sparingly soluble in common organic solvents, but is soluble in nitroglycerine. Its stability is satisfactory, but one of the five NO_2-groups in the molecule is a nitrate rather than a nitro group, so that the compound cannot be as stable as e.g., TNT.

It is prepared by nitration of dinitrophenylaminoethanol, which is in turn formed by condensation of dinitrochlorobenzene with monoethanolamine.

Trinitropyridine

Trinitropyridin; Trinitropyridine, 2,4,6-Trinitropyridine, TNPy

yellow needles
empirical formula: $C_5H_2N_4O_6$
molecular weight: 214.1
energy of formation: +437.9 kJ/kg = +104.6 kcal/kg
enthalpy of formation: +368.5 kJ/kg = +88.0 kcal/kg
oxygen value: −37.4 %
nitrogen content: 26.17 %
specific energy: 129 mt/kg = 1260 kJ/kg
explosion heat (H_2O liq.): 4418 kJ/kg = 1056 kcal/kg
normal volume of gases: 818 l/kg
fusion point: 162 °C (sublimation)
density: 1.77 g/cm^3
detonation rate: 7470 m/s at ρ = 1.66 g/cm^3
impact sensitivity: 4.5−6.5 N m = 0.46−0.66 kp m
sensitivity of friction: at 353 N = 36 kp pin load, no reaction

Trinitropyridine is obtained by means of reduction of → *Trinitropyridine-N-oxide* with sodium nitrite in a solution of sulfuric acid.

Although this compound is a potent explosive, it has yet to gain widespread use.

Trinitropyridine-N-oxide

Trinitropyridin-N-oxid; Trinitropyridine-N-oxide, 2,4,6-Trinitropyridin-1-oxide, TNPyOX

yellow crystals
empirical formula: $C_5H_2N_4O_7$
molecular weight: 230.1

energy of formation: +499.1 kJ/kg = +119.2 kcal/kg
enthalpy of formation: +428.9 kJ/kg = +102.5 kcal/kg
oxygen value: −27.8%
nitrogen content: 24.34%
specific energy: 134 mt/kg = 1315 kJ/kg
explosion heat: (H_2O liq.): 5320 kJ/kg = 1271 kcal/kg
normal volume of gases: 777 l/kg
fusion point: 170 °C (decomposition)
density: 1.86 g/cm^3
detonation rate: 7770 m/s at ρ = 1.72 g/cm^3
impact sensitivity: 1.5−3.0 N m = 0.15−0.31 kp m
sensitivity to friction: 157 N = 16 kp

Trinitropyridine-N-oxide is produced through a cyclical reaction of potassium salt of 2,2-dinitroethanol in diluted phosphoric acid.

The product serves as the basis material for the production of → *Trinitropyridine*, which is not obtainable by means of direct nitration.

2,4,6-Trinitroxylene

Trinitroxylol; trinitrométaxylène; TNX

pale yellowish needles
empirical formula: $C_8H_7N_3O_6$
molecular weight: 241.2
energy of formation: −82.1 kcal/kg = −343.4 kJ/kg
enthalpy of formation: −101.7 kcal/kg = −425.6 kJ/kg
oxygen balance: −89.57%
nitrogen content: 17.42%
volume of explosion gases: 843 l/kg
heat of explosion
 (H_2O liq.): 845 kcal/kg = 3533 kJ/kg
 (H_2O gas): 810 kcal/kg = 3391 kJ/kg
specific energy: 83.5 mt/kg = 819 kJ/kg
melting point: 182 °C = 360 °F

Separation of the xylene isomers is not easy, and nitration to the trinitrate stage is technically difficult.

Tritonal

A castable mixture of 20−40% aluminum and 80−20% TNT.

Trixogen

A pourable mixture of RDX with TNT.

Ullage

Empty volume provided for thermal expansion of propellant in liquid propellant tank.

Unbarricaded

The absence of a natural or artifical barricade around explosive storage areas or facilities.

Unconfined Detonation Velocity

Detonationsgeschwindigkeit ohne Einschluss; velocite de detonation sans enserrement

The detonation velocity of an explosive material without confinement; for example, a charge fired in the open.

Underwater Detonations*)

Unterwasserdetonationen; détonations sous l'eau

Destructive effects of underwater detonations, differ in their distant and proximity effects. The first effect is caused by the action of the pressure shock wave, the latter mainly by the thrust produced by the expanding gas bubble.

Basically the process can be subdivided into three distinct stages:

1. Detonation

The detonation of an explosive charge is triggered off by the fuse. The explosive matter undergoes an extremely rapid deterioration, and the heat developed during this process creates a large amount of gas. This first enters into the cavity previously occupied by the solid explosive and is therefore under a high degree of pressure. This hot compressed gas constitutes the whole of the performance potential.

---

* Extract of a lecture held by *W. E. Nolle* at the annual meeting of the Fraunhofer Institut at Karlsruhe, 1978.

2. Shock wave

The adjacent layer of water is compressed under the influence of this high pressure, which in turn transfers that pressure onto the next layer, and this transfers the pressure onto the next one, and so forth in a chain reaction.

The velocity of propagation increases with pressure, thus creating a steeply ascending pressure front, which imparts the nature of a shock wave to the pressure wave. At the onset, the velocity of propagation exceeds that of the speed of sound, but deteriorates with increasing distance, i.e. to approximately 1450 m/s.

The maximum pressure achieved is directly proportional to the cube root of the charge weight, and inversely proportional to the distance, resulting in the following approximate formula:

$$P_{max} = C \frac{L^{1/3}}{e}$$

p: pressure in bar
L: loading weight in kg
e: distance in m
c: empirical factor; ≈ 500

3. Gas bubble

As stated previously, the gas formed by the underwater explosion first enters the small cavity previously occupied by the explosive, thus creating a gas bubble under a high degree of pressure. The water surrounding the bubble gives away, and the gas bubble expands. This causes the water mass to move radially at great velocity away from the point of explosion. This movement is known as the "thrust".

The maximum amount of kinetic energy imparted to the water during an explosion is called the *thrust energy*.

The increase of expansion of the gas bubble causes a decrease in pressure on the enclosed gases, which slows down expansion to the point where all of the kinetic energy is expended. This causes lowering of pressure in the gas bubble contents, influenced by the static water pressure, and the water mass engulfs it again. The gases are compressed again up to a second minimum, at which point another pressure wave is formed (secondary pressure wave). Oscillation of the gas bubble can be repeated several times, causing a third, and, under favorable conditions, further minima. The gas bubble is propelled upwards towards the surface of the water. The difference in pressure between the top and the bottom layer of the bubble causes the bottom layer to move at greater speed, thus forcing it upwards into the bubble. It is possible for both surfaces to meet. Within a limited area the water receives an upward thrust, creating the so-called waterhammer (water jet).

From the observations it becomes clear that the most effective underwater explosives are those which can produce a high-pressure gas bubble for the formation of the thrust.

Mixtures containing a high percentage of aluminum powder have proved to be most effective (→ *Aluminum Powder;* → *Torpex*; → *Trialenes;* → *Tritonal*).

References:

G. Bjarnholt and *P. Holmberg*, Explosive Expansion Work in Underwater Detonations, Reprints of the Sixth Symposium on Detonation, San Diego, 1976 (from: Office of Naval Research, San Diego, USA).

S. Paterson and *A. H. Begg*, Underwater Explosion, Propellants and Explosives 3,63–89 (1978).

Upsetting Tests

Stauchung; écrasement du crusher

Upsetting tests are used to determinate the → *Brisance* of the explosives. An unconfined cartridge (enveloped in paper or in metal sheet) acts upon a copper or lead crusher; the loss of height of the crusher is a measure for the brisant performance of the tested explosive (→ *Brisance*).

The test according to *Kast* is shown in Fig. 25; the cartridge shock acts by means of a guided pestle onto a copper crusher of 7 mm ⌀ and 10.5 mm height.

The simplified test according to *Hess* is shown in Fig. 26 (see opposite page):

A lead cylinder, 60 mm (2.36 in.) high, 40 mm (1.57 in.) ⌀, protected by two, 6 mm-thick steel disks, is upset by a 100-g (3.53 oz.) cartridge of the same diameter, 40 mm. The cylinder is pressed down into a mushroom shape; in the case of sensitized special gelatins for seismic use, the cylinder can be destroyed completely.

Upsetting Tests

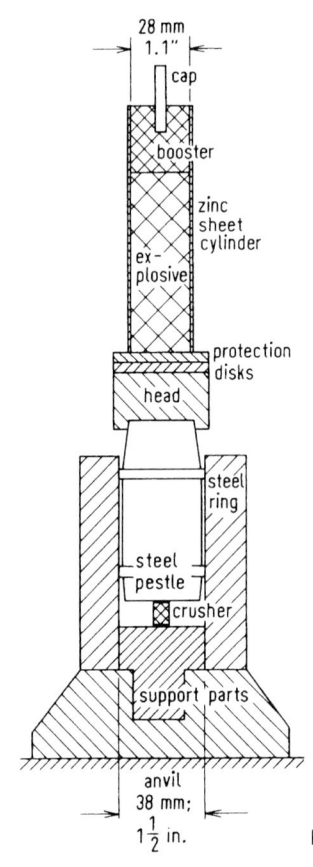

Fig. 25. Upsetting test according to *Kast*.

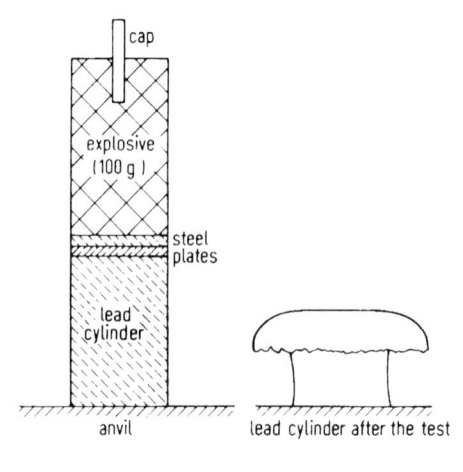

Fig. 26. Upsetting test according to *Hess*.

Urea Nitrate

Harnstoffnitrat; nitrate d'urée

$$O=C\begin{array}{c}NH_2\\ \\ NH_2 \cdot HNO_3\end{array}$$

colorless crystals
empirical formula: $CH_5N_3O_4$
molecular weight: 123.1
energy of formation: -1064 kcal/kg = -4452 kJ/kg
enthalpy of formation: -1093 kcal/kg = -4573 kJ/kg
oxygen balance: -6.5%
nitrogen content: 34.14 %
volume of explosion gases: 910 l/kg
heat of explosion
   (H_2O liq.): 767 kcal/kg = 3211 kJ/kg
   (H_2O gas): 587 kcal/kg = 2455 kJ/kg
specific energy: 77 mt/kg = 755 kJ/kg
density: 1.59 g/cm^3
melting point: 140 °C = 284 °F (decomposition)
lead block test: 270 cm^3/10 g
deflagration point: 186 °C = 367 °F
impact sensitivity: up to 49 N m no reaction
friction sensitivity: up to 353 N
   no reaction
critical diameter of steel sleeve test:
at 1 mm ⌀ no destruction of steel sleeve

Urea nitrate is readily soluble in hot water and sparingly soluble in ethanol. Its thermal stability is satisfactory. The compound is prepared from urea and nitric acid. The salt is strongly acidic. Chemical stability is poor.

U-Zünder

U-detonators are manufactured by ORICA Germany GmbH, Troisdorf, for use in standard situations and where high safety against electrostatic discharges is required. They are safe against 0.45 A and 8 mJ/ohm. All-fire current is 1.5 A, all-fire energy 16 mJ/ohm. U-detonators are available as instantaneous detonators and with 20 ms and 30 ms short period delay, 18 delays each, and 24 delays of 250 ms long period delay.

U-Zündmaschinen: the corresponding blasting machines are produced by WASAGCHEMIE Sythen, Germany.

Vacuum Test

This test, which was developed in the USA and has been adopted by several countries, and is a modification of the → *Taliani Test*, in which the gaseous products of the reaction are determined volumetrically rather than by manometry. The test, which is carried out at 100 °C (212 °F) for single base propellants and at 90 °C (194 °F) for multi-base propellants, is terminated after 40 hours, unlike the *Taliani Test*, which is interrupted after a given pressure or a given volume has been attained.

The vacuum test is used for compatibility testing and applied as a so-called reactivity test. The compatibility between the explosive and the contact material (adhesive, varnish, etc.) is tested by determining the gases liberated by the explosive alone, by the contact material alone, and by the two together. The measure of compatibility (reactivity) is the difference between the sum of the gas volume liberated by each component separately and the gas volume obtained after storing the explosive and the contact material together. If this difference is below 3 ml, the compatibility is considered to be "stable", between 3 and 5 ml, the compatibility is considered „uncertain"; above 5 ml, the two materials are incompatible.

Versuchsstrecke

Berggewerkschaftliche Versuchsstrecke

D-44239 Dortmund-Derne

German institute for research and testing of equiptments and materials for use in gassy coal mines (including → *Permitted Explosives,* → Bridgewire Detonators and → *Blasting Machines*).

Vibrometer

Erschütterungs-Messgerät

Vibrometers are instruments to measure the intensity of shock waves caused by blasting operations. The magnitude of the shock depends on the kind of rocks, underground conditions and distance to the people and buildings to be protected. The regular control of ground shocks caused by blasting is therefore, in the interest of companies active in this field to safeguard friends relations with the neighborhood. Vibrometer records, can also be important in forensic defense against claims in densely populated areas.

The following vibrometers are developed, produced and distributed by WASAGCHEMIE Sythen, Haltern, Federal Republic of Germany:

Vibrometer ZEB/SM 3 and. ZEBI/SM 6 DIN 45669
Indication of the maximum values in alphanumeric display. Registration of the complete recording with the aid of an UV (ultraviolet) recorder.

Vibrometer ZEB/SM 3 DS and ZEBI/SM 6 DS DIN 45669
Indication of the maximum values and frequencies on the screen. Registration of the complete recording with the aid of a four-color plotter, also in graphics form.

Vibrometer ZEBI/SM 6 C DIN 45669
Latest processor technology with hard and floppy disk storage possibilities. Display on a screen and registration of the complete recording on a four-color plotter, both also in graphics form.

Vieille Test

This method for the stability testing of propellants was proposed by *Vieille* in 1896. The sample is heated at 110 °C (230 °F) in the presence of a strip of litmus paper, and is then exposed to air at room temperature overnight, after which the cycle is repeated. This treatment is continued until the litmus paper turns red within one hour. The overall duration of the heating operations thus performed is a measure of the stability.

The advantage of the method is that when the propellant is periodically exposed to the atmosphere, it can reabsorb moisture, which means that the decomposition takes place under realistic conditions. The test is now much less frequently applied ever since a powder manufactured with pentanol as a solvent, which had been tested by this method, had decomposed on board of two warships, which were sunk by the resulting explosion (1911). The *Vieille* test is now used only in France and in Belgium.

Volume of Explosion Gases

fume volume; Normalgasvolumen; volume des produits d'explosion

The volume of the gases (fumes) formed by the explosive reaction, in liters per kg of explosive material, as calculated from the chemical composition of the explosive. The calculation of the number of gas moles of the decomposition products takes the equilibria (e.g. water gas equilibrium and Boudourd equilibrium) at the explosion temperature and during cooling to 1500 K into account. Below 1500 K the equilibria are considered as "frozen".

Conventionally, the volume of explosion gases refers to 0 °C and 1.013 bar. Water is considered to be gaseous.

The volume can be determinated experimentally by test explosion in the → *Bichel Bomb*.

Volume Strength

Same as *Cartridge Strength* or → *Bulk Strength*. Also → *Weight Strength* and → *Strength*.

WASADEX 1

Trade name of a → ANFO explosives marketed in Germany by WASAG-CHEMIE.

Water-driven Injector Transport

Emulsionsförderung; transport par injection d'eau

The liquid nitrate esters – nitroglycerine and nitroglycol – are highly sensitive to impact; handling of these substances in the factory in free unbound condition is dangerous. They are conveyed in the form of emulsions: the explosive oil is sucked up by means of a compressed-water-driven injector, and the emulsion sent through conduit pipes for processing (mixing houses). It is then separated from the carrier water and, if required, is dried by passing through a salt filter.

Water-gel Explosives

Slurries; Sprengschlamm; bouillies

→ *Slurries* and → *Emulsion Slurries*.

Water Resistance

Wasserfestigkeit; résistance a l'eau

In the USA the following method is employed for testing the water resistance of commercial explosives:

Sixteen regularly spaced holes (about 6 mm ⌀) are cut in the cartridge paper (30 mm in diameter, 200 mm long) of the explosive to be tested, and the flaps on the front faces are sealed with tallow. The cartridges thus prepared are placed in a flat, porcelaincoated dish covered with a thin layer of sand, and water at 17–25 °C (63–77 °F) is poured over the sand layer up to a height of about 25 mm. The cartridges are left under water for a certain period of time, are then taken out, the seal is cut off at one end, and the cartridge is tested for detonation and transmission with the aid of a No. 6 blasting cap. The criterion for the water resistance of the explosive is the time of exposure to water, after which it still retains its capacity to detonate the cartridge in three trials, without leaving any non-detonated residual explosive behind.

There is no generally accepted quality classification. Nevertheless, water resistance of an explosive is considered to be satisfactory, acceptable, or poor if the cartridge can still be detonated after 24, 8, or 2 hours respectively.

In Germany, the following method for testing the water resistance of powder-form permissibles has been laid down at the Test Station at Dortmund-Derne.

A train of four cartridges is fixed in a file on a wooden board; the first of the four cartridges is equipped with a detonator No. 8. Five longitudinal, 2 cm long notches, uniformly distributed over the circumference, are cut into each cartridge. The train is immersed for 5 hours in water, in a horizontal position, 20 cm under the water surface, after which they are detonated. The train must detonate in its entirety.

This method, including some additions regarding the preparation of the test sample is standardized as EN 13631-5 as a so-called Harmonized European Standard. The water resistance of partly water – soluble, powder-form explosives (e.g. ammonium nitrate explosives or blasting agents) can be improved by the addition of hydrophobic or gelling agents. If e.g. → *Guar Gum* is added, the water entering immediately forms a gel which blocks the penetration of more water.

Water-resistant Detonator

Unterwasserzünder; détonateur pour tir sous l'eau

Such detonators differ from conventional detonators in being watertight; water cannot penetrate into the detonator even under increased water pressure (→ *Bridgewire Detonator*).

Water Stemming

Wasserbesatz; bourrage à l'eau

Water stemming of explosive-blasted boreholes consists of waterfilled cartridges made of plastic material, which give some protection against firedamp and coal dust explosions.

Web

In a solid propellant grain, the minimum distance which can burn through as measured perpendicular to the burning surface (→ *Burning Rate*).

Weight Strength

The strength of an explosive material per unit of weight expressed as a percentage of the weight of a recognized explosive standard.
→ Strength.

Wetter

Prefix given to all permitted explosives in Austria and in the Germany. The following list gives an overview of all German permitted explosives:

Wetter	Manufacturer	Density g/cm^3	Safety Class	Cartridge mm
Westfalit C	WASAGCHEMIE	1.17	I	30
Westfalit D	WASAGCHEMIE	1.17	I	40
Roburit B	WASAGCHEMIE	1.2	II	30
Securit C	WASAGCHEMIE	1.18	III	30

The powder form explosives are cartridged and inserted in plastic hoses as a loading device.

All class II and class III explosives belong to the group of ion exchanged explosives; for test conditions and applicability restrictions
→ *Permitted Explosives*.

Wetter-Dynacord

Trade name of a detonating cord manufactured by Orica Germany GmbH, with high safety against ignition of methane/air mixtures.

X-Ray Flash

By using special X-ray tubes and very fast high-voltage circuitry, it is possible to generate and trigger ultrashort X-ray flashes down to the nano-/micro-second range.

These X-ray flashes are an important mean of short-time photography because they enable fast occuring phenomena to be recorded by means of X-ray photographs.

In practice, this possibility of short-time radioscopy of test specimens is made use of for shaped charges (→ *Shaped Charges*). Thus, it is possible, during a desired time of detonation, to make a photographic record, in the form of single X-ray photographs, of the penetration *and* streaming behaviour of the sting into a target.

Zinc Peroxide

Zinkperoxid; peroxyde de zinc

$n\ ZnO_2 \cdot Zn(OH)_2$, $n \geq 3$

light yellow amorphous powder
sum formula: ZnO_2
molecular weight: 97.379 g
energy of formation: −344.8 kJ/mole
enthalpy of formation: −347.3 kJ/mole
oxygen balance: 16.43 %
density: 1.57 g/cm^3
melting point: > 150 °C
Fp.: decomposition upwards of 200 °C

Zinc peroxide is not hygroscopic and insoluble in water and organic solvents. The compound results from reaction of an ammoniacal zinc sulfate solution wit 30 % hydrogen peroxide at 80°–95 °C. The bulk density and oxygen value can be varied over a relatively wide range if certain temperature and concentration conditions are maintained.

Zinc peroxide is used in pyrotechnic mixtures and primer compositions whose reaction products should not contain any corrosive and hazardous components (→ *SINTOX Primer Compositions*).

Literature

Books*)

1. Manuals:

Escales, R.: Die Schießbaumwolle, Veit, Leipzig 1905
Escales, R.: Nitroglycerin und Dynamit, Veit, Leipzig 1908
Escales, R.: Ammonsalpetersprengstoffe, Veit, Leipzig 1909
Escales, R.: Chloratsprengstoffe, Veit, Leipzig 1910
Brunswig, H.: Schlagwettersichere Sprengstoffe, W. de Gruyter, Leipzig 1910
Escales, R.: Nitrosprengstoffe, Veit, Leipzig 1915
Escales, R. und *Stettbacher, A.:* Initialsprengstoffe, Veit, Leipzig 1917
Kast, H.: Spreng- und Zündstoffe, Vieweg, Braunschweig 1921
Brunswig, H.: Explosivstoffe, W. de Gruyter, Leipzig 1923
Beyling, C. und *Drehkopf, K.:* Sprengstoffe und Zündmittel, Springer, Berlin 1936
Stettbacher, A.: Spreng- und Schießstoffe, Rascher, Zürich 1948
Naoum, Ph. und *Berthmann, A.:* Explosivstoffe, Hanser, München 1954
Davis, T. L.: The Chemistry of Powder and Explosives, Wiley, New York 1956
Cook, M. A.: The Science of High Explosives, Chapman & Hall, London 1958, korrig. Nachdruck 1971 (Robert E. Krieger Publishing Co. Inc., Huntington, NY, American Chemical Society Monograph Series No. 139); German Translation: Lehrbuch der brisanten Sprengstoffe, MSW-Chemie, Langelsheim 1965
McAdam, R. und *Westwater, R.:* Mining Explosives, Oliver & Boyd, London 1958
Taylor, J. und *Gay, P. F.:* British Coal Mining Explosives, George Newnes, London 1958
Taylor, W.: Modern Explosives, The Royal Institute of Chemistry, London 1959
Berthmann, A.: Explosivstoffe, in: Winnacker-Küchler, Chemische Technologie, 3. Aufl., Hanser, München 1972, Bd. 5, S. 463–527
Urbanski, T.: Chemie und Technologie der Explosivstoffe, VEB Deutscher Verlag für Grundstoffindustrie, Leipzig 1961–1964, (3 Bde); englische erweiterte Auflage: Chemistry and Technology of Explosives, Pergamon Press, Oxford 1964–1967, 1984 (4 Bde)
Kreuter, Th.: Spreng- und Zündmittel, VEB Deutscher Verlag für Grundstoffindustrie, Leipzig 1962

---

* The sequence of listing was made according to the year of publication. The titles published prior to 1970 are historical interest only, they are out of print and only available in a few libraries.

Roth, J. F.: Sprengstoffe, in: Ullmanns Enzyklopädie der technischen Chemie, 3. Aufl., Urband & Schwarzenberg, München und Berlin 1965, Bd. 16, S. 56-109
Calzia, J.: Les Substances Explosives et leurs Nuisances, Dunod, Paris 1969
Newhouser, C. R.: Introduction to Explosives, The National Bomb Data Center, Gaithersburg, USA 1973
Cook, M. A.: The Science of Industrial Explosives, IRECO Chemicals, Salt Lake City, Utah, USA, 1974
Lyle, F.A. and Carl, H.: Industrial and Laboratory Nitrations, ACS Symposium Series No. 22, American Chemical Society, Washington DC, 1976
Oswatitsch, K.: Grundlagen der Gasdynamik, Springer, Wien, New York 1976
Romocki, S. J. von: Geschichte der Explosivstoffe, Bd. 1 und 2, Verlag Gerstenberg, Hildesheim, 1976, Nachdruck der Originalausgabe 1895/96, Berlin
Bartknecht, W.: Explosionen, 2. Aufl., Springer, Berlin 1980
Fordham, S.: High Explosives and Propellants, 2. Aufl., Pergamon Press, Oxford, New York 1980
Biasutti, G. S.: Histoire des Accidents dans l'Industrie des Explosifs, Editor: Mario Biazzi, Vevey 1978, engl. Ausgabe 1985
Lingens, P., Prior, J. und *Brachert, H.:* Sprengstoffe, in: Ullmanns Enzyklopädie der technischen Chemie, 4. Aufl., VCH-Verlagsges., Weinheim 1982, Bd. 21, S. 637-697
Quinchon, J. et al.: Les Poudres, Propergols et Explosifs, Tome 1: Les Explosifs, Technique et Documentation, Paris 1982
Brunisholz, A., Hildebrand, C. und *Leutwyler, H.:* Pulver, Bomben und Granaten. Die Pulvermacher einst und jetzt. Lang Druck AG, Liebefeld, Bern, 1983
Lafferenz, R. und *Lingens, P.:* Explosivstoffe, in: Winnacker-Küchler, Chemische Technologie, Hanser, München 1986, 4. Aufl., Bd. 7
Meyer, R.: Explosives, 3rd Edition, VCH-Verlagsges., Weinheim, New York 1987
Baily, A. und *Murray, S. G.:* Explosives, Propellants & Pyrotechnics, Pergamon Press, Oxford, New York 1988
Ganzer, U.: Gasdynamik, Springer Verlag, Berlin, Heidelberg, New York 1988
Olah, G. A.; Malhotra, R. und *Narang, S. C.:* Nitration, Methods and Mechanisms, VCH-Verlagsges., Weinheim 1989
Nitro Compounds, Recent Advances in Synthesis and Chemistry, Hrsg. *Feuer, H.* und *Nielsen, A. T.,* VCH-Verlagsges., Weinheim 1990
Chemistry of Energetic Materials, Hrsg. *Olah, G. A.* und *Squire, D. R.,* Academic Press, London 1991
Meyer, R. und *Köhler, J.:* Explosives, 4. Edition, VCH-Verlagsges., Weinheim, New York 1993

Structure and Properties of Energetic Materials, Editor: *Liebenberg, D. H., Armstrom, R. W.* und *Gilman, J. J.*, Materials Research Society (MRS), Pittsburgh, USA, 1993 (Symposium Series Vol. 293)
Köhler, J. und *Meyer, R.:* Explosivstoffe, 8. überarbeitete Auflage, VCH-Verlagsges. mbH, Weinheim, 1995
Hazardous Materials Handbook, Editor: *Pohanish, R. B.* und *Greene, S. A.*, Chapman & Hall, London, UK, 1996
Nitration – Recent Laboratory and Industrial Developments, Editor: *Albright, F. L., Carr, R. V. C., Schmitt, R. J.*, American Chemical Society (ACS), Washington, DC, USA, 1996 (ACS Symposium Series Vol. 608)
SIPRI Yearbook 1997 – Armaments, Disarmament and International Security, Stockholm International Peace Research Institute, Oxford University Press, Oxford, UK, 1997
Akhavan, J., The Chemistry of Explosives, Royal Soc Chem, Cambridge, UK, 1998
Köhler, J. und *Meyer, R.*: Explosivstoffe, 9. überarbeitete Auflage, Wiley-VCH Verlagsges. mbH, Weinheim, 1998

2. Application Technique:

Peithner-Jenne: Handbuch des Sprengwesens, ÖGB, Wien 1951
Lathan, W.: Bohr- und Schießarbeiten, VEB Deutscher Verlag für Grundstoffindustrie, Leipzig 1960
Fraenkel, H.: Handbuch für Sprengarbeiten, Atlas Diesel AB, Stockholm 1953–1963
Langefors, U. und *Kihlström, B.:* The Modern Technique of Rock Blasting, Almquist & Wiksell, Stockholm 1963
Biermann, G.: Neuzeitliche Sprengtechnik, Bauverl. Wiesbaden, Berlin 1966
Cole, R. H.: Underwater Explosions, Dover Publ., New York 1965
Wahle M. und *Begrich, K.:* Der Sprengmeister, Heymanns, Köln 1969
Holluba, H.: Sprengtechnik, 3. Aufl., Österreichischer Gewerbeverlag, Wien 1985
Saint-Arroman, Ch.: Pratique des Explosifs, Eyrolles, Paris 1977
Thum, W. und *Hattwer, A.:* Sprengtechnik im Steinbruch und Baubetrieb, Bauverlag GmbH, Wiesbaden und Berlin 1978
Handbuch der Sprengtechnik, Editor: *Heinze, H.*, 2nd Edition, VEB Deutscher Verlag für Grundstoffindustrie, Leipzig 1980
Blasters Handbook, Du Pont de Nemours, Wilmington 1980; laufende Neuauflagen
Weichelt, F.: Handbuch der Sprengtechnik, 6. Aufl., VEB Deutscher Verlag für Grundstoffindustrie, Leipzig 1980
Bodurtha, F. T.: Industrial Explosion, Prevention and Protection, McGraw-Hill, New York 1980
Manual Bickford, Etbls. Davey Bickford, Rouen
Blasting Practice, ICI, Nobel Division, Stevenston, England

Gustafson, R.: Swedish Blasting Technique, SPI, Gothenburg, Schweden 1981
Waffentechnisches Taschenbuch, 6. Aufl., Hrsg. Rheinmetall GmbH, Düsseldorf 1983
Wild, H.W.: Sprengtechnik im Berg-, Tunnel- und Stollenbau, 3. Edition, Verlag Glückauf, Essen 1984
Sprengtechnik. Begriffe, Einheiten, Formelzeichen, DIN 20163, Beuth-Vertrieb GmbH, Köln und Berlin 1985
Shock Waves for Industrial Applications, Editor: *Murr, L.E.,* Noyes Publikations, Park Ridge, New York 1989
Sprengtechnik, Anwendungsgebiete und Verfahren, Editor: *H. Heinze,* 2. Aufl., Deutscher Verlag für Grundstoffindustrie, Leipzig, Stuttgart 1993
Jahrbuch der Wehrtechnik, 1–21, Bernard & Graefe Verlag, Koblenz 1966–1992
Cooper, P.W.: Explosives Engineering, VCH Verlagsgesellschaft, Weinheim, New York, 1996
Introduction to the Technology of Explosives, Editors: *Cooper, P. W., Kurowski, S. R.,* VCH, Weinheim, Germany, 1996
Explosive Effects and Applications, Editors: *Zukas, J.; Walters, W.P.,* Springer, New York, 1998

3. Monographs:

Naoum, Ph.: Nitroglycerin und Nitroglycerinsprengstoffe, Springer, Berlin 1924
Fabel, K.: Nitrocellulose, Enke, Stuttgart 1950
Miles, F.D.: Cellulose Nitrate, Oliver & Boyd, London 1955
Kraus, A.: Handbuch der Nitrocellulose-Lacke, 3 Bde., Pansegrau, Berlin 1955–1961
Nauckhoff, S. und *Bergström, O.:* Nitroglycerin och Dynamit, Nitroglycerin A.B. Gyttorp 1959
Schumacher, J.C.: Perchlorates, their Properties, Manufacture and Use, Reinhold, New York 1960
Feuer, H.: The Chemistry of Nitro and Nitroso Groups, Interscience Publ., New York 1969
Lindemann, E.: Nitrocellulose, in: Ullmanns Enzyklopädie der technischen Chemie, 3. Edition, Urban & Schwarzenberg, München und Berlin 1960, Bd. 12, S. 789–797
Fair, H.D. und *Walker, R.F.:* Energetic Materials, 2 Bde. (over azides), Plenum Press, New York und London 1977
Brachert, H.: Nitrocellulose, in: Ullmanns Enzyklopädie der technischen Chemie, Verlag Chemie, Weinheim 1979, 4. Aufl., Bd. 17, S. 343–354
Cross, J. und *Farrer, D.:* Dust Explosions, Plenum Press, New York 1982

Field, P.: Dust Explosions, Elsevier, Amsterdam 1982
Biasutti, G. S.: History of Accidents in the Explosives Industry, Eigenverlag, Vevey, Schweiz 1985
Bartknecht, W.: Staubexplosionen, Springer, Berlin 1987; engl. Translation: Dust Explosions, 1989
Quinchon, J. und *Tranchant, J.:* Nitrocelluloses, the Materials and their Applications in Propellants, Explosives and other Industries, Ellis Horwood Ltd., Chichester 1989
Toxicity and Metabolism of Explosives, Editor: Yinon, J., CRC Press c/o Wolfe Publishing Ltd., London 1990
Structure and Properties of Energetic Materials, Hrsg.: *Liebenberg, D. H., Armstrong, R. W., Gilman, J. J.,* Materials Research Society (MRS), Pittsburgh, PA, USA, 1993 (Symposium Series Vol. 293)
Handbook of Harzardous Materials, Hrsg.: *Corn, M.,* Academic Press Inc., New York, London, 1993
Nitro Carbons, Editor: *Nielsen, A. T.,* VCH Verlagsgesellschaft mbH, Weinheim, 1995
Liquid Rocket Engine Combustion Instability, Editor: *Yang, V.* und *Anderson, W. E.,* Progress in Astronautics and Aeronautics, Vol. 169, AIAA, Washington, DC, USA, 1995
Introduction in the Technology of Explosives, Hrsg.: *Cooper, P. W.* und *Kurowski, S. R., VCH Verlagsgesellschaft mbH, Weinheim, 1996*
Marinkas, P. L.: Organic Energetic Compounds, Nova Science Publishers, Inc., New York, USA, 1996
Chemical Weapons Destruction and Explosive Waste/Unexploded Ordnance Remediation, Editor: *Noyes, R., Noyes,* Park Ridge, NJ, 1996
Yinon, J.: Forensic and Environmental Detection of Explosives, Wiley, New York, 1999
Teipel, U.: Energetic Materials Particle Processing and Characterization. WILEY-VCH Verlag, Weinheim, 2005

4. Propellants

Brunswig, H.: Das rauchlose Pulver, W. de Gruyter, Berlin und Leipzig 1926
Muraour, J.: Poudres et Explosifs, Vendome 1947
Zähringer, A. J.: Solid Propellant Rockets, Wyandotte 1958
Taylor, J.: Solid Propellants and Exothermic Compositions, George Newnes, London 1959
Kit, B. und *Evered, D. S.:* Rocket Propellant Handbook, Macmillan, New York 1960
Penner, S. S. und *Ducarme, J.:* The Chemistry of Propellants and Combustion, Pergamon Press, London 1960
Summerfield, M.: Solid Propellant Rocket Research, Academic Press, London 1960

Wiech, R. E. und *Strauss, R. F.:* Fundamentals of Rocket Propulsion, Reinhold, New York 1960
Warren, F. A.: Rocket Propellants, Reinhold, New York 1960
Barrère, M., u. a.: Raketenantriebe, Elsevier, Amsterdam 1961 (engl. Edition: Rocket Propulsion, 1960)
Penner, S. S.: Chemical Rocket Propulsion and Combustion Research, Gordon & Breach, New York, London 1962
Hagenmüller, P.: Les Propergols, Gauthlers-Villars, Paris 1966
Pollard, F. B. und *Arnold, J. H.:* Aerospace Ordnance Handbook, Prentice Hall, Englewood, New Jersey 1966
Samer, S. F.: Propellant Chemistry, Reinhold, New York 1966
Dadieu, A., Damm, R. und *Schmidt, E. W.:* Raketentreibstoffe, Springer, Wien 1968
Tavernier, P., Boisson, J. und *Crampel, B.:* Propergols Hautement Energétiques, Agardograph Nr. 141, 1970 (AGARD Publication)
Köhler, H. W.: Feststoff-Raketenantriebe, 2 Bände, Girardet Essen 1971/72
Schmucker, P. H.: Hybrid-Raketenantriebe, Goldmann, München 1972
James, R.: Propellants and Explosives, Noyes Data Corporation, Park Ridge, New Jersey 1974
Sutton, G. P. und *Ross, D. M.:* Rocket Propulsion Elements, 4. Edition, Wiley, New York 1976
Wolff, W.: Raketen und Raketenballistik, 4. Edition, Militärverlag der DDR, Berlin 1976
Baker, D.: The Rocket. The History and Development of Rocket and Missile Technology, New Cavendish Books, London 1978
Cornelisse, J. W., Schöyer, H. F. R. und *Walker, K. F.:* Rocket Propulsion and Spaceflight Dynamics, Pittman, London 1979
Davenas, A. u. a.: Technologie des propergols solides, Série SNPE, Masson, Paris 1988
Gun Propulsion Technology, Hrsg. Stiefel, L., AIAA, New York 1988, Progress in Astronautics and Aeronautics, Vol. 109
Quinchon, J. et al.: Les poudres, propergols et explosifs, Technique et Documentation, Paris:
Tome 1: Les explosifs, 1982, 2. Edition 1987
Tome 2: Les nitrocelluloses et autres matières de base des poudres et propergols, 1984
Tome 3: Les poudres pour armes, 1986
Tome 4: Les propergols, 1991
Davenas, A.: Solid Rocket Propulsion Technology, Pergamon Press New York 1993
Challenges in Propellants and Combustion – 100 Years after Nobel, Editor: *Kuo, K. K.* et al., Begell House, Inc., New York, USA, 1997
Klingenberg, G., Knapton, J. D., Morrison, W. F., Wren, G. P., Liquid Propellant Gun Technology, AIAA, Reston, VA, 1997

Solid Propellant Chemistry, Combustion, and Motor Interior Ballistics, Editors: *Yang, V., Brill, T. B., Ren, Wu-Zhen,* AIAA, Reston VA, 2000

5. *Pyrotechnics:*

Ellern, H.: Modern Pyrotechnics, Fundamentals of Applied Physical Pyrochemistry, Chemical Publishing Comp. Inc., New York 1961
Ellern, H.: Military and Civilian Pyrotechnics, Chemical Publishing Comp. Inc., New York 1968
Brauer, K. O.: Handbook of Pyrotechnics, Chemical Publishing Co, Brooklyn, New York 1974
Shimizu, T.: Feuerwerk (vom physikalischen Standpunkt aus), Hower Verlag, Hamburg 1978
Clark, F. P.: Special Effects in Motion Pictures, Society of Motion Pucture and Television Engineers, Inc., 862 Scarsdale Avenue, Scarsdale, New York 10583, Second Printing 1979
McLain, J. H.: Pyrotechnics, from the Viewpoint of Solid State Chemistry, Franklin Research Center Norristown, PA, 1980
Shimizu, T.: Fireworks, the Art, Science and Technique, Maruzen, Tokio 1981
Barbour, R. T.: Pyrotechnics in Industry, McGraw-Hill International Book Company, New York 1981
Philipp, Ch.: A Bibliography of Firework Books, Works on Recreative Fireworks from the 16th to 20th Century, St. Pauls Biographies, Winchester, Hampshire 1985
Conkling, J.: Chemistry of Pyrotechnics, Basic Principles and Theory, Marcel Dekker Inc., New York 1985
Pyrotechnica I-XV, 1977–1993, Hrsg. *Cardwell, R. G.,* Pyrotechnica Publications, Austin, Texas (irregular)
Lancaster, R.: Fireworks, Principles and Practice, 2. Aufl., Chemical Publishing Comp. Inc., New York 1992
Safety of Reactive Chemicals and Pyrotechnics, Editor: *Yoshida, T., Wada, Y., Foster, N.,* Elsevier Science, Amsterdam, New York, 1995

6. *Theories of Detonation and Combustion:*

Jouquet, E.: Mecanique des Explosifs, Doin et Fils, Paris 1917
Becker, R.: Stoßwelle und Detonation, Z. Physik 9, 321–362 (1922)
Bowden, F. P. und *Yoffe, A. D.:* Initiation and Growth of Explosions in Liquids and Solids, Cambridge University Press, Cambridge 1952, Reprint 1985
Taylor, J.: Detonation in Condensed Explosives, Clarendon Press, Oxford 1952
Bowden, F. P. und *Yoffe, A. D.:* Fast Reactions in Solids, Butterworth London 1958
Penner, S. S. und *Mullins, B. P.:* Explosions, Detonations, Flammability and Ignition, Pergamon Press, London, New York 1959

Zeldovich, J. B. und *Kompaneets, A. S.:* Theory of Detonation (Translation), Academic Press, New York und London 1960
Berger, J. und *Viard, J.:* Physique des explosifs, Dunod, Paris 1962
Andrejev, K. K.: Thermische Zersetzung und Verbrennungsvorgänge bei Explosivstoffen (Translation), Barth, Mannheim 1964
Andrejev, K. K. und *Beljajev, A. F.:* Theorie der Explosivstoffe (Translation), Svenska National Kommittee för Mekanik, Stockholm 1964
Zeldovich, J. B. und *Raizer, J.:* Physics of Shock Waves and High Temperature, Hydrodynamic Phenomena (Translation), Academic Press, New York, London 1966
Lee, J. H., Knystautas, R. und *Bach, G. G.:* Theory of Explosion, McGill University Press, Montreal 1969
Johansson, C. H. und *Persson, P. A.:* Detonics of High Explosives, Academic Press, London, New York 1970
Büchner, E.: Zur Thermodynamik von Verbrennungsvorgängen, 2. Edition, München 1974
Glassmann, I.: Combustion, Academic Press, New York 1977
Turbulent Combustion, Editor: *Kennedy, L. A.,* AIAA, New York 1978
Fickett, W. und *Davis, W. C.:* Detonation, University of California Press, Berkeley 1979
Mader, Ch.: Numerical Modeling of Detonations, University of California Press, Berkeley 1979
Combustion Chemistry, Editor: *Gardiner Jr., W. C.,* Springer, New York 1984
The Chemistry of Combustion Processes, Hrsg.: *Sloane, Th. H.,* ACS Symposium Series No 249, American Chemical Society, Washington 1984
Duguet, J.-R.: Les explosifs primaires et les substances d'initiation, Masson, Paris 1984
Fundamentals of Solid Propellant Combustion, Editor: *Kuo, K. K.* und *Summerfield, M.,* Progress in Astronautics and Aeronautics, Vol. 90, AIAA, New York 1984
Lewis, B. und *Elbe, G.* von: Combustion, Flames and Explosives of Gases, 3. Edition, Academic Press, Orlando, Florida 1987
Energetic Materials: New Synthesis Routes, Ignition, Propagation and Stability of Detonation, Hrsg.: *Field, J. E.* und *Gray, P.,* The Royal Society, London 1992
Nonsteady Burning and Combustion Stability of Solid Propellants, Editor: *De Luca, L., Price, E. W., Summerfield, M.,* Progress in Astronautics and Aeronautics, Vol. 143, AIAA, Washington, DC, USA, 1992
Bartknecht, W. and *Zwahlen, G.:* Explosionsschutz. Grundlagen und Anwendungen, Springer-Verlag Berlin, Heidelberg, New York 1993
Chéret, R.: Detonation of Condensed Explosives, Springer Verlag, Berlin, Heidelberg, New York 1993

Combustion of Boron-Based Solid Propellants and Solid Fuels, Editor: *Kuo, K. K.,* CRC Press Inc., London 1993
S. S. Batsanov: Effects of Explosions on Materials – Modification and Synthesis under High Pressure Shock Compression, Springer-Verlag, Berlin, New York, London, 1994
Decomposition, Combustion, and Detonation Chemistry of Energetic Materials, Editor: *Brill, T. B., Russel, P. B., Tao, W. C., Wardle, R. B.,* Materials Research Society (MRS), Pittsburgh, PA, USA, 1996 (Symposium Series Vol. 418)
Mader, C. L., Numerical Modeling of Explosives and Propellants, Second Edition, CRC, Boca Raton, Fla, 1997
Margolis, S. B., Williams, F. A., Structure and Stability of Deflagrations in Porous Energetic Materials, Technical Report, Sandia Natl. Lab. SAND99-8458, 3–44, Sandia National Laboratories, Livermore, CA, 1999
Dorsett, H., White, A., Overview of Molecular Modelling and ab Initio Molecular Orbital Methods Suitable for Use with Energetic Materials, Technical Report, Def. Sci. Technol. Organ. DSTO-GD-0253, i-iv, 1–35, Weapons Systems Division, Aeronautical and Maritime Research Laboratory, Melbourne, Australia, 2000

7. Military Explosives and Ammunition; Ballistics:

Cranz, C.: Lehrbuch der Ballistik (3 Bde.), Springer, Berlin 1925–1927
Hänert: Geschütz und Schuß, Springer, Berlin 1940
Kutterer, E. K.: Ballistik, Vieweg & Sohn, Braunschwein 1942
Gallwitz, U.: Die Geschützladung, Heereswaffenamt, Berlin 1944
Ohart, T. C.: Elements of Ammunition, Wiley, New York 1952
Hofmann, Fr.: Praktische Sprengstoff- und Munitionskunde, Wehr und Wissen, Darmstadt 1961
Noack, H.: Lehrbuch der militärischen Sprengtechnik, Dt. Militärverlag, Berlin 1964
Ellern, H.: Military and Civilian Pyrotechnics, Chemical Publishing Comp. Inc., New York 1968
Tomlinson, W. R.: Properties of Explosives of Military Interest, Picatinny Arsenal, Dover, N. J. 1971
Gorst, A. G.: Pulver und Sprengstoffe, Militärverlag der Deutschen Demokratischen Republik, Berlin 1977
Interior Ballistics of Guns, Editor: *Krier, R.* u. a., Progress in Astronautics and Aeronautics, Vol. 66, AIAA, New York 1979
Farrar, C. L. und *Leeming, D. W.:* Military Ballistics. A Basic Manual, Brassey's Publishers Ltd., Oxford 1983
Waffentechnisches Taschenbuch, 6. Edition, Editor: *Rheinmetall GmbH,* Düsseldorf 1983

Goad, K. J. W. und *Archer, E.:* Ammunition, Pergamon Press, Oxford, New York 1990
Untersuchung von Rüstungsaltlasten, Editor: *Spyra, W., Lohs, K. H., Preussner, M., Rüden, H., Thomé-Kozmiensky, K. J.,* EF-Verlag für Energie und Umwelttechnik GmbH, Berlin, 1991
Insensitive Munitions, AGARD Conference Proceedings CP511, North Atlantic Treaty Organization, Neuilly sur Seine, France, 1992
Rock Blasting and Explosives Engineering, Editor: *Persson, P. A., Holmberg, R., Lee, J.,* CRC Press Inc., Boca Raton, FL, USA, 1993
Jane's Ammunition Handbook, 2. Edition, Editor: Jane's Information Group, Coulsdon, UK, 1994 (also obtainable on CD-ROM)
Explosivstoffabriken in Deutschland, Editor: *Trimborn, F.,* Verlag Locher, Köln, 1995

8. *Analytical and Testing Methods:*

Berl-Lunge: Chemisch-Technische Untersuchungsmethoden, Bd. 3: Explosivstoffe und Zündwaren, Springer, Berlin 1932
Kast, H. und *Metz, L.:* Chemische Untersuchung der Spreng- und Zündstoffe, Vieweg, Braunschweig 1944 (2. Edition)
Analytical Methods for Powders and Explosives, Bofors A. B., Göteborg 1960
Krien, G. Thermoanalytische Ergebnisse der Untersuchung von Sprengstoffen, Bericht Az. 3.0–3/3960/76, Bundesinstitut für Chemisch-Technische Untersuchungen, Swisttal-Heimerzheim 1976
Malone, H. E.: Analysis of Rocket Propellants, Academic Press, London 1977
Yinon, J. und *Zitrin, S.:* The Analysis of Explosives, Pergamon Press, Oxford, New York 1981
Combustion Measurements, Editor: *Chigier, N.,* Hemisphere Publishing Company, Washington, London, Philadelphia 1991
Determiniation of Thermodynamic Properties, Editor, *Rossiter: B. W.,* John Wiley & Sons, Chichester, England 1992
Yinon, J. und *Zitrin, S.:* Modern Methods and Applications in Analysis of Explosives, John Wiley & Sons, Chichester, England, 1993
Sucéska, M.: Test Methods for Explosives, Springer-Verlag, Berlin, New York, 1995
Modern Methods and Applications in Analysis of Explosives, Editors: *Yinon, J., Zitrin, Sh.,* Wiley, Chichester, UK, 1996

9. *Encyclopedias and Tables:*

Schmidt, A.: Thermochemische Tabellen für die Explosivchemie, Z. ges. Schieß- und Sprengstoffwesen 29 (1934), S. 259 u. 296
Médard, M. L.: Tables Thermochimiques, Mémorial de l'Artillerie Française 28, 415–492 (1954); Imprimerie Nationale, Paris 1954

Selected Values of Chemical Thermodynamics Properties, NBS Technical Note 270, 1968
Stull, D. R., Westrum, E. F. und *Sinke, G. C.:* The Chemical Thermodynamics of Organic Compounds, Wiley, New York 1969
Cox, J. D. and *Pilcher, G.:* Thermochemistry of Organic and Organometallic Compounds, Academic Press, London 1970
Huff, V. N., Gordon, S. and Morell, V. E.: General method and thermodynamic Tables for Computational of Equilibrium Composition and Temperature of Chemical Reactions, NASA Report 1073 (1973)
Gordon, S. and Bride, B. J.: Computer Program for Calculation of Complex Chemical Equilibrium Compositions, Rocket Performance, Incident and Reflected Shocks and Chapman-Jouguet Detonations, NASA Report SP-273 (1971)
Volk, F., Bathelt, H. and *Kuthe, R.:* Thermodynamische Daten von Raketentreibstoffen, Treibladungspulvern und Sprengstoffen, sowie deren Komponenten, Tabellenwerk, Band I und Band II, 1972, 1. Erg. 1981, Selbstdruck des Fraunhofer-Instituts für Chemische Technologie (ICT), Pfinztal-Berghausen
Rossi, B. D. und *Podugnakov, Z. G.:* Commerical Explosives and Initiators, a Handbook, Translation from Russia, NTIS National Technical Information Service, U. S. Department of Commerce, Springfield, VA 1973
Volk, F. and *Bathelt, H.:* ICT Thermochemical Data Base, Disc with Handbook, Fraunhofer-Institut für Chemische Technologie (ICT), Pfinztal-Berghausen, 1994, 3. Update 1997
Kirk-Othmer, Encyclodpedia of Chemical Technology, 2. Aufl., Wiley, New York 1963–1971; Index:
Explosives and Propellants, Vol. 8
Pyrotechnics, Vol. 16
JANAF Thermochemicals Tables, Editor: *Stull, D.* und *Prophet, H.*, National Standard Reference Data Series, National Bureau of Standards, Midland, Michigan, USA; 2. Edition 1971, Updates 1974–1982
Moda, M. C.: Explosive Property Data, University of California Press, Berkeley, California 1980
LASL Explosive Property Data, Editor: *Gibbs, T. R.* und *Popolato, A.*, University of California Press, Berkeley, California 1980
LASL Phermex Vol., Vol. 1–3, Editor: *Mader, Ch. L.*, University of California Press, Berkeley, California 1980
LASL Shock Hugoniot Data, Editor: *Marsh, St. P.*, University of California Press, Berkeley, California 1980
LLNL Explosives Handbook: Properties of Chemical Explosives and Explosive Simulants, Editor: *Brigitta M. Dobratz*, UCLR-52997, Livermore, California 1981; National Technical Information Service, US Department of Commerce, Springfield, VA 22 161, USA
Los Alamos Explosives Performance Data, Editor: *Mader, Ch. L., Johnson, J. N.* und *Crane, Sh. L.*, University of California Press, Berkeley, California 1982

Los Alamos Shock Wave Profile Data, Editor: Morris. Ch. E., University of California Press, Berkeley, Los Angeles, London 1982
Encyclopedia of Explosives and Related Items. PATR 2700, Editor: *Seymour M. Kaye, Dover, N. J.* (USA), Vol. 1–10, 1960–1983; National Technical Information Service, US Department of Commerce, Springfield, Virginia 22161, USA (also obtainable on CD-Rom)
Ullmanns Enzyklopädie der technischen Chemie, 4. Edition, 25 Volumes, Verlag Chemie, Weinheim 1972–1984; Index:
Nitrocellulose, Vol. 17
Pyrotechnik, Vol. 19
Raketentreibstoffe, Vol. 20
Sprengstoffe, Vol. 21
Kirk-Othmer, Encyclopedia of Chemical Technology, 3. Edition, 26 Volumes, Wiley, New York 1978–1984; Index:
Explosives and Propellants, Vol. 9
Pyrotechnics, Vol. 19
DOE Explosives Safety Manuel, US Department of Energy, Springfield, VA 1989
Ullmann's Encyclopedia of Industrial Chemistry, 5. Edition, VCH-Verlagsges. Weinheim, 1985–1996, 37 Volumes, (also obtainable on CD-Rom); Index:
Explosives, Vol. A10
Propellants, Vol. 22A
Pyrotechnics, Vol. 22A
Kirk-Othmer Encyclopedia of Chemical Technology, 4. Edition, 27 Volumes, John Wiley & Sons Ltd., Chichester, New York, 1991–1998; Index:
Explosives and Propellants Vol. 10

10. Governmental Regulations, Acts and Comments

The Handling and Storage of Liquid Propellants. Office of the Director of Defense Research and Engineering, Washington D. C. 1963
Code maritime international des marchandises dangereuses Organisation Intergouvernementale Consultative de la Navigation Maritime, London 1966 ("IMCO"), IMDG-Code english, Storck Verlag, Hamburg, 2001, Editor: International Maritime Organization (IMO), London
Orange Book 11th Edition, Storck Verlag, Hamburg, 1999, Editor: United Nations (UNO), New York
Tests and Criteria 3rd Edition, Storck Verlag, Hamburg, 1999, Editor: United Nations (UNO), New York
"RID", Reglement International concernant le Transport des Marchandises dangereuses par Chemins de Fer, Annexe 1 de la Convention Internationale Concernant le Transport des Marchandises par Chemins de Fer, 1970

Transport of Dangerous Goods, United Nations, New York 1970, Vol, I-N
Recommendations on the Transport of Dangerous Goods. 4th Edition, United Nations, New York 1986
Dangerous Goods Regulations, english 42nd Edition, International Air Transport Association, Montreal – Genf 2001
UVV-Unfallverhütungs-Vorschrift (accident preventing prescription). German Safety Regulations, edited in Jedermann-Verlag, Heidelberg and VCH-Verlagsgesellschaft, Weinheim
Sprengstoffgesetz (German Explosive Law) puplished with all belonging appendices by Apel and Keusgen, Carl Heymanns Verlag, Köln 1978–1990

Periodicals

AIAA-Journal, AIAA, New York
Acta Astronautica, Pergamon Press, New York, Oxford
Bohren, Sprengen, Räumen, Erwin Barth Verlag, Neustadt/Weinstraße
Bundesarbeitsblatt: Beilage Arbeitsschutz, Stuttgart
Combustion, Explosion and Shock Waves, Faraday Press, New York (cover to cover translation of Fizika Goreniya Vzryva)
Combustion and Flame, American Elsevier Publ. Comp., New York
Combustion Science and Technology, Gordon and Breach Science Publ., New York, London, Paris
Explosifs, Edition Commerciales Industrielles, Brüssel
Explosivstoffe, Erwin Barth Verlag, Neustadt/Weinstraße (until 1974)
Explosives Engineer, Wilmington, Delaware (until 1961)
Glückauf, Verlag Glückauf, Essen
Gefährliche Ladung, K.O. Storck Verlag, Hamburg
Industrie der Steine und Erden, herausgegeben von der Steinbruchsberufsgenossenschaft, Verlag Gebr. Janecke, Hannover
Interavia, Luftfahrt-Raumfahrt-Elektronik, Interavia S.A. Genf
International Defense Review, Ineravia S.A., Genf, Schweiz
Internationale Wehrrevue, Interavia S.A., Genf, Schweiz (until Mai 1988)
Journal of Ballistics, Douglas Documentation Systems, Philadelphia
Journal of Energetic Materials, Dowden, Brodman & Devine, Stroudsbury, PE
The Journal of Explosives Engineering, International Society of Explosives Engineers, Cleveland
Journal of Industrial Explosives, Japan, Tokio
Journal of Propulsion and Power, AIAA, New York
Journal of Spacecraft and Rockets, AIAA, New York
Mémorial de l'Artillerie Française, l'Imprimerie Nationale, Paris
Mémorial des Poudres, l'Imprimerie Nationale, Paris (until 1965)

Mining and Minerals Engineering (formerly: Mine and Quarry Engineering) London
Mining Engineer, London
Mining, Engineering, New York
Missiles and Rockets, Washington (until 1966)
Nobelhefte, Sprengtechnischer Dienst der Dynamit Nobel AG, Dortmund
Oxidation and Combustion Reviews, Elsevier Publ. Comp., Amsterdam (until 1973)
Propellants, Explosives, Pyrotechnics, VCH-Verlagsges., Weinheim
Raumfahrtforschung, Deutsche Ges. f. Luft- und Raumfahrt, Ottobrunn
Sprengstoffe, Pyrotechnik, VEB Sprengstoffwerk Schönebeck/Elbe
Sprengtechnik, GEFAS (Gesellsch. f. angewandte Sprengtechnik), Effretikon, Schweiz
Tätigkeitsberichte der Bundesanstalt für Materialprüfung, BAM, Selbstverlag, Berlin
U. S. Bureau of Mines, PB-Reports, Washington
Wehrtechnik, Verlag Wehr und Wissen, Koblenz-Bonn
Wehrwissenschaftliche Rundschau, Verlag Mittler & Sohn, Frankfurt
Zeitschrift für das gesamte Schieß- und Sprengstoffwesen, Verlag August Schrimpff, München (until 1944)

Proceedings of the Fraunhofer-Institut Chemische Technologie (until July 1988: Institut für Treib- und Explosivstoffe), D-76327 Pfinztal-Berghausen; International Annual Conferences of ICT; 1970–2007 (38th):

1970 Wirkungsfaktoren explosionsfähiger Stoffe und deren Dämpfung
1971 Lebensdauer von Raketentreibsätzen, Treib- und Sprengladungen
1972 Probleme und Methoden der Umweltsimulation
1973 Sichere Technologie: Entstehung und Wirkung explosionsfähiger Systeme
1974 Verbrennungsvorgänge bei Treib- und Brennstoffen
1975 Pyrotechnik: Grundlagen, Technologie und Anwendung
1976 Sprengstoffe: Grundlagen, Technologie und Anwendung
1977 Analysenmethoden für Treib- und Explosivstoffe
1978 Moderne Technologie von Treib- und Explosivstoffen
1979 Verbrennungs- und Detonationsvorgänge
1980 Meß- und Prüfmethoden für Treib- und Sprengstoffe
1981 Chemische und Mechanische Technologie von Treib- und Explosivstoffen
1982 Verwendung von Kunststoffen für Treib- und Explosivstoffe
1983 Gütesicherung und Überwachung von Treib- und Sprengmitteln
1984 Technologie von Treib- und Sprengmitteln

1985 Pyrotechnics: Basic Principles, Technology, Application
1986 Analysis of Propellants and Explosives: Chemical and Physical Methods
1987 Technology of Energetic Materials: Manufacturing and Processing – Valuation of Product Properties
1988 Combustion and Detonation Phenomena
1989 Environmental Testing in the 90's
1990 Technology of Polymer Compounds and Energetic Materials
1991 Combustion and Reaction Kinetics
1992 Waste Management of Energetic Materials and Polymers
1993 Energetic Materials – Insensitivity and Environmental Awareness
1994 Energetic Materials – Analysis, Characterization and Test Techniques
1995 Pyrotechnics: Basic Principles, Technology, Application
1996 Energetic Materials – Technology, Manufacturing and Processing
1997 Combustion and Detonation
1998 Energetic Materials – Production, Processing and Characterization
1999 Energetic Materials – Modeling of Phenomena, Experimental Characterization, Environmental Engineering
2000 Energetic Materials – Analysis, Diagnostics and Testing
2001 Energetic Materials – Ignition, Combustion and Detonation
2002 Energetic Materials – Synthesis, Production and Application
2003 Energetic Materials – Reactions of Propellants, Explosives and Pyrotechnics
2004 Energetic Materials – Structure and Properties
2005 Energetic Materials – Performance and Safety
2006 Energetic Materials – Insensitivity, Ageing and Monitoring
2007 Energetic Materials – Characterisation and Performance of Advanced Systems

Further International Conferences (with Proceedings)

1. Symposium (International) on Combustion, every 2nd year, Organizer: The Combustion Institute, Pittsburg, 1928–2006 (31st)
2. Symposium (International) on Detonation, every 4th year, Organizer: Office of Naval Research and others, 1951–2006 (13th)
3. Symposium on Chemical Problems Connected with the Stability of Explosives, every 3rd year, Organizer: Sektionen för Detonik och Förbränning, Schweden, Dr. J. Hansson, 1967–2007 (14th)
4. International Symposium on Ballistics, ca. every 1.5 year, Organizer: ADPA (American Defense Preparedness Association), 1974–2007 (23rd); since Oct. 1997 new name: NDIA (National Defense Industrial Association)

5. Joint International Symposium on Compatibility of Plastics and other Materials with Explosives, Propellants and Ingredients, every year, Organizer: ADPA (American Defense Preparedness Association), 1974–1991; since 1992 new title: International Symposium on Energetic Materials Technology, 1992–1995 (new name since Oct. 1997: NDIA) new title: Insensitive Munitions & Energetic Materials Technology Symposium
6. Symposium on Explosives and Pyrotechnics, Organizer: Franklin Applied Physics, USA, every 3rd year, 1954–2001 (18th)
7. International Symposium on Loss Prevention and Safety Promotion in the Process Industries, Organizer: European Federation of Chemical Engineering, every 3rd year, 1974–2007 (12th)
8. International Pyrotechnics Seminar, Organizer: IPS (The International Pyrotechnics Society, USA), every year 1968–2007 (34th)
9. International Symposium on Analysis and Detection of Explosives, Organizer: changing, every 3rd year, 1983–2004 (8th)
10. Explosives Safety Seminar, Organizer: Department of Defense Explosives Safety Board, every year, since 1974 every 2nd year, obtainable from National Technical Information Service (NTIS), US Department of Commerce), 1958–2006 (32nd)
11. Airbag 2006: International Symposium on Sophisticated Car Occupant Safety Systems, Organizer: Fraunhofer-Institut Chemische Technologie (ICT), Germany, every 2nd year, 1992–2006 (8th)
12. Detection and Remediation Technologies for Mines and Minelike Targets, every year, Organizer: International Society for Optical Eng. – SPIE, 1996–2000 (5th)
13. International Autumn Seminar on Propellants, Explosives and Pyrotechnics: Theory and Practice of Energetic Materials, 1996–2007 (7th)
14. Conference on Explosives and Blasting Technique, every year, Organizer: International Society of Explosives Engineers, 1975–2007 (33rd)
15. Symposium on Explosives and Blasting Research, every year, Organizer: International Society of Explosives Engineers, 1985–1998 (14th)
16. EFEE World Conference on Explosives and Blasting, every second year, Organizer: European Federation of Explosives Engineers, 2001–2007 (4th)

Index

and Technical Dictionary with short information without text reference

A

A → 1-black blasting powder 35
A → bridgewire detonator A (german. now absolete) 41
A → composition A 62
AA = antiaircraft A-IX-2 = RDX/aluminum/wax 73/23/4
abattage par chambre de mine = coyote blasting 65
Abbrand = combustion 60
Abbrandgeschwindigkeit = burning rate 45
abbrennen = to burn down 75
Abel's equation 23
Abel test **1**; 178
abkerben, abspalten → smooth blasting 64
Ablonite = french commercial explosive
abschlagen einer Sprengladung → cut off 67
Absperrzone = blast area 36
Abstand; Sicherheitsabstand = safety distance 136
Abstandsberechnung → scaled distance 276
abstechen → dismantling 75
Abstichladung = jet tapper 193
Acardite → Akardite 9
accessoires pour le sautage = blasting accessoires 36
acceptor 1
Accord Européen relatif au Transport International des Marchandises Dangereuses par Route → A. D. R. **2**, 70
α-cellulose content 222
acétate-dinitrate de glycérine 148
acetone peroxide = tricycloacetone peroxide 249; 343
acétylacétonate de ter 192
Acetyldinitroglycerin 148
acétylsalicylate de plomb 194
Acetylensilber; acétylure d'argent 283

acide picramique 253
acide picrique 254
acide styphnique 300
acide trinitrobenzoique 349
acquisition, handling and storing 136
Acremite 1
Active Binder → Energetic Binder **116**; 151; 205; 257; 258
ADR 2, 70
A.D.C. test = Adreer double cartridge (gap test)
ADN = Ammonium dinitramide **14**; 324
adiabatic 1
adobe charge = mud capping 2; 218
Aeroplex K = solid rocket propellant based on $KClO_4$ and resin
Aeroplex N = solid rocket propellant based on NH_4ClO_4 and resin
Aerozin = hydrazine/dimethylhydrazine 50/50 2
A-E = single base powder
AGARD 2
Airbag **2**; 146
air blast 8
airloader 8
Akardit I, II, III **9**; **10**; 161; 295; 324
Akremit → Acremite 1
Albanite = propellant based on → DINA 108
Alex 10
Alex 20 = → composition B plus 20 % aluminium
alginates 10
aliphatic nitramines 68; 124; 132; 172
all fire 10
allumage spontané = pre-ignition 263
allumer = to inflame 189
allumeur 182
Almatrix = russian trade name for chlorate and perchlorate explosives
Alumatol = AN/TNT/aluminum 77/20/3
aluminum oxide 329

Index

aluminum powder **11**; 12; 215; 216
Amatex 11
Amatol 11
Amilol = diamylphthalate 92
aminoguanidinnitrate → tetracene 309
Ammodyte = powder from commercial explosive (USA)
Ammoksil (Ammokcil; Ammonxyl) = russian name for the mixture AN/trinitroxylene/aluminium 82/12/6
Ammonal = powder form commercial explosive
Ammongelit = german gelatinous commercial explosive 11
ammoniinaya selitra = ammonium nitrate (russian)
ammonium azide 12
ammonium chlorate 58
ammonium chloride **13**; 240; 324
ammonium dichromate 14
Ammonium dinitramide **14**; 324
ammonium nitrate **15**; 16; 20; 36; 124; 161; 216; 239; 316; 324
Ammonium Nitrate Emulsion (ANE) 17
ammonium nitrate explosive **16**; 20
ammonium oxalate 324
ammonium perchlorate **17**; 62; 164; 239; 240; 324
ammonium picrate **19**
Ammonpek = AN/tar 93/5 (russian)
Ammonpulver = ammonium nitrate containing gun powder, now obsolete (german)
Ammonsalpeter = ammonium nitrate 15
Ammonsalpeter-Sprengstoffe = ammonium nitrate based explosives 16; 164; 198
amorçage = priming 265
amorce à pont = birdgewire detonator 41
amorce électrique à l'étincelle = spark detonator 291
amorces **19**; 244; 288
AN = ammonium nitrate 16
analyse thermique différentielle = differential thermal analysis 96
ANC = ANFO (german) 20

Andex = trade name for ANFO (german) 20
ANE → Ammonium Nitrate Emulsion 17
ANF-58 = liquid fuel for rocket motors, say octane (USA)
ANFO = ammonium nitrate fuel oil 16; **20**; 164; 198
angle shot mortar test 267
Anilite = mixture of N_2O_4 and butane (french)
A: nitrocellulose powder
Anlaufstrecke = detonation development distance 88
Antenne → bus wire 47
antigrisouteux → permitted explosives 245
antilueur = muzzle flash damping additive (french)
antimony (in delay compositions) 74
anwürgen = crimping 66
anzünden = to inflame 189
Anzünder = squib 182; 293
Anzündhütchen = amorce 19; 244; 288
Anzündkette = igniter train 183
Anzündlitze = igniter cord 183
Anzündlitzenverbinder = igniter cord connectors 183
Anzündschraube = fuze head 146
Anzündverzug = functioning time 145
APC = ammonium perchlorate 17
APU 20
aquarium test **21**; 298
Arbeitsvermögen = strength 297
area ratio = propellant area ratio ("Klemmung") 267
Argol = crude potassium hydrogen tartrate = muzzle flash damping additive
argon flash 21
armor plate impact test **21**; 266
Armstrong blasting method 22
aromatic nitramines 132
ARRADCOM (Picatinny Arsenal) 22
Arsol = trimethylene trinitrosamine 68
Artillerietreibmittel → gun powder 155
as dimethylhydrazine 97
ASTM = American Society for Testing Materials
Astrolite 22

AT = anti-tank = armor piercing weapon
Athodyd = aerodynamic-thermodynamic = ramjet
Attrappen = mock explosives 216
Audibert tube **22**; 246
Auflegerladung = mud cap 2; 43; 218
Aunt Jemima = 80 % fine ground HMX (→ Octogen) and 20 % plain, old flour. It served for camouflage storage of sabotage explosive in World War II.
Aurol = concentrated hydrogen peroxide (german) 23
Ausbauchung → lead block test 196
aushärten = curing 67
ausschwitzen = exudation 136
Ausströmungsgeschwindigkeit = jet velocity 147
AUSTROGEL G1 und AUSTROGEL G2 = commercial explosives (Austria) 23
average burning rate 23
azides 12; 23; 194; 283; 342
azoture d'ammonium 12
azoture d'argent 283
azoture de plomb 194
azotures 12; 23; 194; 283; 342

B

B → B-black powder 31, 35
B → composition B 62
B → poudre B = nitrocellulose powder (french) 262
B → B-Stoff = german tarning name for methanol
B4 = mixture 60–70 % trinitroanisol and 30–40 % aluminum (italy)
Bachmann process (RDX-Synthesis) 70
ballistic bomb 23
ballistic modfiers 137; 194; 199
ballistic mortar **26**; 37; 297
Ballistite = double base powder with high percentage of nitroglycerine
ball powder **28**; 156
BAM = Bundesanstalt für Materialprüfung (german) **28**; 141; 162; 186

BAM testing methods 141; 162; 186
banc d'essai = rocket test stand 273
Baratols 28
barium chlorate 28
barium nitrate **29**; 161; 324
barium perchlorate 29
barium sulfate 161; 329
Barlow bomb = mixture of liquid oxygen with fuel
Baronal = mixture of barium nitrate with TNT ald aluminum 50/35/15
barricade 29
base bleed propellants 31
base charge 31
bâtiment habite = inhabited building 136
Bazooka 31
B-black blasting powder 31; 35
Bengal fire works → pyrotechnical compositions 269
Benzit = trinitrobenzene (german) 348
benzoyl peroxide 31
Bergbau-Versuchsstrecke = german institution for testing and amittance of permissible explosives, detonators and accessories 363
Bergmann-Junk-test **32**; 178
Bernoulli's equation 290
Berstscheibe = safety diaphragm 274
Berthollet's detonating silver 284
Besatz = stemming 296
Beschußempfindlichkeit = bullet impact sensitivity 43; 266; 273
bewohntes Gebäude = inhabited building 130
BF-122; 151 = polysulfide binder propellant (thiokol)
DGO, BGY, BIC; BID; BIE; BIL; BIM; BIP; BLB; BLC: various castable double base propellants (USA)
Bichel bomb **33**; 364
Bickford's safety fuse 274
BICT = Bundesinstitut für Chemisch-Technische Untersuchungen (german) 33
bilan d'oxygène = oxygen balance 240
Bildungs-Energie-Enthalpie = formation energy (enthalpy) 117; 324

billet 33
binder 34
binary explosives → Astrolite, Hydan 17; 179
Bis-cyclopentadienyl-Eisen = ferrocene 137
BITA = aziridine curing agent 34
bi-trinitroethylnitramine 34
bi-trinitroethylurea 35
BKW = Becker Kistiakowsky Wilson state equation
black powder 31; **35**; 118; 269
Blättchenpulver → gun powder 155
Blasgeräte = air loaders; also → pneumatic placing 8; 257
blast area 36
blast effect 8
blaster; shot firer 36; 283
blasting accessories 36
blasting agent 16; 20; **36**; **37**; 43; 190; 191; 251
blasting galvanometer = circuit tester 58
bullet squib 43
blasting gelatin 26; **37**; 141; 198
blasting machines **37**; 41
blasting mat 38
blasting soluble nitrocotton 222
blasting switch 39
Bleiacetylsalicylat 194
Bleiaethylhexanoat 199
Bleiazid 194
Bleiblockausbauchung = lead block test 196
Bleinitrat 200
Blei-Trizinat 201
BN = barium nitrate 29
Böllerpulver → black powder 35
Bohrlochabstand = spacing 290
Bohrlochpfeife = bootleg 40
Bohrpatrone (german) = 100 gram press-molded cylindrical charge of TNT
Boloron = mixture of concentrated nitric acid and dinitrochlorobenzene (Austria) 203
bomb drop test 39
bombe Bichel 33
bombe Crawford 66
bombe pour essais balistiques 23

Bonit = mixtures of RDX, TNT with and without aluminum (swedish)
boom powder 39
booster **40**; 92; 113
booster sensitivity test 40
bootleg 40
Boronite A, B, C = mixtures of AN, TNT and boron
Borotorpex = castable mixture of RDX, TNT and boron, e.g., 46/44/10 (USA)
boss 40
bouclier contre l'érosion = flame shield 139
Boudouard equilibrium 320; 335
boullies = slurries 287; 365
bourrage = stemming 296
bourrage à l'eau = water stemming 365
bourroir 305
boutefeu = blaster; shot firer 36; 283
BP = russian abbreviation name for → shaped charges
BPZ russian denomination for hollow charges with incendiary effect
branchement en parallèle = parallel connection 241
break 40
break test 247
breech 41
Brenngeschwindigkeit = burnign rate 45
Brennkammer = case; combustion chamber 60
Brennschluß = end of burning 116
Brennschlußgeschwindigkeit = endburning velocity 116
Brennstoff = fuel 143
bridgewire detonators **41**; 92
brisance **42**; 82; 359
Brückenzünder = bridgewire detonator 41
brulage = combustion 60
brûlage regressive = regressive burning 270
B-Stoff = german tarning name for methylalcohol
BSX = 1,7-diacetoxy-2,4,6-tetramethylene-2,4,6-trinitramine
BTM = castable mixture of Tetryl, TNT and aluminum 55/25/20

BTNENA = bi-trinitroethylnitramine 34
BTNEU = bi-trinitroethylurea 35
BTTN = butanetriol trinitrate 48
bulk density 42
bulk mix; bulk mix delivery equipment 42
bulk powder = porous nitrocellulose powder for hunting
bulk strength **43**; 297
bulldoze = mudcap 43
bullet resistant 43
bullet sensitive explosive material 44
bullet hit squib 43
Bundesanstalt für Materialprüfung; BAM 28; 141; 162; 186
burden 44
Bureau of Alcohol, Tobacco and Firearms 44
Bureau of Explosives 44
Bureau of Mines 361
burning chamber → combustion chamber 60
burning rate 23; **45**; 60; 66; 121
bus wire 47
Butarez = carboxy terminated polybutadiene (german)
butanetriol trinitrate 48
BWC = board wood cellulose (UK)
BZ = russian abbreviation for armor piercing charges with incendiary effect

C

C → composition C 62
CA_1 = nitrocellulose 12% N for lacquers 221
CA_2 = nitrocellulose 11–12,5% N for blasting gelatin (french)
calcium carbonate 324; 329
calcium nitrate **48**; 325
calcium stearate 325
calculation of explosives and gun powders 313
calorimetric bomb 159
camphor; camphre **49**; 161; 295; 325
Candelilla wax 161
cannon (testing device in test galleries) 246

caps, detonating 190; 191; 194; 251
cap sensitivity; cap test 49
capacitator (blasting) machines 37
Carbamite denominates al Ethyl-Centralite 56
Carbitol = diethyleneycol monoethyl-ether (USA)
carbazol = tetranitrocarbazol 307
carbon black 161
carbon dioxide 329
carbon oxide 329
carboxyl-terminated polybutadiene 62; 327
cardial medicine 122; 193
Cardox blasting process 52; 146
cartouche = cartridge 52
cartridge 52
cartridge density 53; **75**
cartridge strength 53
case = combustion chamber 60
case bonding **53**; 110; 289
caseless ammunition 54
casting of explosives 55
casting of propellants 55
Catergol = catalytic decomposing rocket propellant (e. g., hydrazine)
cavity effect 280
CBI = clean burning igniter (USA)
CBS = plastic explosive composed of 84% RDX, 16% butyl stearate plus 1,5 stabilizer
CBS-128 K; –162 A = composite rocket propellants (USA)
C. C. = collodion cotton = nitrocellulose 11–12% N (UK) 220
C. C.-propellants = "Cyclonit cannon" – RDX containing gun powders (USA)
CDB-propellants = combined composite and double base rocket propellants (german)
CDT (80) = castable double base propellant (USA)
CE = Tetryl (UK) 310
CEF = tri-β-choroethylphosphate
Cellamite = ammonium nitrate industrial explosive (french)
cellular explosive = foam explosive with closed cavities (USA)
Celluloidwolle = nitrocellulose with about 11% N (german)

Index

Cellulosenitrat → nitrocellulose 220
Centrallt I, II, III = stabilizers (german) 56; 57; 155; 161; 295; 325
Centralite TA = ammonium nitrate industrial explosive (Belgium)
CH_4, methane, fire damp 245
Chakatsuyaku = TNT (Japan) 336
chaleur de combustion = heat of combustion 159
chaleur de formation = energy of formation 117
chaleur d'explosion = heat of explosion 159
chaleur partielle d'explosion = partial heat of explosion 160
chambre de combustion 60
chambre de mine, abattage par = coyote blasting 65
channel effect 58
chantier de tir = blast area 36
Chapman-Jouguet point 83
Charbonniers equation **46**; 289
charcoal 35
charge creuse = hollow charge; shaped charge 67; **280**
charge d'armorcage = primer charge 265
charge de base = base charge 31
charge excitatrice = donor charge 112
charges génératrices de gaz = gas generating units 3, 146
charges nucléaires = nuclear charges
charge réceptrice = acceptor charge 1
chargeurs pneumatiques = air loaders 8
Chauyaku = RDX (Japan) 68
Cheddite = chlorate explosive (swiss)
Chemecol blasting process 145
Chilesalpeter = sodium nitrate (german) 288
chlorate de barium 28
chlorate de potassium 260
chlorate de sodium 288
chlorate explosives 58
chlorodinitrobenzene 100
chlorotrifluoride 239
chlorure d'ammonium 13
chlorure de picryle 350

Chornyi porokh = black powder (russian) 35
chromite de cuivre 64
"cigarette"-burning **58**; 137
circuit en série 280
circuit parallele 280
circuit tester 58
C-J = *Chapman-Jouguet*-condition → detonation 78
CL20 = Hexanitrohexaazaisowurtzitane 172
class A; class B; class C explosives 59
class I, II, III = german safety classes for permitted explosives 247
clearing blasts in oil and gas wells 77
Cloratita = chorate explosive (spanish) 58
closed vessel 23
coal-cement pipe, detonation transmission in 85
coal dust **59**; 325
coated = surface smoothed propellant; → ball powder 28
coefficient of detonation transmission 85
coefficient de selfexitation → sympathetic detonation 85
coefficient d'utilisation pratique c.u.p. (french) → lead block test 196
collodion → nitrocellulose 220
Collodiumwolle = nitrocellulose 12–12,6 % N (german)
column charge 59
comburant = oxidizer 239
combustibility 59
combustible = fuel 143
combustible cartridge cases 60
combustion 60
combustion chamber 60
combustion érosive = erosive burning 121
combustion heat 159
combustion-modifying additive 137; 192; 194; 199
combustion of explosives 75
commande de tir = blasting switch 39
commercial explosives 61
compatibility 61
Compatibility Group **61**; 70
compatibility testing 362

composite propellants 17; **61**; 137; 192; 194; 199
composition A etc. 62
composition B etc. 62
composition C etc. 63
compositions I, II 63
composition lumineuse = illuminant composition 184; 339
compositions pyrotechniques 269
compositions retardatrices = delay compositions 74
confined detonation velocity 63
confinement **64**; 296
consommation specifique d'explosifs = explosive loading factor 130
contained detonating fuse 64
conventional explosive performance data 315
cook off = premature inflammation (→ caseless ammunition) 54
copperchromite 64
cordeau *Bickford* = detonating fuse of lead coated TNT
cordeau détonant = detonating cord 64; 77; 215
cordeau détonant gainé = contained detonating fuse 64
corde d'allumage = igniter cord 183
Cordite = double base gun powder (UK) 65
Corpent = PETN 250
Coruscatives = Delay Compositions **65**; 74
cotton fibres 220
coulée de charges des projectiles = casting of explosives 55
coulée de propergol = casting of propellants 55
coyote blasting 65; 77
CP 1 BFP = nitrocellulose 13 % N 220
CP 2 = nitrocellulose 11,7–12,2 % N
CP SD = nitrocellulose 11,6 % N (french)
crater method 298
Crawford bomb 66
crésylite = mixture of trinitrocresol and picric acid (french)
crimping 66
critical diameter 66

CR-propellants = RDX containing rifle powders (USA)
cross section ratio → propellant area ratio 267
crusher 147; 359
C.T.D. = coefficient of detonation transmission (coefficient de self exitation) (french) 85
CTPB = carboxyl-terminated polybutadiene (USA) 61; 327
cumulative priming 66
c.u.p. = coefficient d'utilisation pratique (french) 197
cuprene 52
cupric salicylate 161
curing 67
cushion blasting 67
cutt off 67
cutting charge **67**; 280
C.W. = nitrocellulose 10–12 % N (german) 220
cyanur triazide 68
Cyclofive = RDX/Fivonite 53/47 (→ page 306) (USA)
Cyclonite = RDX = Hexogen 35; 62; 63; **69**; 114; 132; 141; 164; 187; 198; 206; 252; 325; 337; 339
Cyclops = high energetic rocket propellant
Cyclopentadienyl-Eisen = Ferrocene 137
cyclotetramethylenetetranitramine = HMX = Octogen 132; 141; 206; **237**; 239; 252; 327
Cyclotol = RDX-TNT mixture (USA) 68
cyclotrimethylenetrinitramine = RDX = Cyclonite = Hexogen 174
cyclotrimethylenetrinitrosamine 68
Cylinder Expansion Test 69

D

D-1; D-2 = phlegmatizer for explosives = 84 % paraffin wax, 14 % nitrocellulose and 2 % lecithine
DADE = DADNE = 1,1-Diamino-2,2-dinitroethylene 91
DADNPh = diazodinitrophenol 92
Daisy Cutter = BLU-82 → GSX 152

danger d'explosion en masse = mass explosion risk 208
danger d'inflammation = combustibility 59
dangerous goods regulations **70**
DAP = diamylphthalate 90; 162
DATNB; DATE = diaminotrinitrobenzene
Dautriche method **72**; 77; 88
DBP = dibutylphthalate 161
DBS = dibutylsebacate 161
DBT = mixture of dinitrobenzene and TNT (russian)
DBT = dibutyltartrate 161
DBX = depth bomb explosive (USA) 72
DCDA = dicyanodiamide
DD = mixture of picric acid and dinitrophenol (french)
DDNP = diazodinitrophenol 92
D: DNT added
déblai = muckpile 217
decade counter (chronograph) 88
décapitation de la charge = cut of 67
deflagration **73**; 78; 84; 246
deflagration point 73
DEGDN, DEGN = diethyleneglycol dinitrate 94
degressiver Abbrand = regressive burning 270
Delaborierung = dismantling of ammunition 110
delay; delay compositions 65; **74**
delay fuze 74
delayed initiation; delayed inflammation 74
dense prills 16
densité 75
densité de cartouche = cartridge density 53
densité de chargement = loading density 204
density 42; **75**; 81; 205
Dentex = mixture of RDX/DNT/aluminum 48/34/18 (UK)
dénudation de la charge = cut off 67
DEP = diethylphthalate 161
dépôt = magazine 207
depth charge 357
DES = dimethylsebacate 161
destressing blasting 76

destruction of explosive materials **75**; 110
Detacord = small diameter detonating cord (USA)
Detaflex = detonating cord made of Detasheet
Detasheet = plate shaped flexible explosive consisting of PETN and binders (USA)
détonateur 37; 92
détonateur à fil explosé = mild detonating fuse 216
détonateur instantané = instantaneous detonator 41
détonateur pour tir sou l'eau = water resistant detonator 365
detonating cord, detonating fuse 72; **77**; 113; 191; 251; 364
detonating cord connectors 78
detonation 77 ff.
detonation development distance 89
détonation par influence = sympathetic detonation 85; 139
detonation, selective 84
detonation, sympathetic 85; 139
detonation temperature 317
detonation velocity 42; 72; 75; 87
detonation wave, hydrodynamic theory of 81
détonations dans l'eau = underwater detonations 357
Detonationsgeschwindigkeit = detonation velocity 88
Detonationstemperatur = detonation temperature 317
Detonationsübertragung = sympathetic detonation 85
Detonationsverzögerer = non-electric delay device 236
Detonationswelle = detonation wave 78 ff.
detonator → blasting cap 37; 92
detoninooyuschii shnoor = detonating cord (russian) 77
detonierende Zündschnur, Sprengschnur = detonating cord 77
Deutsch-Französisches Forschungsinstitut St. Louis I. S. L. 192
dextrinated (lead azide) 194
Diaethylendiphenylharnstoff = Centralita 56

Diaethylenglykoldinitrat 94
diamètre critique = critical diameter 66
1,1-Diamino-2,2-dinitroethylene = DADE = DADNE 91
diamylphthalate **92**; 161; 295; 325
diaphragme de sécurité = safety diaphragm 274
diaphragme de potection = environmental seal 118
diazodinitrophenol 92
Diazol = diazodinitrophenol (german) 92
DIBA = diisobutyladipate 161
dibutylphthalate **93**; 161; 295; 325
dibutylsebacate 161
dibutyltartrate 161
dichromate d'ammonium 14
Dichte = density 75
dicyanodiamide 63; 153; 229
diethyldiphenylurea = Centralit I **56**; 155; 161; 295; 325
diethyleneglycol dinitrate **94**; 161; 202; 325
diethylphthalate 161
diethylsebacate 161
differential thermal analysis 96
diglycerol tetranitrate 96
diglycoldinitrate **94**; 161; 217; 356
diisobutyl adipate 161
diisocyanata de toluylène 339
diluent 97
Dimazin = DMH = dimethylhydrazine 97
dimethyldiphenylurea = Centralit II **57**; 155; 161; 295; 325
dimethylhydrazine **97**; 325
dimethylphthalate 161
DINA = dioxyethylnitramine dinitrate 108
Dinal = dinitronaphthalene 102
Dinamon = ammonium nitrate explosive (Italy)
Dinau = dinitroglycolurile (french) 98
dinitrate de butylèneglycol 47
dinitrate de diethylèneglycol 94
dinitrate de dioxyéthyl dinitroxamide 101
dinitrate de dioxyéthylnitramine 108
dinitrate d'éthanolamine 122

dinitrate d'éthylène diamine 123
dinitrate d'éthylnitropropanediol 223
dinitrate de glycérine 149
dinitrate de glycérine-dinitro phényl-éther 150
dinitrate de glycérine nitrolactate 150
dinitrate de glycol 227
dinitrate de méthylnitropropanediol = dinitrate de nitrométhylpropanediol 232
dinitrate de propylèneglycol 267
dinitrate de triéthylèneglycol 343
dinitrate de triméthylèneglycol 345
dinitrate de trinitrophénylglycérine-éther 150
dinitrate d'hexaméthalènetétramine 166
dinitrate d'isosorbitol 192
dinitroaminophenol = picramic acid 252
dinitrobenzene → metadinitrobenzene 210
dinitrobenzofuroxane 99
Dinitrobenzol 210
Dinitrochlorhydrin 148
dinitrochlorobenzene 100
dinitrocresol 103
Dinitrodiaethanoloxamiddinitrat 101
Dinitrodiglykol 94
dinitrodimethyloxamide 101
dinitro-dinitrosobenzene → dinitrobenzofuroxane 99
dinitrodioxyethyloxamide dinitrate 101
dinitrodiphenylamine 102
Dinitroethylendiamin 124
dinitroethylene diamine 123
Dinitroglycerin 149
Dinitroglycerin nitrolactat 150
Dinitrokresol 103
dinitronaphthalene 102
dinitroorthocresol 103
dinitrophenoxyethylnitrate 104
Dinitrophenylglycerinaetherdinitrat 150
Dinitrophenylglykolaethernitrat 104
dinitrophenylhydrazine 104
dinitrosobenzene 106
dinitrotoluene **106**; 161; 164; 325
Dinitryl 150

Index

Dinol = diazodinitrophenol (german) 92
dioctylphthalate 161
dioxyethylnitramine dinitrate 107
DIPAM = diaminohexanitrodiphenyle
DIPEHN = dipentaerythritol hexanitrate 108
diphenylamine **109**; 161; 295; 325
Diphenylharnstoff 9
diphenylphthalate 161
diphenylurea; diphénylurée 9; 161
diphenylurethane **108**; 161; 295; 325
dipicrylamine 169
dipicrylsulfide 170
dipicrylsulfone 171
dipicrylurea 110
dismantling of ammunition 75; **110**
distance d'évolution de détonation = detonation development distance 88
Di-(trinitroäthyl)-Harnstoff; di-trinitroéthylurée 35
di-trinitroéthylnitramine 34
ditching dynamite 111
Dithekite = liquid explosive consisting of nitric acid, nitrobenzene and water (USA) 112; 203
Diver's solution = high concentrated solution of AN and NH_3 in water; proposed as monergol
DMEDNA = dimethylethylene dinitramine (USA)
DMNA = dimethyldinitramine (USA)
DMS = dimethylphthalate 161
DMSO = dimethylsulfoxide
DNAG= Dynamit Nobel AG, Troisdorf, Germany
DNAP = dinitrodiazephenol 92
DNB = dinitrobenzene 210
DNBA = dinitrobenzaldehyde
DNCB = dinitrochlorobenzene 100
DNDMOxm = dinitrodimethyloxamide 101
DNDMSA = dinitrodimethylsulfamide
DNDPhA = dinitrodiphenylamine
DNEtB = dinitroethylbenzene
DNEU = dinitroethylurea
DNF = dinitrofurane
DNG = diglycerol dinitrate
DNMA = dinitromethylaniline

D.N.N. = dinitronaphthalene (british) 102
DNPA = 2,2-dinitropropyl acrylate
DNPh = dinitrophenol
DNPT = dinitrosopentamethylene tetramine
DNR = dinitroresorcinol
DNT = dinitrotoluene 106
DNX = dinitroxylene
donor charge 112
DOP = dioctylphthalate 161
DOS = dioctylsebacate 161
double base propellants 9; 56; 95; 112; 155; 194; 199; 205; 214
douilles combustibles = combustible cartridge cases 60
DOV = distilled of vitriol = H_2SO_4 96% (UK)
downline 78
DPA = diphenylamine 102
DPEHN = dipentaerythrol hexanitrate 108
DPhA = diphenylamine 102
DPP = diphenylphthalate 161
Drehgriffmaschine = twist knob blasting machine 38
dreibasiges Pulver = triple-base gun powder
drop test → bomb drop test 39
Druck, spezifischer = specific energy 291
Druckexponent = pressure exponent 46; 290
Druckkochen = boiling under pressure (→ nitrocellulose) 221
Druckluft-Ladeverfahren = pneumatic placing 257
Druckluft-Sprengverfahren = Armstrong blasting process 22
Drucksprung → hydrodynamic theory of detonation 78 ff.
Druckstoßwirkung = air blast 8
Drucktränksprengen = pulsed infusion shotfiring 268
Druckwelle = shock wave; air blast 8; 81
DTA = differential thermal analysis 96
Düse = nozzle 237
Dunnit = ammonium picrate (USA) 14

dutch test 112
Duxita = RDX phlegmatized with 3 % castor oil (Italy)
dwell time 113
Dyno Boost 113
Dynacord = detonating fuse (german) 113
Dyneschoc 113
Dynatronic 114
dynamic vivacity 24
Dynamit-Collodiumwolle = gelatinizing nitrocellulose for dynamites 222
Dynamite LVD = low velocity explosive 114
Dynamite MVD = medium velocity explosive 114
Dynamite gomme = blasting gelatin (french) 37
Dynamites 114

E

E → E-process (RDX synthesis) 70
E → E-Wolle = ester soluble nitrocotton (german) 222
EBW = exploding bridgewire detonator (USA) 128
écran = barricade 29
Ecrasit; Ekrasit = picric acid; in France: = ammonium trinitrocresolate
EDA = ethylene diamine
EDADN; EDD = etyhlendiamine dinitrate 123
Ednafive = EDNA/Fivonite 50/50 124; 306
EDNA = ethylene dinitramine 124
Ednatoal = Ednatol + 20% aluminum
Ednatol = pourable mixtue of EDNA and TNT **115**; 124
EED = electro-explosive device 115
effet Neumann = shaped charge effect (french) 280
effet du souffle = air blast 8
EF poudre = blank cartridge powder (french)
EFI = Exploding Foil Imitiator 128

EGDN = ethyleneglycol dinitrate 227
Einbasige Pulver = nitrocellulose gun powder 284
Einschluß = confinement 64
Eisenacetylacetonat 192
Ekrasit = picric acid 253
EL-506 = plate shaped explosve ("Detasheet") (USA)
EL 387 A; B = slurry consisting of water, starch and aluminum (Belgium)
EL-763 = permitted explosive (USA)
elektrische Zünder → bridgewire detonators 41
E.L.F. = extra low freezing (of nitroglycerine explosives)
EMMET= ethyltrimethylol methane trinitrate
Empfängerladung = acceptor charge 1
Empfindlichkeit = sensitivity 280
emulsion explosives 16; 17; 20; **115**; 193; 236; 365
end burning velocity 116
endothermal 116
energetic binders **116**; 151; 205; 257; 258
energy, enthalpy of formation **115**; 324 ff.
energy, specific → specific energy 291
Entzündungstemperatur = deflagration point 73
environmental seal 118
environmental testing → rocket test stand 273
épreuve de chaleur → hot storage test 178
épreuve de fracture = fragmentation toot 139
E-process → RDX 70
éprouvette 118
éprouveur = circuit tester 58
equation de *Bernoulli* 290
equation de *Charbonnier* 46; 290
equilibrium constants 335
equation of state 118
Erla = epoxy compound (USA) 120
erosion 121
erosive burning 120
ErTeN = erythritol tetranitrate 121

Index

essai au bloc de plomb = lead block test 196
Estane = polyester consisting of urethane, adipic acid, 1,4-butane-diol and diphenylmethane diisocyanate (USA)
EtDP = ethyl 4,4-dinitropentanoate
Ethanolamindinitrat 122
ethanolamine dinitrate 122
Ethriol trinitrate **122**, 325
Ethyldiphenylharnstoff 10
Ethylendiamindinitrat 123
Ethylendinitramin 124
Ethylenglykoldinitrat 227
Ethylglykoldinitrag = propyleneglycol dinitrate 267
Ethylnitrat 125
Ethylphenylurethan 125
Ethylpikrat 126
Ethyl -Centralit 56; 155; 325
ethyldiphenylurea 10; 155; 295; 324
ethylene diamine dinitrate 123
ethylene dinitramine 115; 124
ethylene glycol dinitrate = nitroglycol 227
ethylglycoldinitrate = propyleneglycol dinitrate 267
ethylhexanoate de plomb 199
ethyl nitrate 125
ethylphenylurethane **125**; 155; 295; 325
ethylpicrate = trinitrophenetol 126
ethyl-tetryl 126
EUROOYN 2000 127
European Committee for the Standardization of Test of Explosive Materials; now → International Study Group for the Standardization of the Methods of Testing Explosives 85; 198
European test fuze 51
E.V.O. = Eisenbahn-Verkehrsordnung (german) → RID 271
E-Wolle → nitrocellulose, estersoluble (german) 222
EWALIDW 127
exchanged salt pairs 14; 245
exothermal 127
explode 128
exploding bridgewire 128

exploseur = blasting machine 37
explosif en vrac = bulk mix 42
explosif-liant plastique = plastic explosive 255
explosifs 131
explosifs allégés = low density commercial explosives containing foamed additives (french)
explosifs à l'oxygène liquide = liquid oxygen explosives 203
explosifs antigrisouteux = permitted explosives 245
explosifs au nitrate d'ammonium; explosifs nitratés = ammonium nitrate explosives 15
explosifs chloratés = chlorate explosives 58
explosifs chlorurés = permitted explosives containing sodium chloride 245
explosifs d'amorcage = initiating explosives 190
explosifs gainés = sheathed explosives 282
explosifs liquides 202
explosifs nitratés = ammonium nitrate explosives 15
explosifs perchloratés = perchlorate explosives 244
explosifs primaires = initiating explosives 190
explosifs pulvérulents = powder form explosives 140; 263
explosifs secondaires = base charge 31; 277
explosifs sismiques 277
explosif S.G.P. = permitted explosive (Belgium) 245
explosion 128
explosion heat → heat of explosion 159
Explosionstemperatur = explosion temperature 129
Explosionswärme, partielle = partial heat of explosion 160
explosion tardive (long feu) = hangfire 158
explosion temperature 129
explosive bol 130
explosive casting → casting of explosives 55

explosive D = ammonium picrate 15
explosive forming 129
explosive loading factor 130
explosive materials 130
explosive train 131
explosives 131
explosives equal sheathed 248
"Extra" = ammonium nitrate containing commercial explosives
Extragummidynamit = gelatinous commercial explosive (Norway)
extrudeuse à vis = screw extruder 277
exudation 136

F

F 8 = mixture of aluminum and bariumnitrate (USA)
F (φ) = secondary explosive with high performance (russian)
FA = furfuryl alcohol
FA/AN = mixture furfuryl alcohol/aniline/hydrazine 46/47/7
face burning 136
Fackel = flare 139
factices = Mock explosives 216
FAE = Fuel Air Explosives 8; **142**; 312
fallhammer 137; 184
farine de guar = guar gum 151
Favier explosives = ammonium nitrate explosives (France; Belgium; now obsolete)
Federzugmaschine → blasting machines 37
fendage préliminaire = pre-splitting 263
fertilizer gerade ammonium nitrate; FGAN 16
Ferrocene (combustion moderating additive) 137
Feststoff-Raketen = solid propellant rockets 289
Feuergefährlichkeit = combustibility 59
Feuerwerkssätze = pyrotechnical compositions 269
Feuerwerkszündschnüre = pyrotechnical fuses (→ also: quick-match) 269

FGAN = fertilizer gerade ammonium nitrate (USA) 16
fils du détonateur = leg wires 202
Filmeffektzünder = Bullet Hit
fin de cumbustion = end of burning 115
firedamp 137; 245
firing current 137
firing line 138
first fire 138
Fivolite = tetramethylolpentanol pentanitrate
Fivonite = tetramethylolpentanone tetranitrate 305
flamebeau = flare 139
flame shield 138
flare 138
flash over 85; 138
flash point 139
flegmatiser 252
FLOX = mixture of liquid oxygen and liquid fluorine
flüssig-Luft-Sprengstoffe = liquid oxygen explosives 203
flüssig-Treibstoff-Raketen = liquid propellant rockets 204
flüssiges Ammoniak = liquid ammonia 204
flüssige Luft = liquid air 203
flüssiges N_2O_4 204
flüssiger Sauerstoff = liquid oxygen 203
flüssiger Wasserstoff = liquid hydrogen 204
flüssiges Fluor = liquid fluorine 204
flüssige Sprengstoffe = liquid explosives 202
fly rock 139
F-M = double base powders
FM = titanium tetrachloride (USA)
F = nitrocellulose nitroglycerine
FNR = tetrafluoroethylene trifluoronitrosomethan copolymer
force = strength 24; 291; **297**
Formfunktion → burning rate 24
FOX-7 = DADNE = 1,1-Diamino-2,2-dinitroethylene 91
FOX-12 = GUDN = Guarnyl-ureadinitramide 155
Fp 02 = TNT (german) 336
Fp 60/40 = TNT/AN 60/40 (german)

fracture, épreuve de 139
fragmentation test 139
fragment velocity 140
Fraunhofer-Institut für Chemische Technologie, ICT 182
free flowing explosives 20; **140**; 193
freezing of nitroglycerine 140
Frühzündung = premature firing 263
friction sensitivity 141
FTS = solid rocket propellant (german)
fuel 143
Fuel Air Explosives = FAE 8, **142**; 312
Füllpulver 02 = TNT (german) 336
fugasnost = lead block test (russian) 196
fulmicotone = nitrocellulose (italy) 220
fulminate d'argent = silver fulminate 284
fulminate de mercure = mercury fulminate 209
fumes; fumées 143
fume volume 364
functioning time 145
Furaline I-III = fuel mixture consisting of furfuryl alcohol, xylidine and methanol
fuse; fuze; fusée 145
fusée pyrotechnique = pyrotechnical fuse 269
fusée retardatrice = delay fuze 74
Fuß-Vorgabe = toe 339
fuze head 145
FV = Fivonite = tetramethylolpentanone tetranitrate 305
FV/EDNA = Ednative = mixture of Fivonite and EDNA
FV/PETN = Pentative = mixture of Fivonite and PETN
FV/RDX = Cyclofive = mixture of Fivonite and RDX (USA)

G

Galcit = solid rocket propellant consisting of $KMnO_4$ and asphalt pitch (now obsolete) (USA)

galerie d'essai = test gallery 246
galette = paste (for double base powders) 242
galvanometer 58
GAP = → glycidyl azide polymere 151
gap test → sympathetic detonation 85 ff.
Gasdruck = gas pressure 23; 45; **147**
gas-erzeugende Ladungen = gas generators 2; 146
gas generators 2; 146
gas jet velocity 147
gasless delay compositions 65; 74
gas pressure 23; 45; 147
gas volume 316; 364
GC = gun cotton = nitrocellulose ca. 13 % N (UK) 220
GcTNB = glycol trinitrobutyrate
GDN = glycol dinitrate 227
Geberladung = donor charge 112
gelatin explosives; gelatins 147
Gelatine Donarit S = gelatinous commercial explosive (Austria) 147
gelatinizer 49; 56; 295
Gelex = semigelatinous commercial explosive
generator machines → blasting machines 37
Geomit = powder form commercial explosive (Norway)
Geosit = gelatinous seismic explosive (german) 148
Geschützpulver = gun powder 155
Gesteinsprengstoffe = commercial rock explosives
gestreckte Ladung = column charge 59
gewerbliche Sprengstoffe = commercial explosives (german) 61
GGVE = german transport regulations 70; **271**
Gheksogen = Hexogen = Cyclonite = RDX (russian) 68
Gießen von Sprengladungen = casting of explosives 55
Gießen von Treibladungen = casting of propellants 55
Globular-Pulver = ball powder 28
glossary → Preface
GLTN = dinitroglycerinnitrolactate 150

Glühbrücke → bridgewire detonators 41
Glycerindinitrat 149
glycerol 226
glycerol acetate dinitrate 148
glycerol dinitrate 149
glycerol-dinitrophenylether dinitrate 150
glycerol nitrolactata dinitrate 150
glycerol-trinitrate 224
glycerol-trinitrophenylether dinitrate 149
glycide azide polymere 151
glycide nitrate 224
glycol 162; 228; 325
glycolurile dinitramine; tetramine 98
GND = pressure proof (seismic) explosive (german)
G = nitrocellulose-diglycol dinitrate
Goma pura = blasting gelatin (spanish)
Gomma A = blasting gelatin (Italy)
Gomme A etc. = blasting gelatin (french) 37
Grade A Nc = nitrocellulose 12.6–12.7 % N
Grade B Nc = nitrocellulose 13.35 % N
Grade C Nc = nitrocellulose blend of A and B
Grade D Nc = nitrocellulose 12.2 % N, also Grade E (USA)
grade strength 27; 297
grain 152
Granatfüllung 88 = picric acid (german) 253
granulation 152
graphite **152**; 162
Grenzstromstärke = no-fire current 236
grisou = fire damp 138; 245
Grisoudynamite chlorurée No. 1 etc. = permitted explosive (french)
Großbohrlochsprengverfahren = large hole blasting 193
group P 1 etc. = british safety classes for permitted explosives 248
Groupe d'Etude International pour la Normalisation des Essais d'Explosifs Secretary; Der. Per Anders

Persson, Swedish Detonic Research Foundation, Box 32 058, S 12 611 Stockholm, Sweden 85; 198
Grubengas = firedamp 138; 245
G-Salz = nitroguanidine (german) 228
GSX 152
guanidine nitrate 63; **153**; 198; 325
guanidine perchlorate 154
guanidine picrate 154
Guanite = nitroguanidine 228
guar gum; Guarmehl **154**; 325
Guarnylureadinitramide = GUDN 155
Gudolpulver = gun powder consisting of nitrocellulose, nitroglycerine and Picrite (german)
guhr dynamite 114; 198
Gummidynamit = gelatinous commercial explosive (Norway)
GUNI = guanidine nitrate (german) 153
gun cotton 37; 220
gun powder 23; **155**
Gurney energy 157

H

H-6 = mixture RDX/TNT/aluminum/ wax 45/30/20/5 (USA)
H-16 = 2-acetyl-4,6,8-trinitro-2,4,6,8-tetrazanonane-1,9-diol-diacetat
HADN = hexamethylenediamine dinitrate
härten; aushäurten = curing 67
Haftvermögen = case bonding 53
Halbsekundenzünder = half-second-step-delayed bridgewire detonators 41
Haleite, Halite = ethylenedinitramine (USA) 124
halogen fluoride 239
hangfire 158
Hansen test 155
Harnstoffnitrat = urea nitrate 361
Haufwerk = muckpile 217
HBX-1 = mixture of RDX, TNT and aluminum 155

HC = mixture of hexachloroethane and zinc (fume generator) (USA)
HE = high explosive (USA)
HEAP = armorpiercing
HEAT = antitank; hollow charge
heat of combustion 155
heat of explosion 117; 155; 315
heat of formation → energy of formation 117; 324 ff.
heat sensitivity 162
HEATT = hollow charge with tracer
HEBD = base detonating
HEDA = delayed action
HEF = high energy fuel, e. g., boranes (USA)
HE/SH = squashhead (UK)
HEF-2 = propylpentaborane
HEF-3 = triethyldekaborane
HEF-5 = butyldekaborane
HEH = heavy projectile
HEI = brisant incendiary ammunition with tracer
HEIA = Immediate action (USA)
heineiyaku = trinitrophenetol (Japan) 126
Heizsatz = heating charge for gas generating units 146
heinelyaku = trinitrophenetol (Japan)
HELC = long case (USA)
Hellhoffit = liquid mixtures of nitric acid and nitrocompounds
HeNBu = hexanitrobutane
HEPT = brisant ammunition with squashhead and tracer
Heptanitrophenylglycerin
Heptryl = 2,4,6-trinitrophenyl trinitromethyltrimethylnitramine trinitrate 164
HES = shell (USA)
Hess, upsetting test according to 359
HETRO = cast-granulated mixtures of RDX, TNT and additives (swiss)
HEX = high energy explosive (USA) 165
Hex-24; −48 = mixtures of $KClO_4$, aluminum, RDX and asphalt 32/48/16/4 (USA)
Hexa = hexanitrodiphenylamine (german) 165
Hexal = RDX/aluminum (german) 165

hexamethylene diisocyanate 165
hexamethylenetetramine dinitrate 70; 166
hexamethylentriperoxide diamine 166
Hexanit; Hexamit = mixture of hexanitrodiphenylamine and TNT (german)
hexanitrate de dipentaérythrite 108
hexanitrate de mannitol 207
hexanitroazobenzene 167
hexanitrobiphenyl 168
hexanitrocarbanilide 110
hexanitrodipentaerythritol 108
hexanitrodiphenylamine 100; **168**; 325
hexanitrodiphenylaminoethyl nitrate 169
hexanitrodiphenylglycerol mononitrate 170
hexanitrodiphenyloxamid 173
hexanitrodiphenyloxide 170
hexanitrodiphenylsulfide 171
hexanitrodiphenylsulfone 171
hexanitroethane 172
Hexanitrohexaazaisowurtzitane **172**; 326
hexanitromannitol 207
hexanitrooxanilide 173
hexanitrostilbene 55; **174**
hexanitrosulfobenzide 171
Hexastit = RDX phlegatized with 5 % paraffine
Hexite = hexanitrodiphenylamine (german) 168
Hexogen = Cyclonite = RDX 37; 131; 141; 164; **174** ff; 187; 198; 206; 242; 252; 255; 325; 339
Hexotol = mixtures RDX-TNT (swedish)
Hexotonal = mixture of TNT, RDX and aluminum (german)
Hexyl = hexanitrodiphenylamine (german) 168
HiCal = high-energetic rocket fuel
HITP = High Ignition Temperature Propellants 54; 258
HMTA = hexamethylenetetramine
HMTD, HMTPDA = hexamethylene triperoxide diamine 166
HMX = Homocyclonite = Octogen 132; 206; 206; **237**; 252; 327

HN = hydrazine nitrate 180
HNAB = hexanitroazobenzene 168
HNB = hexanitrosobenzene 342
HNCb 1 = hexanitrocarbanilide 110
HNDP, HNDPhA = hexanitrodiphenylamine 168
HNDPA = hexanitrodiphenyl 168
HNDPhA; HNDP = hexanitrodiphenylamine 168
HNDPhAEN = hexanitrodiphenylaminoethyl nitrate
HNDPhBzi = hexanitrodiphenylbenzyl
HNDPhGu = hexanitrodiphenylguanide
HNDPhSfi = hexanitrodiphenylsulfide 170
HNDPhSfo = hexanitrodiphenylsulfone 171
HNDPhU = hexanitrodiphenylurea 110
HNEt = hexanitroethane 171
HNF = Hydrazine nitroform = salt of Trinitromethane
HNG = hydrine-nitroglycerine
HNH = hexanitroheptane
HNIW = Hexanitrohexaazaisowurtzitane 172
HNO = hexanitrooxanilide 173
HNS = hexanitrostilbene 173
Hohlladung = hollow charge; shaped charge 280
Hohlraumsprengen = cushion blasting 67
Holland test → Dutch test 112
hollow charge = shaped charge 42; 62; **280**
Holzkohle = charcoal 3
Homocyclonite = HMX = Octogen 132, 206; 206; **237**; 252; 327
hot spots 177
hot storage tests 173
HOX = high oxygen explosive 34
Hoxonit = plastic explosive consisting of RDX, nitroglycerine and nitrocellulose (swiss)
H. P. 266 = powder form permitted explosive (USA)
HT = RDX and TNT (french)
HTA = RDX/TNT/aluminum 40/40/20 (USA)
HTA-3 = HMX/TNT/aluminum 49/29/22 (USA)

HTP = hydrogen peroxide (UK)
HTBP = hydroxol-terminated polybutadiene (USA) 62
HU-detonators = highly unsensitive bridgewire detonators (german) 38; 179
Hugoniot's equation; *Hugoniot* curve 83
hybrids 1; **179**; 289
Hydan 179
hydrazine 2; **180**; 204; 326
hydrazine nitrate 180; 326
hydrazine perchlorate 181
hydrazoic acid 23
hydrodynamic theory of detonation 78 ff.
hydrogen peroxide 20
Hydrox blasting 146
HYDYNE = UDMH/diethylene triamine 60/40 (USA)
hygroscopicity 181
Hyman = nitromethylglycolamide nitrate (USA)
Hypergols 75; 182; 204

I

IBEN = Incendiary bomb with explosive nose (USA)
ICBM = Intercontinental ballistic missile (USA)
ICT = Fraunhofer-Institut für Chemische Technologie, D-7507 Pfinztal-Berghausen 182
Idrolita = AD/RDX/paraffine/water 70/20/3/7 (Italy)
Ifzanites = slurries (russian)
Igdanites = ANFO (russian)
Igniter 182
igniter cord/igniter cord connector 183
igniter safety mechanism 183
igniter train 131; 183
ignitibility 183
ignition system 183
illuminant composition 184
Imatrex = swedish trade name for Miedziankit 58
IMO = Intergovernmental Maritime Consultative Organization 70; 184

IME = Institute of Makers of Explosives 184
immobilization 184
impact sensitivity 44; **184**; 266
important explosives 132
impulse 189
impulse spécifique 292
incendiary 189
incompatibility 61; 362
Industrial explosvies → commercial explosives 61
inert 189
to inflame 189
inflammabilité = ignitibility 183
inflammer = to inflame 189
infusion blasting → water infusion blasting 268
Ingolin = concentrated hydrogen peroxide (german) 23
inhabited building 136
inhibited propellant 190; 301
Initialsprengstoffe = initiating explosives 133; 190
initiation 190
initiator 92; 190
injector nitration 225
injector transport → water driven injector transport of nitroglycerine 365
instantaneous detonators 41; 191
insulation 191
internal energy 331
internal enthalpy 333
International Study Group for the Standardization of the Methods of Testing Explosives, cretary: Dr. Per-Anders Persson. Swedish Detonic Research Foundation, Box 32058, S 12611, Stockholm, Sweden 85; 198
inverse salt pair: ion exchanged (salt pair) permitted explosives
ion propellants 191
IPN = Isopropyl nitrate (USA) 268
iron acetylacetonate **192**; 326
I.S.L. = Deutsch-Französisches Forschungsinstitut St. Louis 192
isobaric; isochoric 2; 317; 322
Isocyanate 257
isolement; Isolierung = insulation 191

isopropyl nitrate 268
isosorbitol dinitrate 192

J

JATO = jet assisted take of charge (USA)
jet perforator = → hollow charge for oil and gas well stimulation
jet tapper 192
JP = jet propellant (USA)
JP-1; -2; -3; -4; -5; = rocket fuels; hydrocarbons (USA)
JPN = improved JP
JPT = double base propellant in tube elements for → Bazooka 31
JP-X = JP-4/UDMH 60/40 hypergolized fuel (USA)
Julnite = ethylendiurethane (french)
jumping mortar test 298

K

K-1; -2 Splav = mixture of TNT and DNB or dinitronaphthalene (russian)
K1F = chlorotrifluoroethylene polymer (USA)
KA-process (RDX-synthesis) 70
Kalisalpeter = Kaliumnitrat 260
Kaliumbitartrat 219
Kaliumchlorat 260
Kaliumsulfat 219
Kaliumperchlorat 261
Kalksalpeter = calcium nitrate 48
kalorimetrische Bombe = calorimetric bomb 159
Kammerminensprengung = coyote blasting 65
Kampfer = camphor 49
Kanaleffekt = channel effect 58
Kantenmörser = angle shot mortar 248
Karben = carbene (polyacetylene) 52
Karitto = black powder (Japan) 35
KA-Salz = RDX (Hexogen) by synthesis according to *Knöfler*-Apel 70
K = ball powder

Kast's upsetting Test 359
Kcilil = trinitroxylene (russian) 356
KDNBF = potassiumdinitrobenzofuroxane (USA)
Kel-F = chlorotrifluoroethylene polymer 9010 = RDX/Kel-F 90/10
Kelly 192
kerosine 326
Kerosole = metal dispersion in kerosine (USA)
Keyneyaku = trinitrophenetol (Japan) 126
Kibakuyaku = explosive for primers (Japan)
Kieselgur = guhr 114; 329
Klasse I; Klasse II; Klasse III = german safety classes for permitted explosives 249
Klemmung = propellant area ratio 267
knäppern = modcapping 218
Knallquecksiler = mercury fulminate 209
Knallsilber = silver fulminate 284
Kochsalz = rock salt 330
Kohlenstaub = coal dust 59
Kohlenstaubsicherheit = coal dust safety → permitted explosives 245
Kokoshokuyaku = black powder (Japan) 35
Kollergang = heavy roller mill for black powder mixing
Kollodiumwolle = nitrocellulose 12–12.6% N (german) 220
Kolloksilin = nitrocellulose 11–12% N (russian) 220
Kondensatoranzündmaschinen = capacitator blasting machines 37
Kontur-Sprengen = smooth blasting 288
Koomooliativuyye = shaped charge (russian) 280
Koruskativs 65
Krater-Methode = crater method 298
Kresylit = trinitrocresol (german) 351
kritischer Durchmesser = critical diameter 66
krut = gun powder (swedish) 155
K-Salz = RDX (Hexogen) by synthesis according to *Knoefler* (german) 70

Kugelpulver = ball powder (german) 28
kugelsicher = bullet resistant 43
kumulative Zündung = cumulative priming 66
Kumullativuyye = cumulatively acting charge (hollow charge) (russian)
Kunkeln = press-molded perforated clinders of B-black powder (german)
kunststoffgebundene Sprengstoffe = plastic explosives 255
Kupferchromit 64

L

laboratory combustion chamber 60
Ladegeräte = buld mix delivery equipment 42
Ladedichte = loading density 205
Ladeschlauch = plastic hose as loading device for permitted explosives
Ladestock = tamping pole 304
Lager = magazine 44; 136; 207; 296
Lambrex = slurry (Austria) 192
Lambrit = ANFO blasting agent (Austria) 193
Langzeit-Teste = long time test; shelf life tests 112; 283
large hole blasting 193
LASL = Los Alamos, National Scientific Laboratory (USA)
LDNR = lead dinitroresorcinate
LE = low explosive; propellant (UK)
lead acetyl salicylate 162; **194**; 326
lead azide 23; 43, 141; **194**; 326
lead block test **196**; 298; 340
lead ethylhexanoate 162; **199**; 326
lead nitrate **199**; 326
lead picrate 200
lead salicylate 162
lead stearate 162
lead styphnate 29; 43, **201**; 326
lead sulfate 162; 170
lead trinitroresorcinate = lead styphnate 29; 43, **201**; 326
leading lines 201
Lebensdauerteste = shelf life tests 112; 283

Index

Lebhaftigkeitsfaktor = vivacity → burning rate 45
leg wires 202
Leitungsprüfer = circuit tester 58
Leit-Sprengschnur = trunkline 78
Leuchtsätze = illuminant compositions 184
Leuchtspur = tracer 339
L.F. = low freezing
LH_2 = liquid hydrogen (USA)
Ligamita 1; 2; 3 = nitroglycerine base industrial explosives (Spain)
ligne de tir = firing line 138
ligne de cordeau détonant 78
linear burning 202
line-wave generator = linear detonation front generating charge
lined cavities → shaped charges 280
linters 222
liquid ammonia 204
liquid explosives 202
liquid fluorine 204
liquid hydrogen 204
liquid N_2O_4 204
liquid oxygen (LOX) 203
liquid oxygen explosives 203
liquid propellant rockets 203
lithergoles = hybrids 179
lithium nitrate 204
lithium perchlorate 204
LJD = Lennard-Jones-Devonshire state equation 119
LN = lead nitrate 200
LN_2 = liquid nitrogen
loading density 42; 53; 204
long time tests 112; 283
low explosive = propellants 155
LOVA = low-vulnerability ammunition 206
LOVA propellant 205
LOX = liquid oxygen 239 → also liquid oxygen explosives 203
LOZ = liquid ozone
L-Splav = TNT with 5% trinitroxylene (russian)
Lp = liquid propellant 204
lueur à la bouche = muzzle flash 218
LX = octogen (HMX)-formulations 206
Lyddite = picric acid 253

M

M-1; M-6; M-15 etc. are short names for american gun powders
M3 = kerosine hypergolized witz 17% UDMH (USA)
MABT = mixture of TNT, picric acid and dinitrophenol (Italy)
Macarite = mixture of TNT and lead nitrate (Belgium)
MAF-40: − X = mixed amine fuels (Hydine) (USA)
magazine 44; 136; **207**; 296
magnesium carbonate 326
magnesium trinitroresorcinate; magnesium styphnate 201
MAN = methylamine nitrate 326
mannitol hexanitrate 207; 326
Manöverpulver → porous powder 259
Mantelsprengstoffe = sheathed explosives 281
MAPO = methylaziridine phosphinoxide 208
Marschsatz = sustainer charge in rocketry
mass explosion risk 208
mass flow 46
mass ratio 209
Massen-Durchsatz = mass flow 46
Massen-Explosionsgefährlichkeit; Massen-Explosionsfähigkeit = mass explosion risk 208
Massenverhältnis = mass ratio 209
MAT = mixture of TNT and picric acid (France; Italy)
Matagnite = Industrial explosive (Belgium)
match cord 270
Matsu = blasting gelatin (Japan) 37
maturer = curing 67
MBT = mixture of picric acid and dinitrophenol (France; Italy)
MDF = mild detonating fuse (0.2−0.4 gram PETN/m = 10−20 grain/ft) USA
MDN = mixture of picric acid and dinitronaphthalene (France)
MeAN = MAN = methylamine nitrate 326
mèches = safety fuses 274

MeDINA; MeEDNA = methylethylene-dinitramine
Mehlpulver = ungranulated black powder 35
Mehrlochpulver = multiperforated grain 155
Meiaku = Tetryl (Japan) 310
Mélinite D = picric acid (french) 253
Mélinite/O ("ordinaire") (french) = picric acid with 0.3 % trinitrocresol
Mélinite/P = picric acid with 12 % paraffine (french)
MeN = methylnitrate 212
MeNENA = 1-nitroxytrimethylene-3-nitramine (USA)
Menkayaku = nitrocellulose (Japan) 220
mercury fulminate 133; 141; **209**
merlos = barricade 29
Mesa burning → burning rate 45
mesurage de la poussée = thrust determination 336
metadinitrobenzene 210
Metallbearbeitung durch Sprengstoffe = explosive forming 129
metatelnyi zariad = Treibladung (russian)
methane 138; 245
methoxytrinitrobenzene = trinitroanisol 346
methylamine nitrate **212**, 326
methyldiphenylurea; Methyldiphenylharnstoff = Akardit II **9**; 161; 295; 324
methylenedinitrotetrazacyclooctane (precursor product for the Octogen (HMX) synthesis) 238
Methylethyldiphenylharnstoff = Centralit III (german) 57
methylethyldiphenylurea = Centralit III (german) **57**; 161; 295; 319
methyl mehacrylate (MMA) 162
methyl nitrate 212
Methylnitroglykol = propyleneglycol dinitrate 267
methylphenylurethane **213**; 295
methyl picrate = trinitroanisol 346
methyltrimethylolmethane trinitrate = Metriol trinitrate 214
methyl violet test 213
Metriol trinitrate 162; **214**; 326

MetrTN = Metrioltrinitrate 214
MF = mercury fulminate (USA) 209
MHF = mixed hydrazine fuel containing hydrazine nitrate (USA)
MHN = mannitol hexanitrate (USA) 207
Microballoons 215
Micrograin = solid fuel consisting of zinc powder and sulfur (USA)
Miedziankit = one site mixed chlorate explosive (german) (now obsolete) 58
mild detonating fuse 216
military explosives 135
millisecond delay blasting 215
millisecond delay detonator 215
Millisekundenverzögerer = detonating cord MS connector 78
mineral jelly 162
minex = mixture of RDX, TNT, AN and aluminum powder (UK)
miniaturized detonating cord 215
Minimal-Zündimpuls = all fire 11
minol = mixture of AN, TNT and aluminum powder (UK)
Minolex = mixed explosive consisting of AN, RDX, TNT and aluminum (UK)
Minurex = commercial explosive (french)
mise à feu 190
misfire 216
missile 216
M. J. = mineral jelly = vaseline (british) 162
MltON = maltose octanitate (USA)
MMeA = mononitromethylaniline (USA)
MMA = methyl methacrylate 162
MN = mononitrotoluene (France)
MNA = mononitroaniline (USA)
MNAns = mononitroanisol (USA)
M.N.B. = mononitrobenzene (USA; UK)
MNBA = mononitrobenzaldehyde (USA)
MNBAc = mononitrobenzoic acid (USA)
MNCrs = mononitrocresol (USA)
MNDT = mixture consisting of AN, dinitronaphthalene and TNT (France; Italy)

MNM = mononitromethane 231
M.N.N. = mononitronaphthalene
MnnHN = mannose hexanitrate
MNO = dinitrodimethyloxamide 101
M.N.T. = mononitrotoluene
MntHN = mannitol hexanitrate 207
MNX = mononitroxylene
mock explosives 216
Mörser; ballistischer = ballistic mortar 26
Mörser; Kantenmörser; Bohrlochmörser = mortar; cannon 247
Momentzünder = instantaneous bridgewire detonator 41
Monergol 23; 97; 203; **216**; 269
Monobel = permitted explosive (USA)
Monoethanolamindinitrat 122
mononitrate d'hexanitrodephénylyglycérine 170
mortier ballistique 26
morteur fusée = rocket motor 273
motor 217; 273
moulage des propellants = pressing of propellants 264
moulage d'explosifs = pressing of explosives 264
mouton de choc = fallhammer 137
MOX = metal oxidizer explosives 217
MP = picric acid with 12% paraffin (France)
MP 14 = catalyst for decomposition of hydrogen peroxide (permanganates) (german)
MTN = Metrioltrinitrate 162; **214**; 326
muckpile 217
mud cap 3; 43; 217
Mündungsfeuer = muzzle flash 218
Muenyaku = smokeless powder (Japan)
Multiprime 236
Munroe effect = hollow charge effect **218**; 280
muzzle flash 218

N

NAC = nitroacetylcellulose (USA; Italy)
Nachdetonation = hang fire 158
nachflammen = post-combustion 260
Nafolit = tetranitronaphthalene (France)
NAGu = nitrominoguanidine (USA) 219; 326
Nano Energetic Materials 218
Naphthit = trinitronaphthalene (german) 353
NATO advisory group for aeronautical research and development → AGARD 3
Natriumchlorat 288
Natriumnitrat Natronsalpeter = sodium nitrate 288
Natriumperchlorat 289
NBSX = 1,7-dinitroxy 2,4,7-trinitro-2,4,6-triazaheptane
NBYA = di-trinitroethylurea
Nc; NC = nitrocellulose (german) 220
NCN = Nitrocarbonitrate 36; 219
NDNT = AN/dinitronaphthalene/TNT 85/10/5 (french)
NENA = N-(2-nitroxy)-nitramine-ethane (USA)
NENO = dinitrodioxyethyl-oxamide dinitrate 103
NEO = diglycol dinitrate (french) 94
Neonite = surface smoothet nitrocellulose gun powder (UK)
NEPD = nitroetyhlpropanediol dinitrate
Neumann effect = hollow charge effect 280
neutral burning 219
Neuvalin = concentrated hydrogen peroxide
New Fortex = ammonium nitrate explosive with Tetryl (UK)
NG; N/G = nitroglycerine 224
NGc = nitroglycol (USA) 227
Ngl = nitroglycerine (german) 224
nib-glycerine trinitrate; NIBTN = nitroisobutylglycerine trinitrate 230
nicht-sprengkräftige detonierende Zündschnur = mild detonating cord; miniaturized detonating cord 216
Nigotanyaku = RDX/TNT mixtures (Japan)
Nigu = nitroguanidine (german) 228
Nilite = powder form blasting agent (USA)

NIMPHE = nuclear isotope mono propellant hydrazine engine (temperature maintenance device for decomposition catalysts in monergol driven rockets)
NIP = nitroindene polymer (USA)
Niperit = PETN (german) 250
Nisalit = stoechiometric mixture of nitric acid and acetonitrile (german)
nitrate d'amidon 233
nitrate d'ammonium 15
nitrate de barium 29
nitrate de calcium 48
nitrate de dinitrophénoxyéthyle 104
nitrate d'éthyle 125
nitrate de glycide 226
nitrate de guanidine 153
nitrate de lithium 204
nitrate de méthylamine
nitrate de méthyle 212
nitrate de plomb 199
nitrate de polyvinyle 259
nitrate de potassium 260
nitrate de propyle 268
nitrate de sodium = nitrate de soude 288
nitrate de strontium 300
nitrate de sucre 234
nitrate de tétraméthylammonium 305
nitrate de triaminoguanidine 340
nitrate de triméthylamine 344
nitrate de trinitrophényl-nitramine-éthyle 354
nitrate de trinitrophényloxéthyle 353
nitrate d'hexanitrodiphénylglycérine 169
nitrate d'hydrazine 180
nitrate d'urée 361
nitre = salpetro 260
nitric acid esters 133
Nitroaminoguanidine **219**; 326
nitrocarbonitrate 219
nitrocellulose 133; 155; 162; 164; 187; **220**; 326; 327
nitrocellulose propellants → single base propellants 284
nitrocompounds 132
Nitrodiethanolamindinitrat 108
nitrodiphenylamine 327
Nitroethan 223

nitroethane 222
Nitroethylpropandioldinitrat 223
nitroethylpropanediol dinitrate 223
nitroform = trinitromethane 352
nitrogen 329
nitrogen dioxide 329
nitrogen oxide 329
nitrogen tetroxide 203
nitroglycerin 37; 114; 133; 162; 164; 187; 198; 203; **224**; 240; 327
nitroglycerin powders → double base propellants 112; 155
nitroglycide 226
nitroglycol = Nitroglykol 17; 133; 162; 164; 187; 198; 203; **227**; 327
nitroguanidine (picrit) 132; 162; **228**; 240; 327
nitroguanidine powders 156
nitrogurisen = nitroglycerin (Japan) 224
Nitroharnstoff 235
Nitzrohydren 234
nitroisobutantriol trinitrate = nitroisobutylglycerol 230
nitrokletchatka = nitrocellulose (russian) 220
Nitrolit = trinitroanisol 346
Nitromannit = nitromannitol 207
nitromethane 198; 203; **231**; 327
nitromethylpropanediol dinitrate 232
nitroparaffins 233
Nitropenta = PETN (german) 250
nitropentanone 305
Nitropentaglycerin = Metriol trinitrate 214
nitrostarch = Nitrostärke 233
nitro-sugar 233
Nitrotetryl = tetranitrophenylmethylnitramin
nitrotoluene 234
Nitrotriazolone **234**, 327
nitrourea **235**, 327
Nitrozucker 234
NM = nitromethane 231
Nobelit = emulsion explosive and blasting agent (german) 236
non-allumage, intensité de courant de = no-fire current 236
non-electric delay device 236
Nonel – non electric firing device (swedish) 236

Normalgasvolumen = Normalvolumen = volume of explosion gasses 364
Norm-Brennkammer = standard combustion chamber 60
Novit = mixture of hexanitrodiphenylamine, TNT and aluminum (swedish)
nozzle 237
N-P = triple base powders
Np 5; Np 10 = PETN phlegmatized with 5 resp. 10 % wax (german)
NQ = nitroguanidine 228
NS = nitrostarch 233
NSP = black powder – smokeless powder compound (german)
NSug = nitrosugar (USA) 233
NTNT = AN/TNT 78/20 (french)
NTO = Nitrotriazolone 234
NT = TNT/AN 30/70 (french)
N2N = AN/SN/TNT 50/30/20 (french)
NX = AN/trinitroxylene (french)

O

Oberflächenbehandlung = surface treatment; also → inhibited propellant; → restricted propellant 190; 271; 301
Obturate 237
Octogen = Homocyclonite = HMX 132; 141; **237**; 252, 327
Octol = micture HMX/TNT 77/23 239
Octyl = Bitetryl = N,N'-dinitro-N,N'-bis (2,4,6-trinitrophenyl)-ethylenediamine (UK)
ohmmeter 58
Oktogen = Octogen (german) 237
ON 70/30 = N_2O_4/NO_2 70/30 oxidator (german)
onayaku = mixture of picric acid and dinitronaphthalene (Japan)
onde de choc = shock wave 8; **78 ff.**
onde de détonation = detonation wave 8, 78 ff.
ONERA = Office Nationale d'études et de Recherches, Paris
Optolene = liquid rocket fuel consisting of vinylethylether, aniline, tar, benzene and xylene (german)

oshokuyaku = picric acid, press molded (Japan)
ouvreuses explosives de percée = jet tapper 193
Ox = carborane-fluorocarbon copolymer (USA)
oxidizer 143; **239**
oxygen 329
oxygen balance 240
Olyliquit 203
Oxypikrinsäure = styphnic acid (german) 300
Oxytetryl = trinitromethylnitraminophenol
Ozo-benzene = benzene triozonide

P

P1 = trimethyleneglycol dinitrate 345
P2 = methylene dioxidimethanol dinitrate (USA)
P (salt) = piperazine dinitrate (USA)
P.A. = picric acid (french) 253
PAC = ANFO (german) 20
Palatinol = phthalic acid ester (german)
PAN = explosive consisting of PETN; pentaerythrol tetraacetate and AN
PANA = same mixture plus aluminum (Italy)
Panclastit = mixture of nitrobenzene and N_2O_4 30/70
Panzerfaust → Bazooka 31
paraffin **241**; 327
parallel connection; also bus wire 47; 241
Parazol = dinitrochlorobenzene (USA)
partial heat of explosion 160
paste 242
Patrone = cartridge 52
Patronendichte = cartridge density 53
PBAN = polybutadiene acrylonitrile
PB-RDX = polystyrene bonded RDX, consisting of 90 % RDX, 8.5 % polystyrene and 1.5 % dioctylphthalate
PBTC = carboxylterminated polybutadiene (USA)

PBU = phenylbenzylurethane (UK)
PBX = plastic bonded explosive 242; 255
PBX 9010, 9011, 9404·03; various compositions with 90 % RDX and 94 % MHX, grain sized 242
PCX = 3,5-dinitro-3,5,-diazopiperidine nitrate (USA)
PDNA = propylenedinitramine
PE 1; 3A = plastizised RDX (USA)
PE-Wolle = low nitrogen percentage nitrocellulose (german)
PEG = polyethylene glycol 162
pellet powder 243
pellets 140; 243
pendulum test → ballistic mortar 26
Penta = PETN (german) 250
pentaerythritol 251
pentaerythritol tetranitrate 37; 77; 133; 141; 162; 164; 187; 198; 244; 244; **250**; 252; 327
pentaerythritol trinitrate 162; 244
Pentastit = phlegmatized PETN 242
Pentol = mixture of PETN and TNT (german)
Pentolite = pourable mixture of TNT and PETN (USA) 244
Pentrit = PETN (german) 250
PENTRO = mixture of PETN, TNT and paraffin 49/49/2 (Swiss)
Pentryl = trinitrophenylnitramine ethylnitrate 354
PEP-2; PEP-3; PIPE = mixtures of PETN and Gulf Crown Oil (USA)
perchlorate d'ammonium 17
perchlorate de barium 29
perchlorate explosives 242
perchlorate de guanidine 154
perchlorate d'hydrazine 181
perchlorate de lithium 204
perchlorate de potassium 261
perchlorate de sodium 289
percussion cap; percussion primer 93; 200; 201; **242**; 265; 285; 287
perforation of oil and gas wells 245
perle d'allumage = squib 43, 234
Perlit = picric acid (german) 253
permissibles; permitted explosives 14; 73; 84; 86; **245**; 367
peroxides 31; 167; **250**; 343; 368
peroxide de benzoyle 31

peroxide de tricycloacétone 343
peroxide de zinc = zinc peroxide 368
Perspex = acrylic acid methylester polymer (same as Plexiglas; Lucite) (USA) 87
Pertite = picric acid (Italy) 253
pétardage = mud cap 2; 43; 217
PETN = pentaerythrol tetranitrate 37; 77; 133; 141; 162; 164; 187; 198; 244; 244; **250**; 252; 327
PETRIN = pentaerythrol trinitrate (USA) 162; **250**
petroleum jelly **252**; 327
PETS = pentaerythrol tetrastearate (USA)
PGTN = pentaglycerine trinitrate
pH test → Hansen test 158
PHE = plastic high explosive (UK)
phlegmatization 252
Ph-Salz = ethylenediamine dinitrate (german) 123
phthalate diamylique 92
phthalate dibutylique 93
Picatinny Arsenal 21; 185; 197
picramic acid 253
picramide = trinitroaniline 346
picrate d'ammonium 19
picrate d'éthyle 126
picrate de guanidine 154
Picratol = pourable mixture of TNT and mmmonium picrate 253
picric adic 164, 198, **254**, 327
Picrinata = picric acid (spanish) 253
picrylchloride = trinitrochlorobenzene 350
picryl sulfide = hexanitrodiphenylsulfide
picryl sulfone = hexanitrodiphenylsulfone
Picurinsan = picric acid (Japan) 253
piezo-quartz 24
Pikramid = trinitroaniline 346
Pinkrinsäure = picric acid 253
Pikrinsäuremethylester = trinitroanisol 346
Pirosilinovyye porokha = single base nitrocellulose propellant (russian)
Piroksilins N° 1 = nitrocellulose 12–13 % N; N° 2 = > 13 % N (russian) 220
plane charge (shaped charge) 280

plane wave generators = charges generating plane detonation fronts
plastic explosives 237; 255
Plastisol = solid rocket propellant consisting of APC or AN, PVC and plasticizers
plate dent test 255
plateau combustion 46; 62
Platzpatrone = blank cartridge
Plumbaton = lead nitrate/TNT 70/30 (USA)
PLX = Picatinny liquid explosive = → nitromethane or 95% NM and 5% ethylenediamine (USA)
PMA; PMMA = polymer acrylic methylester (Plexiglas Lucite, Perspex) 87
PN = poudre noir (french) = black powder 35
PNA = pentanitroaniline
PNDPhEtl = pentanitrodiphenyleethanol
PNDPhEth = pentanitrodiphenylether
PNDPhSfo = pentanitrodiphenylsulfon
pneumatic placing 257
PNL = D – 1 = phlegmatizing mixture consisting of 84% paraffin wax, 14% nitrocellulose (12% N) and 2% Lecithine (USA)
PNP = polynitropolyphenylene 257
poiont d'inflammation = flasch point 139
polyacetylene → carbene 52
poly-3-azidomethyl-3-methyl-oxetane 256
poly-3,3-bis-(azidomethyl)-oxetane 257
polybutadiene acrylic acid; polybutadiene acrylonitrile → composite propellants 61
polybutadiene carboxyterminated 62; 327
polyethyleneglycol 162
polyisobutylene 327
polymethacrylate 162
polynitropolyphenylene 258
polypropyleneglycol 162; **255**; 327
polysulfide 61
polytrop exponent 82
polyurethane 61; 258

polyvinyl nitrate 133; 162; **259**; 327
porous AN-prills 17
porous powder 259
post-combustion 218; 260
post-heating 11
potassium bitartrate 219
potassium chlorate 58; 240; **260**
potassium chloride 329
potassium nitrate 35; 162; 240; **260**; 327
potassium perchlorate 162; **261**
potassium sulfate 162; 218
poudre = gun powder 155
poudre à base de la nitroglycérine = double base propellant 112
poudre à simple base = poudre B = single base nitrocellulose powder (french) = single base powder 259; 285
poudre composite = composite propellant 61
poudre d'aluminium = aluminum powder 11
poudre noir = black powder 35
poudre noir au nitrate de soude = B-black powder 31
poudre poreux = porous powder 259
poudre progressive = progressive burning propellant 265
poudre sans fumée 284
poudre sphérique = ball powder 28
pousée = thrust 336
poussière = coal dust 59; 245
powder (gun powder) 155
powder form explosives 263
PPG = polypropyleneglycol (USA) **258**; 327
pre-ignition 263
premature firing 263
prequalification test 263
pre-splitting 264
pressing of rocket propellant charges 264
press-molding of explosives 264
pression de gaz = gas pressure 147
pressure cartridge 264
pressure exponent 25; 46
prills 16; 264
Primacord = detonating fuse (USA) 77

Primadet = mild detonating fuse (USA) 216
primary blast 265
primary explosives 190; 265
primer 92; 265
primer charge 265
produits de détonation = fumes 143
progressive burnig 155; 265
projectile forming (shaped) charge 280
projectile impact sensitivity 266
propanediol dinitrate 267
propellant 155; 266
propellant-actuated power device 267
propellant area ratio ("Klemmung") 267; 290
propellant casting 55
propellant grain 46
propellant traité de surface → surface treatment 301
propergols 267
propulseur = rocket motor 273
propyleneglycol dinitrate 267
propylnitrate **268**; 327
protection contre les courants vagabonds = stray current protection 296
Protivotankovaya roochnaya zazhigatelnaya granata = Molotow cocktail (russian)
Prüfkapsel = test cap 305
PS = polysulfide 61
PSAN = Phase Stabilized Ammonium Nitrate 16
PTX-1 = Picatinny ternary explosive = RDX/Tetryl/TNT 30/50/20 (USA)
PTX-2 = RDX/PETN/TNT 44/28/28 (USA)
PTX-3 = mixture of EDN, Tetryl and TNT (USA)
PTX-4 = mixture of EDA, PETN and TNT (USA)
PU = polyurethane 61
pulsed infusion shotfiring 268
Pulver, Schießpulver = gun powder 155
pulverförmige Sprengstoffe = powder form explosives, mostly AN based 262
Pulverrohmasse = paste 260

PVN = polyvinyl nitrate 133; 162; **259**; 327
Pyrocore = detonating fuse for rocket ignition
Pyronite = Tetryl; also: mixture of Tetryl and trinitrophenylmethylaniline 310
pyrophoric 269
Pyropowder = single base nitrocellulose propellant (UK)
pyrotechnical compositions 28; 29; 30; 43; 64; 204; 260; 261; **269**; 270; 368
pyrotechnical fuse 269
PYX = 2,6-bis-bis-(pikrylamino)-3,5-dinitropyridine

Q

quality requirements for Industrial and military explosives 133, 135
quantitiy-distance table 136; 269
quartz 147
QDX = SEX = 1-acetyloctahydro-3,5,7-trinitro-1,3,5,7-tetrazocine (USA)
Quecksilberfulminat = mercury fulminate 209
quick-match; match cord; cambric 269
quickness 46

R

R-Salz = cyclotrimethylenetrinitrosamine (german) 68
radicals, free radicals 322
Rakete = rocket 273
Raketenflugkörper = missile 216
Raketenmotor = rocket motor 273
Raketenprüfstand = rocket test stand 273
Raketentreibmittel = rocket propellant 112; 322
Raketentriebwerk = rocket motor 273
Ramjet = air breathing rocket system
rapport d'expansion = propellant area ratio 267

raté = misfire 216
RATO = rocket assisted take off
rauchloses; rauchschwaches Pulver = smokeless powder 284
Rayleigh line 83
RDX = Cyclonite; Hexogen 37; 62; 63; **68**; 114; 132; 141; 164; 187; 198; 206; 252; 255; 325; 339
RDX class A-H = various grain sized qualities of Hexogen (USA) 68
RDX type A = product nitrated by nitric acid 68
RDX type B = product fabricated by the acetic anhydride (Bachmann) process; contains 3–12% → Octogen 68; 237
RDX Polar PE = plastizised RDX; containing 12% Gulf 297 process oil and lecithine 90/10 = RDX/KelF 90/10
reactivity test 362
recommended firing current 270
recommended test current 270
recul = setback 280
Reduced Sensitivity 270
Règlement International concernant le transport des marchandises dangereuses 271
regressive burning 270
Reibempflindlichkeit = friction sensitivity 141
relais = booster 40
relation des masses = mass ratio 209
relative weight strengt 26; 37; 297; 366
relay 271
reliability 271
requirements on industrial and military explosives 133
résistance à l'eau = water resistance 364
résistant au balles = bullet resistant
resonance = erosive burning 271
resserrement = propellant area ratio 267
restricted propellant 271
retart = delay 74
retard d'allumage = delayed inflammation; funtioning time 74; 145
RF-208 = organic phosphorus compound for hypergolizing rocket fuels
RFG = rifle fine grain powder (UK)
RFNA = red fuming nitric acid (UK)
RID = Réglement International concernant le transport des marchandises dangereuses 70; 162; 184; **271**
rifle bullet impact test 273
RIPE = plastizided RDX with 15% Gulf Crown Oil (USA)
rocket 273
rocket motor 273
rocket propellants 61; 112; 322
rocket test stand 273
Röhrenpulver; Röhrchenpulver = tube shaped gun powder 155
Rohmasse; Pulverrohmasse = paste 242
roquette à propergol liquide = liquid propellant rocket 204
roquette à propergol solide = solid propellant rocket 289
Rossite = guanylnitrourea (USA)
rotational firing 274
Round Robin Test 274
RP-1 = kerosin type as fuel in liquid rocket motor systems
rubberlike propellant = polysulfide-, polyurethane- or Plastisolbonded → compostie propellants
Rückstoß = setback 280
russian method = fibre length evaluation by pulp sedimentation 221
Russkii Koktel = russian cocktail = potassium chlorate and nitrotoluene in glass containers; ignition by concentrated sulfuric acid
Russkii Splav = mixture of pricric acid and dinitronaphthalene
RX-05-AA = RDX/polystyrene/dioctylphthalate 90/8/2 (USA)
RX-09-AA = HMX/dinitropropylacrylate/ethyldinitropentanoate 93.7/5/0.6 (USA)
RX-04-AV = HMX/polyethylene 92/8 (USA)
RX-04-BY = HMX/FNR 86/14 (USA)
RZ-04-AT = HMX/carborane-fluorocarbon polymer 88/12 (USA)
RZ-04-PI = HMX/Viton 80/20 (USA)

S

safe & arm 274
safety classes (of permitted explosives) 245
safety diaphragm 274
safety distance of buildings 136
safety fuse 269; **274**
SAFEX INTERNATIONAL 275
Saint-Venant and *Wantzel* formula 147
Salpeter = salpetre; potassium nitrate 259
Salpetersäure = nitric acid
salt pair (ion exchanged) permitted explosives 245
sand test 275
Sanshokitoruoru = TNT (Japan) 336
Sauerstoffträger = oxidizer 239
Sauerstoffwert; Sauerstoffbilanz 240
sautage à gran trou = large hole blasting 193
sautage par grands fourneaus de mines = coyote blasting 65
SBA = slurry blasting agent (USA)
scaled distance 276
SCAN = Spay Crystallized Ammonium Nitrate 16
SCB → Semiconductor Bridge Igniter 278
Schießpulver = gun powder 165; 220
Schießwolle = underwater explosives consisting of TNT, hexanitrodiphenylamine and aluminum, now obsolete (german)
Schlagempfindlichkeit – impact sensitivity 184
Schlagwetter = firedamp 138; 245
Schneckenpresse = screw extruder 277
Schneiderit = mixture of AN and nitrocompounds, mostly dinitronaphthalene (french)
Schneidladungen = cutting charges 67; 280
schonendes Sprengen = smooth blasting
Schub = Thrust 336
Schubmessung = thrust determination 336
Schuttdichte = bulk density 42
Schutzwall = barricade 29
Schwaden = fumes 143
Schwarzpulver = black powder 35
Schwarzpulverzündschnüre = safety fuses 274
Schwefel = sulfur 301
screw extruder 277
SD = double base propellant fabricated without solvents (french)
SDMH = symmetric dimethylhydrazine (USA)
SE = slurry explosive (USA) 287
secondary blasting 277
secondary explosives 277
Securit → Wetter-securit = permitted explosive (german) 367
seismic explosives; seismic gelatins 277
seismic shots 77; 277
Seismo-Gelit = gelatinous special gelatin for seismic use 278
seismograph; seismometer 278
Seismoplast = special explosives for seismic shots (german) 278
selective detonation 84
self forging fragment; SFF 281
Sekundärladung = base charge 31
Semiconductor Bridge Igniter (SCB) 278
semigelatin dynamites 279
Semtex 279
sensibilité 305
sensibilité à l'impact = impact sensitivity 184
sensibilié à l'impact projectiles = projectile impact sensitivity 266
sensibilité au chauffage – heat sensitivity 162
sensibilité au choc de détonateur = cap sensitivity 49
sensibilité au frottement = friction sensitivity 140
sensitivity 141; 162; 266; 280
Serienschaltung = series electric blasting cap circuit 278
Serien-Parallel-Schaltung = series in parallel electric balsting cap circuit 278
setback 278

Index

SFF = EFP 281
S.G.P. = denomination of → permitted explosives in Belgium
S-H-process (RDX-synthesis) 68
shaped charge 42; 67; 280
sheathed explosives 282
shelf life (storage life) 112; 283
Shellite (UK) = Tridite (USA) = mixtures of picric acid and hexanitrodiphenylamide
Shimose = picric acid (Japan) 253
shock pass heat filter 87
shock wave 3; 78 ff.; 283
Shoeiyaku = PETN (Japan) 250
Shotoyaku = An/TNT 50/50 (Japan)
Shouyaku-koshitsu = plastizised RDX
short delay blasting (→ also: millisecond delay blasting) 215
shot anchor 283
shot firer, blaster 36; 283
SH-Salz = RDX fabricated by decomposing nitration according to *Schnurr-Henning* 70
Sicherheitsabstands-Tabelle (USA) 136
Sicherheitszündschnüre = safety fuses 274
Silberacetylid 283
Silberazid 283
Silbercarbid 283
Silberfulminat 284
silver acetylide 283
silver azide 283
silver carbide 283
silvered vessel test 284
silver fulminate 284
SINCO = gas generating propellant (german) 285
single bases powders 285
Single-Event FAE = Thermobaric Explosives (TBX) 312
Sinoxid = non eroding primer composition (german) 286
SINTOX = lead free priming composition (german) 287
Sixolite = tetramethylolcyclohexanol pentanitrate 306
Sixonite = tetramethylolcyclohexanone tetranitrate 306
skid test 287
slotted mortar (test gallery) 247

slurries 10; 17; 36; 49; 115; 155; 193; 203; 212; **284**
slurry casting 55
small arms ammunition primers 285
SN = sodium nitrate 288
snake hole 285
Sodatol = sodium nitrate/TNT 50/50 288
sodium chlorate 285
sodium chloride; salt 245; 246; 330
sodium hydrogen carbonate 327
sodium nitrate 63; 239; 240; **285**; 327
sodium perchlorate 289
soft grain powder (black powder) 35; 289
Sohlenbohrloch = snake hole 288
solid propellant rocket 289
soluble gun cotton = nitrocellulose 12.2 % 221
solventless processing for propellants 155
Sorguyl = tetranitroglycolurile (french) 98
soufre 301
spacing 291
Spätzündung = hangfire 158
Spaltzünder = spark detonator 291
Span = sorbitane monooleate (USA)
spark detonators 291
special effects 291
specific energy **291**; 298; 317
specific impulse 292
spezifischer Sprengstoffverbrauch = explosive loading factor 130
spinner = spin stabilized rocket
Splittertest = fragmentation test 139
Spränggrummi = blasting gelatin (Norway)
Sprengbereich = blast area 36
Sprenggelatine = blasting gelatin 37
Sprengkapselempfindlichkeit = cap sensitivity 49
Sprengkapseln = blasting caps; detonators 37
Sprengkraft = strength 297
Sprengluft-Verfahren → liquid oxygen explosives 203
Sprengmeister = shot firer; blaster 36; 283

Sprengmittel = explosive materials 130
Sprengoel = nitroglycerine, nitroglycol and mixtures thereof 224; 227
Sprengupulver = black powder for blastings 31; 35
Sprengriegel = explosive bolt 130
Sprengsalpeter = B-black powder 31
Sprengschlamm = slurry 10; 17; 36; 49; 115; 155; 193; 203; 212; **287**
Sprengschnur 77
Sprengstoff-Attrappen = Mock Explosives 216
Sprengstoff-Ladegeräte = loading devices
Sprengstoff-Lager = magazine 207
Sprengstoff-Prüfstrecke = test gallery 246
Sprengzubehör = blasting accessories 36
Sprengzünder = blasting cap 37
springing 293
squib 43; 293
Stabilität; stabilité; stability 1; 32; 112; 155; 213; 280; **294**; 362; 363
stabilizer 5; 10; 56; 57; 102; 109; 125; 155; 213; 295
stable detonation 83
Stärkenitrat = nitrostarch 233
Stahlhülsenverfahren = steel sleeve test (heat sensitivity) 163
standard combustion chamber 60
Standardization of tests: International Study Group for the Standardization of the Methods of Testing Explosives Secretary: Dr. Per-Anders Persson, Swedish Detonic Research Foundation, Box 32058, S 12611 Stockholm, Sweden 85
Start-Rakete = booster 40
state equation 83; 118
Stauchprobe = upsetting test 359
Steinflug = fly rock 139
steel sleeve test 162
stemming 64; 296
Stickstofftetroxid = nitrogen tetroxide 203
Stirnabbrand = face burning 58; 137
Stoppine = quick-match; cambric 270
storage of explosives 136; 136; 296

Stoßgriffmaschinen = blasting machines wiht impact knob 37
Stoßtränkungssprengen = pulsed infusion shotfiring 268
Stoßwelle = shock wave 78
straight dynamites 27
strands 66
stray current protection 296
strength 27; 196; 291; **297**
Streustromsicherheit = stray current protection 296
strontium nitrate 300
styphnic acid = trinitroresorcinol 300; 328
styphnyldichloride = dichloro-2,4,6-trinitrobenzene
sulfur 35; 240; **298**
Supercord = detonating fuse with ca. 40 g/m PETN (german) 301
superconic propagation → shock wave; detonation 78 ff.

T

T4 = RDX (Italy) 68
T 4 plastico = RDX plastizised with 17,3 % diglycol dinitrate or with 11 % petroleum (Italy)
T-Stoff = concentrated hydrogen peroxide (german)
TA = triacetine 162
Tacot = tetranitrodibenzole-trazapentalente 303
tableau des distances de sécurite: → table 13, page 136
TAGN = triaminoguanidine nitrate 340
talc 329
Taliani test 304
tamping = stemming 296, 305
tamping pole 305
Tanoyaku = mixtures of RDX, TNT and Tetryl (Japan)
TAT = 1,3,5,7-tetracetyl-1,3,5,7-octahydroazocine, precursor-product for the → Octogensynthesis 238
TATB = triaminotrinitrobenzene 341
TATNB = triazidotrinitrobenzene 342
TAX = acetylhexahydrodinitrotriazine (USA)
TBX = Thermobaric Explosives 312

TDI = toluene diisocyanate 339
TEDFO = bis-(trinitroethyl)-difluorformal
TEGMN = triethyleneglycolmononitrate
TEGN = triethyleneglycoldinitrate 343
temperature coefficient → burning rate 45
température de décomposition; température d'inflammation = deflagration point 73
température d'explosion = explosion temperature 129
TEN = PETN (russian) 250
TeNA = tetranitroaniline 306
TeNAns = tetranitroanisol
TeNAzxB = tetranitroazoxybenzene
TeNB = tetranitrobenzene
TeNBPh = tetranitrodiphenylamine
TeNBu = tetranitrobutane
TeNCB = tetranitrochlorobenzene
TeNCbl = tetranitrocarbanilide
TeNCbz = tetranitrocarbazol 307
TeNDG = tetranitrodiglycerine 96
TeNDMBDNA = tetranitrodimethylbenzidinedinitramine
TeNDPhETa = tetranitrophenylethane
TeNDPhEtla = tetranitrodiphenylethanolamine
TeNHzB = tetranitrohydrazobenzene
TeNMA = tetranitromethylaniline (Tetryl) 310
TeNMe = tetranitromethane 307
TeNN = tetranitronaphthalene 308
TeNOx = tetranitrooxanilide
TeNPhMNA = tetranitrophenylmethylnitramine
TeNT = tetranitrotoluene
TeNTMB = 3,5',5'-tetranitro-4,4'-tetramethyldiaminobiphenyl (USA)
TePhUR = tetraphenylurea (USA)
Territ = plastic explosive consisting of nitroglycerine, APC, DNT, TNT, SN and nitrocellulose (Sweden)
test cap 305
test préliminaire = prequalification test 263
testing galleries 246
Tetralit = Tetryl (german) 311
Tetralita = Tetryl (italian, spanish) 311

tetramethylammonium nitrate 305
tetramethylenetetranitramine = Octogen; HMX 132; 141; 206; **237**; 252; 327
tetramethylolcyclohexanol pentanitrate = Hexolite 306
tetramethylolcyclohexanone tetranitrate = Hexonite 306
tetramethylolcyclopentanol pentanitrate = Fivolite 306
tetramethylolcyclopentanone tetranitrate = Fivonite 306
tétranitrate de diglycérine 96
tétranitrate d'erythrile 121
tétranitrate de tétraméthylolpentanone 305
tétranitrate de pentaérythrite = PETN 250
tetranitroaniline 307
tetranitrocarbazol 307
tetranitrodibenzotetrazapentalene = Tacot 303
tetranitrodiglycerol 96
tetranitroerythritol 121
tetranitroglycolurile 98
tetranitromethylaniline = Tetryl 310
tetranitromethane 203; **308**; 328
tetranitronaphthalene
Tetrasin = tetrazene (russian) 309
Tetratetryl = tetra-(trinitrophenylnitraminoethyl)methane (USA)
tetrazene 133; 141; 187; 201; 285; 287, **310**
tetrazolylguanyltetrazene hydrate = tetrazene 309
Tetril = Tetryl (russian) 311
Tetritol-Cyclonite = Tetryl/TNT/RDX 11.7/16.4/71.9 (russian)
Tetroxyl = trinitrophenylmethoxynitramine (USA)
Tetryl 37; 132; 164; 187; 198; **311**; 328
Tetrytol = mixture Tetryl/TNT 312
TFENA = trifluoroethylnitramine
TG = thermogravimetry 96
TG = Trotil-Gheksogen = TNT/RDX mixtures (russian)
théorie hydrodynamique de la détonation = thermohydrodynamic theory of detonation 78 ff.
thermic differential analysis 96

thermische Sensibilität = heat sensitivity 162
thermites 313
Thermobaric Explosives (TBX) 312
thermodynamic calculation of decomposition reactions of industrial explosives, gun powders, rocket propellants, tables 313 ff.
thermogravimetry 96
thermohydrodynamic theory of detonation 78 ff.
Thional = pentanitrodiphenylsulfon (USA)
thrust 290; **334**
tir à microretard = millisecond delay blasting 215
tir d'impregnation = pulsed infusion shotfiring 268
tir primaire = primary blast 265
tir sou pression d'eau = underwater detonation 357
TLP = Treibladungspulver = gun propellant; the following suffix (e. g., TLP/A) indicates;
TMEMT = trimethylenetrinitrosamine 68
TNA trinitroaniline 307
TNAmPH = trinitroaminophenol
TNAZ = Trinitroazetidin 328; 347
TNnd = trinitroanilide
TNAns = trinitroanisole 346
TNB = trinitrobenzene 348
TNBA = trinitrobenzaldehyde
TNBAc = trinitrobenzoic acid 349
TNBxN = trinitrobenzol nitrate
TNC = tetranitrocarbazol 307
TNCB = trinitrochlorobenzene 350
TNCrs = trinitrocresol 351
TNDCB = trinitrodichlorobenzene
TNDMA = trinitrodimethylaniline
TNDPhA = trinitrodiphenylamine
TNEB = trinitroethylbenzene
TNEDV = trinitroethyldinitrovalerate
TNETB = 2,2,2-trinitroethyl-4,4,4,-trinitrobutyrate
TNG = trinitroglycerine 224
TNM = tetranitromethane 308
TNMA = trinitromethylaniline
TNMeL = trinitromelamine
TNMes = trinitromesitylene
TNN = trinitronaphthalene 353

TNPE = PETN (spanish) 250
TNO = tetranitrooxanilide
TNPh = trinitrophenol = picric acid 253
TNPhBuNA = trinitrophenylbutylnitramine
TNPhDA = trinitrophenylendiamine
TNPhENA = trinitrophenylethyl-nitramine 126
TNPhlGl = trinitrophloroglucine
TNPhMNA = trinitrophenylmethyl-nitramine = Tetryl 310
TNPhMNAPh = trinitrophenyl-methyl-nitraminophenol
TNPht = trinitrophenetol 126
TNPy = trinitropyridine 355
TNPyOx = trinitropyridine-1-oxide 355
TNR = trinitroresorcine (styphnic acid) 300
TNRS = lead styphnate 201
TNStl = trinitrostilbene
TNT = trinitrotoluene 16; 27; 55; 62; 68; 114; 115; 132; 162; 164; 187; 215; 239; 253; 288; 328; **336**; 339; 340
TNTAB = trinitrotriazidobenzene 342
TNTClB = trinitrochlorobenzene 350
TNTMNA = trinitrotolylmethyl-nitramine
TNX = trinitroxylene 356
toe 339
TOFLOX = solution of ozonefluoride in liquid oxygen
Tolita = TNT (spanish) 336
Tolite = TNT (french)
Tolite/D = TNT, solidification point 80.6 °C = 177.1 °F
Tolite/M = the same, 78 °C = 172.4 °F
Tolite/O = 79 °C = 174.2 °F
Tolite/T = 80.1 °C = 176.2 °F (french)
totuylene diisocyanate 133; **339**
Tonka = aniline and dimethylaniline as liquid rocket fuel together with nitric acid as oxidator (german)
Torpex = mixture of RDX, TNT and aluminum 11; 70; **339**
Totalit = AN with ca. 6 % paraffin (swiss)
toxic fumes 8; **143**

Index

TPEON = tripentaerythroloctanitrate tracers 339
traitement des métaux par explosion 129
traitement de surface = surface treatment 301
transmission of detonation → detonation, sympathetic detonation; gap test 85; 146
transport regulations
→ ADR (road) 2
→ RID (rail) 271
→ IATA (air) 182
→ IMDG (shipping) 184
transport par injection d'eau 365
Tauzl test → lead block test 196
Treibstoff = propellant 266
triacetin 162
Triacetonperoxid; triacétoneperoxide 249; **343**
Trialen = mixture of RDX, TNT and aluminum (german) 11; 70; **340**
triaminoguanidine nitrate 6; 206; 328; **340**
triaminotrinitrobenzene 328; **341**
triazide cyanurique 68
triazidotrinitrobenzene 342
Tribride = 3-component rocket system, e. g., metal fuels suspended in liquid fuels and oxidators (USA)
Tricinat = salt of styphnic acid (german)
tricycloacetone peroxide 249; **343**
Triergol = same as Tribride
triethyleneglycol 344
triethyleneglycol dinitrate = Triglykoldinitrat (german) 162, **343**
Trilita = TNT (spanish) 336
trimethylamine nitrate 328; **344**
trimethyleneglycol dinitrate 345
trimethylenetrinitrosamine 68
trimethylenetrinitramine = RDX = Cyclonite 68
trimethylolethylmethane trinitrate 122
trimethylolnitromethane trinitrate 230
trimethylolpropane trinitrate 122
Trinal = trinitronaphthalene 353
trinitrate de butanetriol 48
trinitrate de glycérine 224
trinitrate de nitroisobutylglycérine 230
trinitrate de pentaérythrite 244

trinirate de triméthyloléthylméthane 122
trinitrate de triméthylolméthylméthane 214
Trinitril = glyceroltrinitrophenylether dinitrate
trinitroaniline 198; **346**
trinitroanisole 346
Trinitroazetidin 328; **347**
trinitrobenzene 328; 348
trinitrobenzoic acid 349
trinitrochlorobenzene 100; 328; **350**
trinitrocresol 351
trinitrodioxybenzene 300
trinitroethylalcohol 35
trinitrometacresol 351
trinitrometaxylene 356
trinitromethane 352
trinitronaphthalene 353
trinitrophenetol 126
trinitrophenol; picric acid 164; 198; **253**; 327
trinitrophenoxy-ethyl nitrate 353
trinitrophenylethylnitramine = ethyltetryl 126
Trinitrophenylglycerinetherdinitrat 150
Trinitrophenylglykolethernitrat 353
trinitrophenylmethylaniline = Tetryl 310
trinitrophenylmethylether = trinitroanisol 346
trinitrophenylmethylnitramine = Tetryl 37; 132; 141; 164; 187; 198; 240; **311**; 328
trinitrophenylnitramino ethylnitrate 354
trinitropyridine 328; **355**
trinitropyridine-N-oxide 328; **355**
trinitrorésorcinate de plomb 201
trinitroresorcinol = styphnic acid **300**; 328
trinitrosotrimethylenetriamine 75
trinitrotoluene = TNT 16; 27; 55; 62; 68; 114; 115; 132; 162; 164; 187; 215; 239; 253; 288; 328; **336**; 339; 340
trinitroxylene 356
Trinol = trinitroanisol 346
Triogen = trimethylenetrinitrosamine (USA) 75

triple base propellants 155
Trisol = trinitroanisol 346
Tritonal = mixture of RDX, TNT and aluminum (german) 11; 356
Trixogen = mixture of RDX and TNT (german) 357
Trizin = trinitroresorcinol (german) 300
Trojel = slurry blasting agent
Trotyl = TNT (german) 336
trunkline (of detonating cord) 78
tuyère = nozzle 237

U

U = bridgewirde detonator U (unsensitive), (german) 41
UDMH = unsymmetrical dimethylhydrazine (USA) 97
Übertragung = sympathetic detonation 85; 139
ullage 357
ummantelte Sprengstoffe = sheathed explosives 281
unbarricated 357
unconfined detonation velocity 88; 357
undersonic propagation (= deflagration) 45; 73
underwater detonations 21; **357**
underwater explosives → DBX; → Torpex 72; **339**; 357
Unterwasserzünder = water resistant detonator 366
upsetting tests 42; **359**
urea nitrate 108; 328, **343**
urethane → stabilizer 295
US Bureau of Mines 361

V

vacuum test 61; **362**
vaseline 252
velocite de combustion = burning rate 45
Veltex = mixture of HMX, nitrocellulose, nitroglycerine, nitrodiphenylamine and triacetine 362

Verbrennung = combustion 60
Verbrennungswärme = heat of combustion 159
Verbundtreibsätze = composite propellants 61
verdämmen = stemming 296
Vernichten von Explosivstoffen = destruction of explosive materials 75
Verpuffungspunkt = deflagration point 73
Versager = misfire 216
Verstärkungsladung = booster 40
Versuchsstrecken = test galleries 246
Verträglichkeit = compatibility 61
Verzögerungszünder = delay fuze 74
vessel mortar test 298
Viatra = Inert stemming cartridge (swiss)
Vibrometer 362
Vieille Test 178; **363**
Virial equation 119
viscosity of nitrocellulose 222
Visol; Visol-1; -4; -6 = vinylethylether and mixtures with isopropylalcohol and vinylbutylether; liquid rocket fuel (german)
Vistac N° 1 = low molecular weight polybutane (USA)
vitesse de combustion = burning rate 23; 45; 121
vitesse de détonation = detonation velocity 42; 72, 75; 88
vitesse moyenne de combustion = average burning rate 23
Viton A = perfluoropropylene-vinylidine fluoride copolymer
vivacity factor 46
VNP = polyvinyl nitrate (USA) 259
volume of explosion gases 363
volume strength 42; 317; 364
Vorgabe = burden 44
vorkesseln = springing 293
vorspalten = pre-splitting 263
VV = Vzryvchatoiye veschestvo = explosive (russian)
V-W = porous powders (german)

W

W I; W II; W III = permitted explosives belonging to the german safety classes I, II and III 367
warm storage tests; Warmlagerteste → hot storage tests 178
Wasadex 1 = ANFO-explosive 364
Wasserbesatz = water stemming 366
water driven injector transport of nitroglycerine 365
water 328; 329
water gels 10; 287; 365
water hammer; water jet 359
water infusion blasting 359
water resistance 364
water resistant detonator 365
water stemming 365
web 366
Weichkornpulver = soft grain black powder 35; 289
weight strength → relative weight strength 26; 37; 297; 366
Wetter- = german prefix given to all trade name of permitted explosives 367
Wetter-Sprengstoffe = permittet explosives **245**; 367
Wetter-Roburit B; Wetter-Securit C; Wetter-Westfalit = permitted explosives (german) 367
WFNA; WFN = white fuming nitric acid (UK)
WhC = white compound = 1.9-dicarboxy 2,4,6,8-tetranitrophenazin-N-oxide (USA)
wood dust 328
Wood's metal 73
W-Salz = RDX according *Wolfram* (german) 70

X

X-306 A = igniter for mild detonating fuse (USA)
X28M = Plasticizer containing → Polyvinyl Nitrate 259
X-Ray-Flash 367
X-Stoff = tetranitromethane (german) 307

Xilit = trinitroxylene (russian) 356
XTX = RDX coated with Sylgard, a silicon resin
Xyloidine = nitrostarch 233

Y

Yonkite = industrial explosive (Belgium)
Yuenyaku = black powder (Japan) 35

Z

Zazhigateinaya = Molotow-cocktail (russian)
Zeitzünder → bridgewire detonators 41
Zentralit → Centralite 56; 57
Zhirov = mixtures of tetryl with APC or potassium perchlorate (russian)
Zigarettenabbrand = face burning 137
zinc peroxide 287; 368
Z-Salz; Z-Stoff C; Z-Stoff N = permanganates for decomposition catalysis of hydrogen peroxide
Zuckernitrat 233
Zündabzweigung = downline
Zündanlage = ignition system 183
Zündblättche → amorces 19
Zünder = fuze 145
Zünder, elektrische → bridgewire detonators 41
Zünderdrähte = leg wires 202
Zündhütchen = amorce 19
Zündimpuls = electric pulse for bridgewire detonators 41
Zündkabel = firing line 138; 202
Zündkreis = circuit 58; 280
Zündkreisprüfer = circuit tester 58
Zündladung = primer charge 265
Zündmaschine = blasting machine 37
Zündpille = squib 293
Zündschalter = blasting switch 38
Zündschnur = safety fuse 269, **274**
Zündschraube = fuze head 146
Zündsicherung = igniter safety mechanism 183

Zündstrom = firing current 138
Zündung = initiation 190
Zündverzug = functioning time 75; **145**
Zündwilligkeit = ignitibility 183

Zustandsgleichung = equation of state 83; **118**
zweibasige Pulver = double base propellants 112